Microwave and RF Semiconductor Control Device Modeling

For a listing of recent titles in the
Artech House Microwave Library,
turn to the back of this book.

Microwave and RF Semiconductor Control Device Modeling

Robert H. Caverly

ARTECH HOUSE
BOSTON | LONDON
artechhouse.com

Library of Congress Cataloging-in-Publication Data
A catalog record for this book is available from the U.S. Library of Congress

British Library Cataloguing in Publication Data
A catalog record for this book is available from the British Library.

ISBN-13: 978-1-63081-021-4

Cover design by John Gomes

685 Canton Street
Norwood, MA

10 9 8 7 6 5 4 3 2 1

To my wife,
Maggie

Contents

Preface

The microwave and RF design engineer always seeks to develop a design that will meet specifications the first time that the circuit is fabricated. To do so requires that as many elements and phenomena as possible associated with the control devices and circuit be accurately modeled. In the case of the microwave and RF semiconductor control circuits, accurate modeling of the solid-state control components over frequency, voltage, current, and power is key to successful control system design. This book was written to provide the RF and microwave design engineer insight into the physical operation and modeling of PIN diodes and field-effect transistors (FETs) as control components and their use in microwave and RF control circuits. This insight I hope will be of some aid to design engineers to help them wisely choose and adapt device and circuit parameters during design optimization for best circuit performance.

The book is organized in the following manner. Chapters 1 and 2 cover fundamental material that provide the foundation for better understanding of the control device models introduced in later chapters. These two chapters cover the basics of control circuits, noise theory, device packaging issues, thermal behavior, and nonlinear device theory. Forward and reverse bias operation of the PIN diode in both switch states is the subject of Chapters 3 and 4. Chapter 3 focuses on the theory of operation of the PIN diode as an RF and microwave control element. The linear modeling of this operation is extensively covered. Chapter 4 expands on this modeling of the linear operation of the PIN diode and presents modeling of the device's nonlinear behavior and the subsequent introduction of unwanted signals into the circuit. Chapters 5 and 6 present the theory, operation, and modeling of FET-based control devices, with the metal-oxide-semiconductor FET (MOSFET) and metal-semiconductor FET (MESFET) being the two main focus technologies. Because the operation of the MOSFET and MESFET is fundamentally different, the operation and modeling of each device is covered in separate chapters. Chapter 5 covers the linear and nonlinear on- and off-state operation of the silicon MOSFET, focusing on the n-channel device. Chapter 6 covers the linear and nonlinear operation of the MESFET: traditional MESFETs and high electron mobility MESFET (HEMTs) in both operational states is discussed. The final two chapters, Chapters 7 and 8, present example control circuit simulations based on the PIN diode and FET models presented in the earlier chapters and provide a basis for further exploration of control device operation. The author, with the gracious agreement of Artech House, has placed simulation files and other support resources and files described in more detail in Chapter 1 on the web at a SourceForge repository: http://sourceforge.net/projects/pindiodemodel/files/

Acknowledgments

I have spent most of my engineering career working on microwave and RF control devices, and there have been many people I have collaborated with and discussed details of the operation of these components. If you are not mentioned in these acknowledgments, please be assured that I appreciate all the insights that I gained through our discussions.

I would first like to thank Wesley Fields and Rockford Curby at Cobham Metilics for their help in obtaining the photographs of various PIN diode packages and micrographs in Chapters 2 and 3. They explained in one picture what would have taken me many words to present. My Artech House editors, Marissa Koors and Molly Klemarczyk, provided invaluable assistance in keeping the book on track, keeping me aware of upcoming deadlines, and generally guiding me through all aspects of the development and production process. I also wish to thank Mark Walsh of Artech House for his assistance during the preliminary stages of the book development in helping me with understanding the entire publishing process and for being supportive of this endeavor over the many years prior to the book actually being written. Dr. Steve Maas was also extremely helpful in the proposal stages of the book, providing valuable suggestions that helped strengthen the presentation of the material. In addition to the editors, the technical reviewer gave me many suggestions in improving the overall manuscript; I used practically all of them.

In my first book, I acknowledged two individuals that gave me general but invaluable advice in the mechanics of book writing, advising me to stick to a strict and detailed outline and to schedule and keep plugging away on a daily basis: Dr. Steve Maas and Dr. William Bushong. Their advice stood the test of time and was still valuable in the writing of this book. I collaborated with Gerald (Jerry) Hiller for almost 20 years on modeling of microwave and RF control devices, and I am honored to have had the opportunity to work with him. I also wish to thank William E. Doherty, Jr. and Ronald Watkins for a collaboration that spanned almost 10 years and who first introduced me to the MRI field. Other acknowledgments go to Dr. Peter Rizzi, who was a mentor early in my career, and to the Amateur Radio community at large for using microwaves and RF as a hobby.

Putting together a book like this takes many hours of work a day. I thank my family for putting up with the long hours spent at my computer during the book's preparation. I'd like to thank my two sons, William and Matthew, for their interest in the progress of the book. Finally, I want to thank my wife, Maggie, for her patience and support during the writing of the book.

Robert H. Caverly, PhD
January 2016

CHAPTER 1

Introduction

1.1 Historical Perspective and Background

The modern era of semiconductor radio frequency (RF) and microwave control began more than a half-century ago, soon after the invention of the transistor. For the previous 50 years, semiconductor devices were used at RF and microwave frequencies, using such fragile devices as "cat whiskers," and point contact diode technologies [1–5]. With the advent of more reliable commercial semiconductor diodes in the 1950s, theoretical and applied research for RF and microwave control applications increased dramatically. At the time, both germanium and silicon diodes were available, with the most famous being the n-type germanium 1N263 and the p-type silicon 1N23B [6]. Interestingly, the 1N263 exhibited superior performance when used as a waveguide switch compared with the 1N23B [7]. This superior performance was attributed to the difference in the hole and electron mobilities between the two semiconductor and doping types [6, 7]. Garver published a series of papers in the 1950s and 1960s that described the theory and operation of these waveguide switching diodes in great detail and provided a solid foundation for further work on the subject [7–9]. Uhlir described the concept of using a PIN diode, a relatively new type of diode structure at the time, as a variable resistor at RF and microwave frequencies in 1958 [10, 11]. The PIN diode was first introduced as a high-voltage rectifier but was found to be a poor rectifier above a few megahertz [11]. Further investigation, however, showed the diode to have controllable impedance at very high frequencies [11]. This opened up the possibility of accurate control of RF and microwave signals with variable dc forward bias; this ability to vary the amplitude gave rise to the term variolosser [12]. White wrote the first and what is still considered to be one of the standard microwave semiconductor control texts on applications of PIN diodes [13]. Research on microwave and RF control devices continued during this time on not just discrete PIN diodes. One of the first silicon monolithic microwave integrated circuits (MMIC) was a PIN diode transmit-receive switch [14]. Since that time, control circuit designs using PIN diodes in both discrete and monolithic form for RF and microwave control applications have expanded from these early radar applications to wide-ranging applications in personal and infrastructure communication systems and even medical applications, such as in magnetic resonance imaging (MRI) scanners.

From the 1950s onward, transistors were being intensely investigated for many uses, including RF and microwave control. The first usable microwave bipolar transistor was introduced in 1965, and steady progress in frequency and performance has continued to this day. The use of field effect transistors (FETs) for microwave applications started in 1971; Liechti published an excellent review of the development history of FETs up to 1976 [15]. Garver, who published extensively in the 1950s and 1960s on RF and microwave diode control, in 1979 discussed using FETs instead of diodes for RF and microwave control [9]. The next year, Ayasli and colleagues developed a gallium arsenide-based (GaAs) monolithic FET transmit-receive switch, followed up two years later by a switch with higher power handling [16, 17]. The first RF and microwave control FETs were of the metal-semiconductor (MESFET) type, but as digital complementary metal-oxide-semiconductor (CMOS) FETs improved in performance due to decreasing feature size, MOSFETs started to play a larger role in the control area. Since the MESFET and MOSFET are easy to include with other digital and analog circuitry on an integrated circuit, their use has important impact on fully monolithic integrated communication and other system-on-chip (SoC) solutions.

Increase in circuit and system complexity requires the designer to use computer tools to reduce design time and increase the probability of first-pass design success. Computer design tools are only as good as the models they use, and modeling of semiconductor control devices is no exception. This book seeks to provide the theoretical background governing the operation of these semiconductor control devices and to link this theoretical understanding to the computer models that aid in simulating the device's behavior at the circuit level. It is hoped that the reader will gain better insight into the physical operation of these devices during the design process and understand the interplay between device model parameters and their impact on the control circuit performance. The next sections in this chapter cover fundamental control concepts using a simple switch and other control circuits as examples for a more general overview of device modeling. The later sections in the chapter generalize the discussion to include other control circuits such as attenuators and phase shifters.

1.1.1 Simplified Switch Concepts

The basic concept of the switch depends on the location of the control element and its impedance Z_{CTL} in the control circuit. For a series-connected switch element, low impedance Z_{low} is required for the low loss or on-state, whereas a high impedance Z_{high} is required for the high loss or off-state. For a shunt-connected switch element, the opposite is true: a high impedance is required for the on-state and a low impedance for the off-state. In an ideal series-connected mechanical switch, these two impedance states would correspond to the switch contacts touching with very small resistance $Z_{low} = R_{ON}$ between the contacts; the open switch would then have infinite resistance. However, while this may be true at dc, in the open-switch condition, the two air-separated contacts (or vacuum or other gas, depending on the switch type) would create a capacitance C_{OFF} that is a function of the area and the contact separation distance. This capacitance exhibits a reactance of $1/j\omega C_{OFF}$ (ω is the radian frequency) and would show a frequency dependent loss in the off-state, with the loss decreasing with increasing frequency (although still infinite at dc, unless the dc bias voltage were large enough to cause breakdown in the medium between the contacts). In the off-state, there will be a small resistance in series with this capacitance, R_{OFF}, because of the finite conductivity and dimensions of the

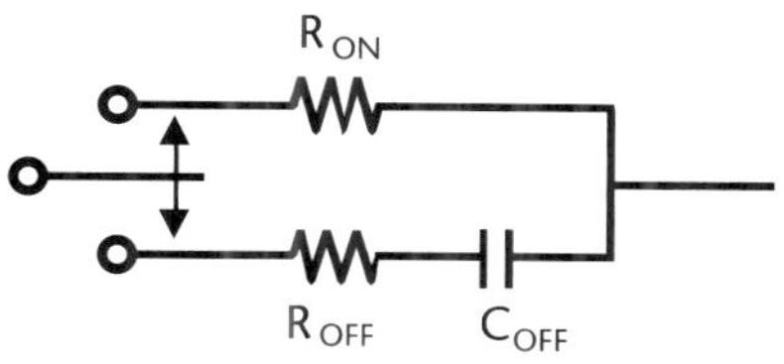

Figure 1.1 Simplified switch model showing the low and high impedance switch states with actuator.

contacts. In the case of the mechanical switch, R_{OFF} and R_{ON} are approximately equal but may not be in actual practice. The high impedance condition of the switch state Z_{high} is the series sum of these two components, $Z_{high} = R_{OFF} + 1/j\omega C_{OFF}$. The switch actuator then automatically controls which state is in play (Figure 1.1).

In this book, the focus will not be on mechanical switches but on common semiconductor devices that make up modern RF and microwave control circuitry. In the case of each one of these semiconductor control devices, the impedance concepts are exactly the same: the control elements are required to exhibit low or high impedance values, depending on their circuit connections and the control state (on or off). The actuator that controls the impedance state is no longer mechanical but will be some electrical signal from one or more control points that governs the semiconductor device operating behavior as it toggles between the two impedance states. In the case of the PIN diode, the "actuator" is the application of a dc forward current to create the low impedance state and a reverse bias voltage to create the high impedance state. For FET-based circuits, the "actuator" is voltage-based and the application or removal of a voltage on the gate terminal controls the drain-source channel impedance state. In any case, the equivalent switch circuit shown in Figure 1.1 can be applied to any of these control circuit components when studying their ideal operation.

1.2 General Control Circuit Terminology and Operation

The design engineer will encounter various terms associated with RF and microwave control components and modules as part of the search for the best device and subsequent circuit design choices. This section covers the major terminology seen in the control environment and presents fundamentals of RF and microwave control circuits and relationships that describe ideal operation. The discussion of these fundamental control circuits is general and will eventually be used later in the chapters to derive more detailed expressions for specific control circuit topologies, such as single- and multiple-control device switches, attenuators, and phase shifters.

1.2.1 Switching Quality Factor (*Q*)

Kurokawa and Schlosser presented an elegant theory on switch design based on the impedance of the switching elements (in their original case, switching diodes), Z_{low} and Z_{high} [18]. The theory is based on the switching *Q*-factor (quality factor) and relates the losses in the on/off-state (or the associated impedances) by showing the tradeoffs between losses in the two switch states. The two switch states (States 1 and 2) are defined by impedances based on the equivalent circuit of the switch element, $Z_1 = R_1 + jX_1$ and $Z_2 = R_2 + jX_2$ when placed at the output of a lossless two-port network described by S-parameters S_{ij}. The switching quality factor is then defined in terms of these impedances, Z_i, or the corresponding reflection coefficients, Γ_i:

$$Q = \frac{\sqrt{(R_1 - R_2)^2 + (X_1 - X_2)^2}}{\sqrt{R_1 R_2}} = \frac{2|\Gamma_1 - \Gamma_2|}{\sqrt{\left(1 - |\Gamma_1|^2\right)\left(1 - |\Gamma_2|^2\right)}} \tag{1.1}$$

In (1.1), the switching Q expression in the case of equal resistances (but not reactances) reduces to the traditional form for the quality factor as a reactance-resistance ratio. If the reflection coefficients only differ in phase ($\Gamma_2 = \Gamma_1 e^{j\Theta}$) as might be seen in a matched switch, then the switching Q can be written as

$$Q = \frac{2|\Gamma||1 - e^{j\Theta}|}{\left(1 - |\Gamma|^2\right)} \tag{1.2}$$

Figure 1.2 shows the switching Q in (1.2) plotted as a function of loss (defined as $-20\log_{10}|\Gamma|$ dB) using the reflection coefficient phase angle Θ as a parameter. High switching Q occurs at low loss with further improvement at a fixed loss as the phase angle between the two switch state reflection coefficients approaches 180°.
Figure 1.3 shows the relationship between the two switching states using the switching Q as a parameter assuming zero phase ($\Theta = 0°$) of the reflection coefficients. The reflection coefficients could be either on- or off-state; the curves are interchangeable. For a given switching Q as defined by the switching element, changing Γ_1 by varying the parameters of the associated lossless switch circuit changes the corresponding Γ_2. For the $Q^2 = 25$ circuit, a peak in Γ_2 occurs for Γ_1 of approximately 0.5 dB, implying large reflections and hence loss in the opposite switch state. A lower or higher Γ_1 would cause a corresponding reduction in the reflection loss in the other

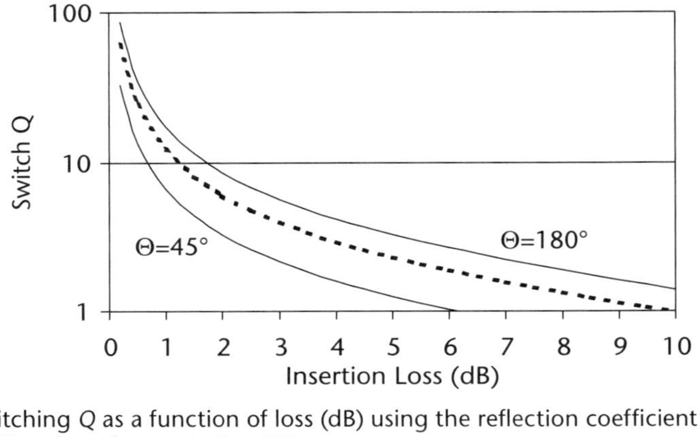

Figure 1.2 Switching Q as a function of loss (dB) using the reflection coefficient phase angle Θ as a parameter. The dotted curve is $\Theta = 90°$.

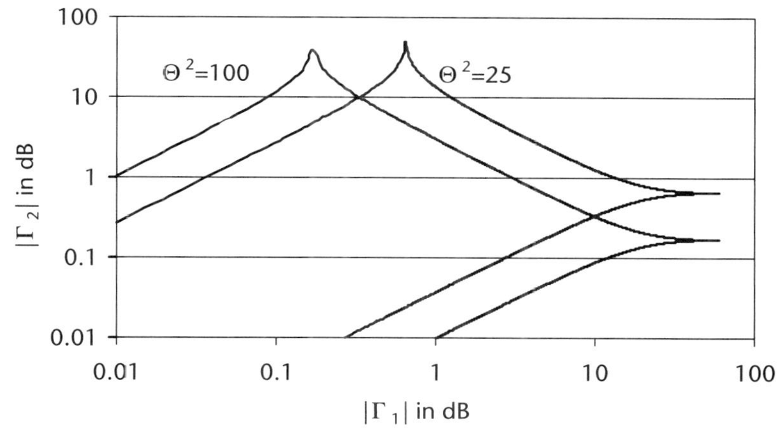

Figure 1.3 Switch State 2 reflection coefficient ($|\Gamma_2|$) as a function of Switch State 1 reflection coefficient ($|\Gamma_1|$) for two different values of switching Q.

switch state. A higher switching Q circuit (such as $Q^2 = 100$) requires lower losses, and hence the resulting smaller reflection coefficient Γ_1 achieves the same peak in the opposite switching state.

Using the general switch equivalent circuit shown in Figure 1.1, where State 1 is the on-state and State 2 is the off-state, and further, if the difference between R_{ON} and R_{OFF} is significantly less than the reactance X_2, the switching Q can be written as

$$Q = \frac{X_2}{R_{ON}} = \frac{1}{\omega\sqrt{R_{ON}R_{OFF}}C_{OFF}} \tag{1.3}$$

where ω is the radian frequency. The RC product in (1.3) can be used to define a cutoff frequency called the switch cutoff frequency [19], F_C, which allows the switching Q under this scenario to be written as

$$F_C = \frac{1}{2\pi\sqrt{R_{ON}R_{OFF}}C_{OFF}} \tag{1.4a}$$

$$Q = \frac{F_C}{F} \tag{1.4b}$$

where F is the frequency ($\omega = 2\pi F$) and the loss factor K in Hines' paper is assumed to be unity [19]. A high switch cutoff frequency F_C will keep the switching Q high, with the Q decreasing as the operation frequency increases. The two cases shown in Figure 1.3 correspond to operation at frequencies at 20% and 10% of the cutoff frequency, F_C. The concept of switching Q is so general that it can be used in the design of both PIN diode and FET-based semiconductor switching circuits [19–21]. The two impedance states Z_1 and Z_2 can be determined by measurement and may include other elements along with the device [20, 21].

1.2.2 Circuit Analysis

The circuit analysis approach in this section is a study of the voltage and current characteristics of two common control circuits and includes the impact on these circuits of the control element. This section discusses the voltage and current characteristics of the control element in Figure 1.1 in series and shunt connections. The study is a general circuit analysis and so will apply to any type of control circuit using the equivalent circuit shown in Figure 1.1, switch or attenuator.

Figure 1.4(a) shows a series-connected control element in a series configuration in a Z_0 system with the generator matched to the load. Assuming a series element impedance Z_{CTL}, the peak voltage and current at the load can be computed using traditional circuit analysis as

$$V_{load} = V_g \frac{Z_0}{2Z_0 + Z_{CTL}}; \qquad I_{load} = V_g \frac{1}{2Z_0 + Z_{CTL}} = \frac{V_{load}}{Z_0} \tag{1.5a}$$

The power available from the generator P_A is the load power in the absence of the control element and is defined as

$$P_A = \frac{V_g V_g^*}{8Z_0} \tag{1.5b}$$

where the generator voltage is assumed to be a peak voltage and the superscript * denotes the complex conjugate. The corresponding load power with Z_{CTL} in the circuit can then be computed using (1.5b) as

$$P_{load} = \frac{1}{2} RE\left(V_{load} I_{load}^*\right) = \frac{1}{2} RE\left[\left(V_g \frac{Z_0}{2Z_0 + Z_{CTL}}\right)\left(\frac{V_g}{Z_0}\frac{Z_0}{2Z_0 + Z_{CTL}}\right)^*\right] = P_A \left|\frac{2Z_0}{2Z_0 + Z_{CTL}}\right|^2 \tag{1.6}$$

where RE() denotes taking the real part. For Z_{CTL} much less than Z_0, the load and available power are approximately the same; a high impedance value for Z_{CTL} corresponds to a significant reduction in the load power compared with the available power from the generator.

For the shunt-connected control element (Figure 1.4(b)), the load voltage and current can be derived in a similar fashion, yielding the following expressions:

$$V_{load} = V_g \frac{Z_{CTL}}{2Z_{CTL} + Z_0}; \qquad I_{load} = \frac{V_g}{Z_0} \frac{Z_{CTL}}{2Z_{CTL} + Z_0} = \frac{V_{load}}{Z_0} \tag{1.7}$$

Using the expression for the power available from the source, P_A in (1.5b), the load power can be written as

$$P_{load} = \frac{1}{2} RE\left(V_{load} I_{load}^*\right) = \frac{1}{2} RE\left[\left(V_g \frac{Z_{CTL}}{2Z_{CTL} + Z_0}\right)\frac{V_g^*}{Z_0}\left(\frac{Z_{CTL}}{2Z_{CTL} + Z_0}\right)^*\right] = P_A \left|\frac{2Z_{CTL}}{2Z_{CTL} + Z_0}\right|^2 \tag{1.8}$$

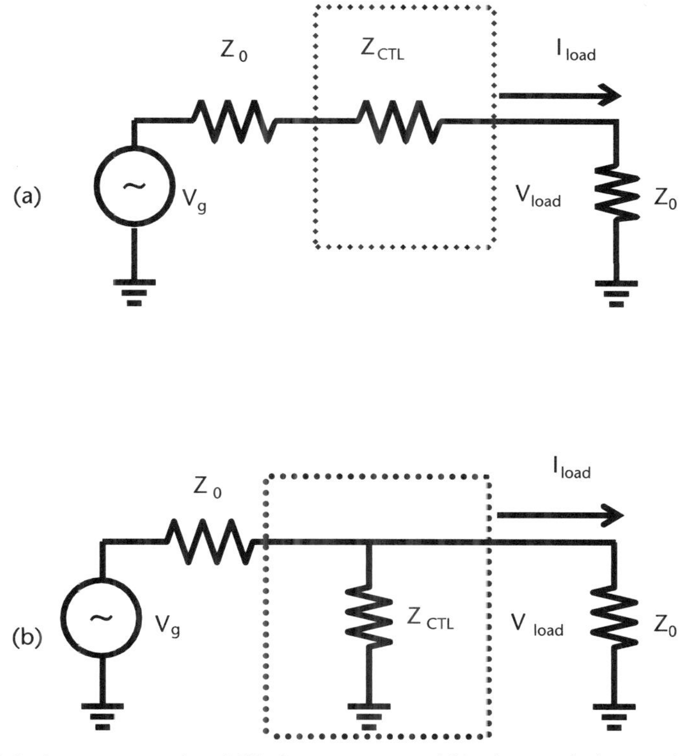

Figure 1.4 (a) Series-connected and (b) shunt-connected ideal control element in a circuit. The dotted box represents the control device as an added two-port network to the circuit.

For the shunt-connected case, the control behavior is complementary to that of the series-connected case. For Z_{CTL} much greater than Z_0, the load and available power are approximately the same; a low impedance value for Z_{CTL} corresponds to a significant reduction in the load power compared with the available power from the generator.

Figure 1.5 shows power loss ratio P_A/P_{load} as a function of the controlling impedance Z_{CTL} for series and shunt-connected circuits in a 50Ω system using (1.6) and (1.8). Z_{CTL} is assumed purely resistive, and the power loss ratio P_A/P_{load} is presented

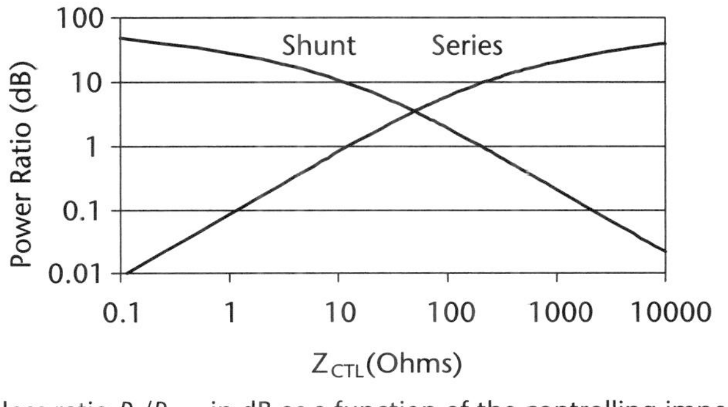

Figure 1.5 Power loss ratio P_A/P_{load} in dB as a function of the controlling impedance Z_{CTL} for series and shunt-connected circuits in a 50Ω system.

in decibels (dB) as $10\log_{10}(P_A/P_{\text{load}})$. The figure shows that for the series-connected control device case, power loss ratios less than 1 dB occur for impedances less than approximately 10Ω, with the power loss ratio increasing with increasing impedance. The converse is true for the shunt-connected device; power loss ratios greater than 10 dB occur for impedances less than approximately 10Ω, with the ratio decreasing as the impedance increases.

1.2.3 Control Circuit Power Handling

The previous section introduced the voltage, current, and power at various locations in the simple control circuits shown in Figure 1.4. Eventually, the general impedance Z_{CTL} will be the impedance exhibited by a semiconductor control device such as a PIN diode or an FET. These semiconductor control elements are limited in the level of voltage across or current through them, and if either one is exceeded, device destruction will occur and the control module will fail. An example of the voltage and current properties of a purely resistive control element in both device states will be presented in this section and used as an introduction to a more general discussion of voltage, current, and power handling of microwave and RF control devices.

For example, assume the control element toggles between one of two impedance states $Z_{\text{CTL}} = R_{\text{CTL}}$, where the low impedance state is $Z_0/20$ and the high impedance state is $20Z_0$. The voltage across and current through the series and shunt-connected devices in both switch states can be computed with the aid of (1.5) and (1.7) and are shown in (1.9) and (1.10) for both series and shunt-connected devices. Figure 1.6 shows the control element voltage and current for the two resistance states calculated using (1.9) and (1.10), as the generator voltage V_g is varied between 1 and 100 V.

Series:

$$V_{\text{load}} = V_g \frac{Z_0}{2Z_0 + R_{\text{CTL}}}; \qquad I_R = \frac{V_{\text{load}}}{Z_0}; \qquad V_R = R_{\text{CTL}} I_{\text{load}} \tag{1.9}$$

Shunt:

$$V_{\text{load}} = V_R = V_g \frac{R_{\text{CTL}}}{2R_{\text{CTL}} + Z_0}; \qquad I_R = \frac{V_{\text{load}}}{R_{\text{CTL}}} \tag{1.10}$$

Figure 1.6(a) shows that in the device off-state, the control device voltage drop is larger than the on-state, although the voltage drop is higher in the series-connected case. This indicates that the higher voltage stress occurs during the control device off-state, which is the series-connected control module off-state and the shunt-connected control module on-state. Figure 1.6(b) shows the device on-state current is higher than the off-state current, with the current higher in the shunt-connected device on-state. Higher current stress occurs during the control device on-state, which occurs in the series-connected control module on-state and the shunt-connected control module off-state. The higher voltage stress occurs in the off-state series-connected device, and the higher current stress occurs in the on-state shunt-connected device.

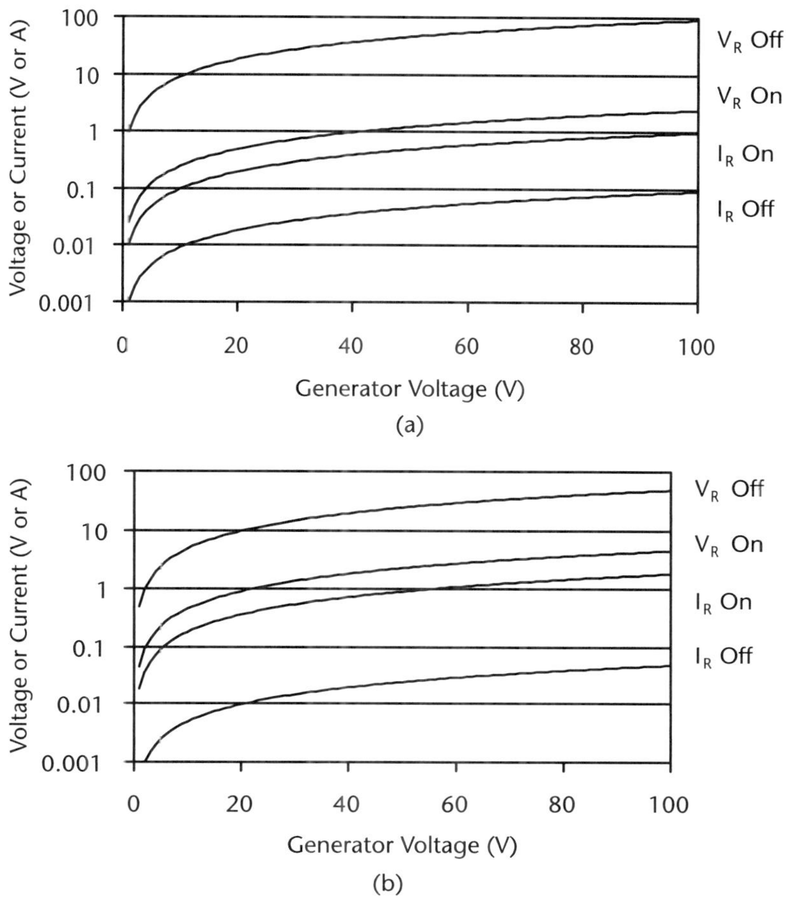

Figure 1.6 (a) Series and (b) shunt-connected control device voltage V_R and current I_R as a function of generator voltage, V_g. ON and OFF refer to the state of the control device, not the switch state.

A general theory on the fundamental limits of RF and microwave control using semiconductor diodes was first advanced by Hines [19] and later for FET-based switches by Atwater and Sudbury [20] (based on Hines' theory). The power handling capability of control devices was derived by observing the difference between the control module's on-state and off-state reflection coefficients and the resulting voltages and currents in the circuit:

$$\left|\Gamma_{\mathrm{OFF}} - \Gamma_{\mathrm{ON}}\right| \leq \frac{I_{sc} V_{oc}}{8P_A} \tag{1.11}$$

where I_{sc} is the peak value of the current in the switch when closed and V_{oc} is the peak value of the voltage across the switch when open. This theoretical approach explains the example scenario illustrated in Figure 1.6 but extends the example to show the resulting power limitations of the control device. Assuming that the reflection coefficient in the ideal on-state switch is zero and the reflection coefficient in the ideal off-state is unity, (1.11) yields the factors limiting the power handling in semiconductor control devices:

$$P_{\text{MAX}} \le \frac{I_{sc}V_{oc}}{8} \tag{1.12}$$

1.2.4 Definition of Control Circuit Terms

The previous section introduced the term power loss ratio as the ratio of the power available from the source, P_A, to the power delivered to the load, P_{load}: P_A/P_{load}. This ratio is not just applicable to the simple control circuit structures shown in Figure 1.4 but applies to all RF and microwave control circuitry whether PIN diode or FET-based and no matter how simple or complex the circuit topology. The power loss ratio is a fundamental control circuit term and is directly related to the fundamental term insertion loss that RF and microwave engineers widely use. The fundamental definition of insertion loss is the ratio of the power delivered to the load before insertion of a network to the power delivered to the load after network insertion [22, 23]. The power loss ratio P_A/P_{load} directly fits this definition because before the control network is added, P_A defines the power delivered to the load, and P_{load} is the power after the control network is introduced (dotted blocks in Figure 1.4). The insertion loss, IL, can be written in terms of the power loss ratio in dB as:

$$IL = 10\log_{10}\frac{P_A}{P_{\text{load}}}\ \text{dB} \tag{1.13}$$

Note that the definition of the term is power loss ratio and so the IL, being a loss measure, will have positive dB values. If for some reason the control circuit could increase the power delivered to the load, then IL would be negative since a negative loss (in dB) represents gain (insertion gain). This never occurs under typical operating conditions for RF and microwave control circuits.

The insertion loss is often calculated from the feed-forward S-parameter, S_{21} (assuming a two-port network), in which case the insertion loss can be computed as:

$$IL = -10\log_{10}|S_{21}|^2\ \text{dB} \tag{1.14}$$

Note the negative sign in (1.14) when using S-parameters; for passive networks such as RF and microwave control circuits, $|S_{21}|$ will always be less than unity, and this negative sign is required to maintain the concept of a loss mechanism. It is frequently seen in the literature that $|S_{21}|$ is presented as a negative number and is calculated as $10\log_{10}|S_{21}|^2$. Confusion can arise when $|S_{21}|$ is termed the insertion loss; while it is another parameter that describes power loss, the use of the negative sign is contradictory to the insertion loss definition [22, 23] and so is to be avoided. Always ensure positive dB values for insertion loss IL for RF and microwave control circuitry built with PIN diodes and FETs.

For the RF and microwave engineer, the insertion loss is a fundamental parameter based on first principles derived from circuit theory. For the control circuit in the on-state, insertion loss is used to define the level of signal loss between the input of the control module and its output. For the RF and microwave engineer involved with control circuitry, however, there are two other definitions that are mathematically based on insertion loss but have different interpretations: isolation and attenuation. Isolation and attenuation are both calculated using (1.13) and are described in terms of a positive dB value, but the control circuit engineer uses these terms for different purposes. From the dictionary definition, the term isolation implies separation [24]; for the engineer involved with control circuitry, the isolation would then be a measure of the load-source electrical separation when the control circuit is in its off-state. For attenuation, the dictionary definition implies a reduction or a weakening [25]; for the engineer involved with control circuitry, this term is used to describe the amount of weakening or reduction of power to the load in a controlled manner. This reduction is typically performed in the control circuit on-state by varying component values to control the degree of loss introduced into the system. Early in the history of RF and microwave engineering, attenuation usually only referred to dissipative losses by the network, but it now includes losses from all sources, dissipative and reflective alike [24]. No matter the term, attenuation (for attenuators), insertion loss (for the on-state), or isolation (for the off-state), all are discussed and presented in terms of positive dB.

1.3 Circuits

This section covers the fundamental operation of a number of widely used RF and microwave control circuits. The presentation is not meant to be an exhaustive coverage of circuits but rather to show the theory of operation of fundamental circuits so that more elaborate circuits can be better understood. The chapter covers reflective switches and attenuators, matched attenuators, simple transmit-receive switches, and other circuit elements often seen in control modules.

1.3.1 Reflective Switches and Attenuators

The simplest of the RF switching circuits is the single-pole, single-throw (SPST) switch, in which a single control element replaces the more general Z_{CTL} impedance shown in Figure 1.4 The SPST switches in this case are often called series or shunt-reflective switches, since they operate by creating an impedance mismatch between the matched source and load impedances, Z_0.

1.3.1.1 Series Reflective Control Circuits

The control impedance Z_{CTL} of Figure 1.4(a) creates an impedance mismatch between the source and load and, in a Z_0 system, the associated reflection coefficient as seen by the source is defined as:

$$\Gamma = \frac{\left(Z_{CTL} + Z_0\right) - Z_0}{\left(Z_{CTL} + Z_0\right) + Z_0} = \frac{Z_{CTL}}{Z_{CTL} + 2Z_0} \tag{1.15}$$

A certain portion of the power available from the source, P_A, will be reflected due to the impedance mismatch and is defined as $P_{REF} = P_A\Gamma\Gamma^* = P_A|\Gamma|^2$. The total power delivered to the input of the switching module is the difference between the power available from the source and the reflected power, $P_{IN} = P_A(1 - |\Gamma|^2)$. The power dissipated by the load is then a fraction of the power available, P_A:

$$P_{load} = P_A \left| \frac{2Z_0}{2Z_0 + Z_{CTL}} \right|^2 \tag{1.16}$$

Using the equivalent circuit for the control element shown in Figure 1.1 and (1.15), the reflection coefficient in the two switch states for the series-connected control element can be written as

$$\Gamma = \begin{cases} \dfrac{R_{ON}}{R_{ON} + 2Z_0} & \text{ON} \\[2ex] \dfrac{\left(R_{OFF} + 1/j\omega C_{OFF}\right)}{\left(R_{OFF} + 1/j\omega C_{OFF}\right) + 2Z_0} & \text{OFF} \end{cases} \tag{1.17}$$

In the switch on-state, application of (1.17) defines the insertion loss of the switch as

$$IL = 20\log_{10}\left(1 + \frac{R_{ON}}{2Z_0}\right) \text{ dB} \tag{1.18}$$

In a similar manner, the off-state isolation (ISO) can be defined, again using (1.17), as

$$ISO = 10\log_{10}\left|1 + \frac{\left(R_{OFF} + 1/j\omega C_{OFF}\right)}{2Z_0}\right|^2 \text{ dB} \tag{1.19}$$

For operation well below the switch cutoff frequency F_C, the reactance of the off-state capacitance is much greater than R_{OFF}, and so the isolation expression can be simplified:

$$ISO = 10\log_{10}\left|1 + \frac{1}{2j\omega C_{OFF}Z_0}\right|^2 = 10\log_{10}\left(1 + \frac{1}{\left(2\omega C_{OFF}Z_0\right)^2}\right) \text{ dB} \tag{1.20}$$

Ideal operation of the series-connected control device shows no frequency dependence in the low insertion loss on-state, but the isolation is frequency dependent because of the reactive nature of Z_{CTL} in this state. Figure 1.7(a) plots the insertion loss for the on-state series-connected control device as a function of R_{ON}, indicating

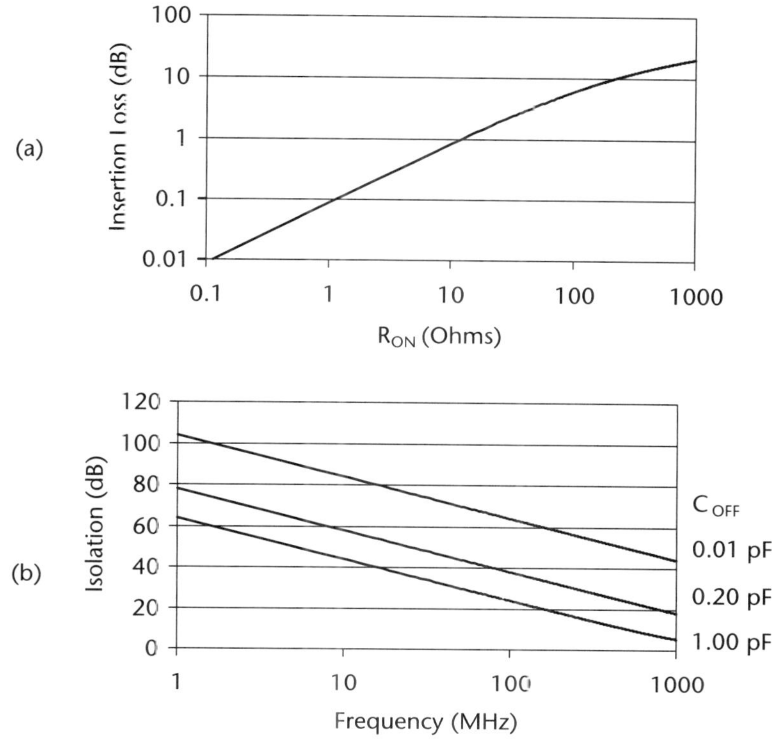

Figure 1.7 (a) Series-reflective SPST switch on-state insertion loss as a function of resistance R_{ON}. (b) Series-reflective SPST switch off-state isolation as a function of frequency with capacitance C_{OFF} as a parameter.

that smaller values of insertion loss correspond to low values of R_{ON}. As an example, a 1.0Ω on-state resistance corresponds to an insertion loss of 0.1 dB. Figure 1.7(b) shows the isolation for the off-state series-connected control device as a function of frequency using off-state capacitance C_{OFF} as a parameter; $R_{OFF} \ll 1/\omega C_{OFF}$ has been assumed in this figure. The isolation results show decreasing isolation as a function of frequency for any capacitance value, with smaller capacitance values yielding significantly higher isolation. At 100 MHz in the off-state, an 0.01 pF off-state capacitance provides more than 60 dB isolation, whereas a 1.0 pF capacitance only provides about 20 dB isolation. In both switch states, small R_{ON} and C_{OFF} provide the best insertion loss and isolation, a fact consistent with the concept of maximizing switch cutoff frequency ($F_C = 1/2\pi\sqrt{R_{ON}R_{OFF}}C_{OFF}$) and switching Q.

The peak RF current through the control component in its on-state (low switch insertion loss) is $I_{R_{ON}} = V_g/(2Z_0 + R_{ON})$, and so the power dissipated in the control component can be computed as

$$P = \frac{1}{2} I_{R_{ON}}^2 R_{ON} = \frac{4R_{ON}Z_0}{\left(2Z_0 + R_{ON}\right)^2} P_A \cong \frac{R_{ON}}{Z_0} P_A \tag{1.21a}$$

with an alternate form of the control component current in terms of P_A:

$$I_{R_{ON}} = \frac{\sqrt{8Z_0P_A}}{\left(2Z_0 + R_{ON}\right)} \cong \sqrt{2P_A/Z_0} \tag{1.21b}$$

In the off-state with the control component at a high impedance for good isolation (reactance much greater than Z_0), the majority of the generator voltage V_g is dropped across the control component:

$$V = V_g = \sqrt{8Z_0P_A} \tag{1.21c}$$

1.3.1.2 Shunt Reflective Control Circuits

An approach similar to that used in the series-reflective switch case can be used for analysis of the shunt-connected or shunt-reflective SPST switch. The control impedance Z_{CTL} creates an impedance mismatch (Figure 1.4(b)) between the source and load, and in a Z_0 system, the associated reflection coefficient as seen by the source can be written as

$$\Gamma = \frac{-Z_0}{2Z_{CTL} + Z_0} \tag{1.22}$$

Using the equivalent circuit for the control element shown in Figure 1.1 and (1.15), the reflection coefficient in the two switch states can be written as

$$\Gamma = \begin{cases} \dfrac{-Z_0}{2\left(R_{OFF} + 1/j\omega C_{OFF}\right) + Z_0} & \text{ON} \\ \dfrac{-Z_0}{2R_{ON} + Z_0} & \text{OFF} \end{cases} \tag{1.23}$$

In the switch on-state, (1.13) is used to define the insertion loss of the switch as

$$IL = 10\log_{10}\left|1 + \frac{Z_0}{2\left(R_{OFF} + 1/j\omega C_{OFF}\right)}\right|^2 \text{ dB} \tag{1.24a}$$

For normal operation, well below the switch cutoff frequency F_C, the reactance of the off-state capacitance is much greater than R_{OFF}, and so the isolation expression can be simplified:

$$IL = 10\log_{10}\left|1 + \frac{j\omega C_{OFF}Z_0}{2}\right|^2 = 10\log_{10}\left(1 + \left(\frac{\omega C_{OFF}Z_0}{2}\right)^2\right) \text{ dB} \tag{1.24b}$$

In a similar manner, the off-state isolation (ISO) can be defined, again with (1.13):

$$ISO = 10\log_{10}\left(1 + \frac{Z_0}{2R_{ON}}\right)^2 \text{ dB} \tag{1.25}$$

Ideal operation of the shunt-reflective SPST switch shows a low insertion loss state frequency dependence due to the reactive nature of the shunt capacitance, C_{OFF}. Figure 1.8(a) plots the insertion loss for the on-state SPST shunt-reflective switch as a function of frequency using capacitance C_{OFF} as a parameter; $R_{OFF} \ll 1/\omega C_{OFF}$ has been assumed in this figure. At 1000 MHz in the switch on-state, a 0.01 pF capacitance exhibits a very low insertion loss, and a 1.0 pF capacitance shows about 0.25 dB insertion loss; for all capacitances, the insertion loss increases with increasing frequency. The ideal off-state isolation does not show any frequency dependence because the shunting element controlling the isolation is a pure resistance, R_{ON}. Figure 1.8(b) shows the isolation for the off-state SPST shunt-reflective switch as a function of R_{ON}, indicating that high isolation corresponds to small values of R_{ON}. As an example, a 1.0Ω on-state resistance provides approximately 30 dB of generator-load isolation. Similar to the case of the series-reflective switch,

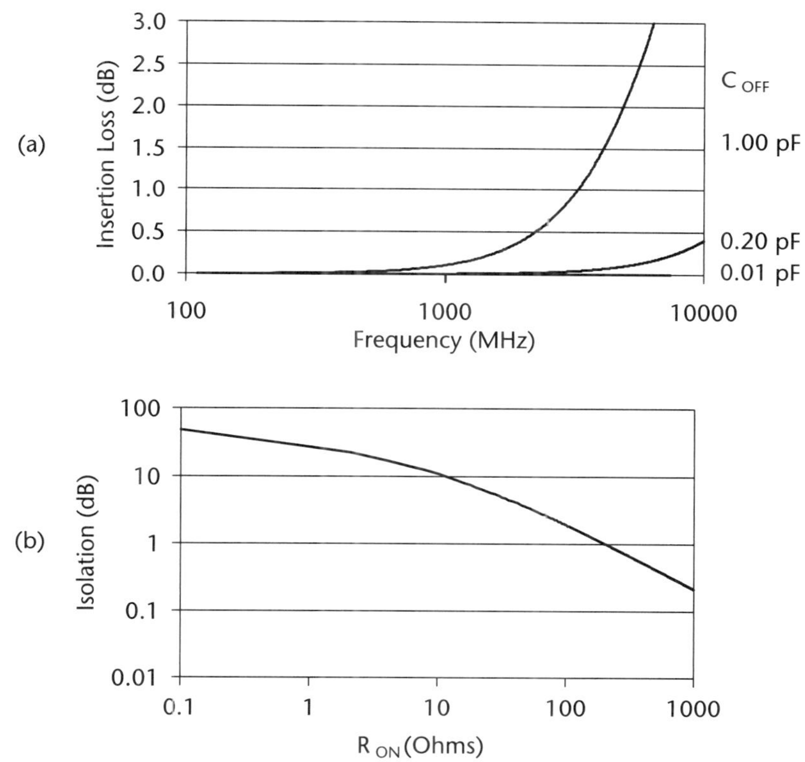

Figure 1.8 (a) Shunt-reflective SPST switch on-state insertion loss as a function of frequency with capacitance C_{OFF} as a parameter (note the change in frequency scale from Figure 1.7). (b) Shunt-reflective SPST switch off-state isolation as a function of resistance R_{ON}.

small R_{ON} and C_{OFF} provide the best insertion loss and isolation, a result consistent with the concept of maximizing switch cutoff frequency ($F_C = 1/2\pi\sqrt{R_{ON}R_{OFF}}C_{OFF}$) and switching Q. This result shows the utility of using switching Q and/or F_C in describing switch performance, since these terms are independent of switch circuit topology.

The peak RF current through the control component in its on-state (high switch isolation) is $I_{R_{ON}} = V_g/(2R_{ON} + Z_0)$, and so the power dissipated in the control component can be computed:

$$P_{PIN} = \frac{1}{2} I_{R_{ON}}^2 R_{ON} = \frac{4R_{ON}Z_0}{\left(2R_{ON} + Z_0\right)^2} P_A \cong \frac{4R_{ON}}{Z_0} P_A \tag{1.26a}$$

with an alternative form of the control component current in terms of P_A:

$$I_{R_{ON}} = \frac{\sqrt{8Z_0P_A}}{\left(2R_{ON} + Z_0\right)} \cong \sqrt{8P_A/Z_0} \tag{1.26b}$$

In the control component off-state with a high shunt impedance providing low switch insertion loss, half of the generator voltage V_g is dropped across the device:

$$V_{PIN} = V_g/2 = \sqrt{2Z_0P_A} \tag{1.26c}$$

The voltage and current forms for the series and shunt switch cases in (1.21) and (1.26) show the on-state shunt connected control element has higher current stress for a given generator power P_A than in the series connected case. Conversely, the off-state control element has higher voltage stress in the series connected case compared with the shunt connection.

1.3.1.3 Reflective Attenuators

The insertion loss and isolation expressions in (1.18) and (1.25) for the series and shunt-connected device controlled by purely resistive control devices show that the control devices may be used as reflective attenuators since the on-state resistance in semiconductor control devices is often controllable. For the sake of completeness, (1.18) and (1.25) are repeated as attenuation expressions in (1.27) and (1.28), and Figure 1.9 shows the circuit attenuation achieved as a function of the control device resistance, R_{ON}.

$$ATT = 10\log_{10}\left(1 + \frac{R_{ON}}{2Z_0}\right)^2 \text{ dB} \quad \text{Series} \tag{1.27}$$

$$ATT = 10\log_{10}\left(1 + \frac{Z_0}{2R_{ON}}\right)^2 \text{ dB} \quad \text{Shunt} \tag{1.28}$$

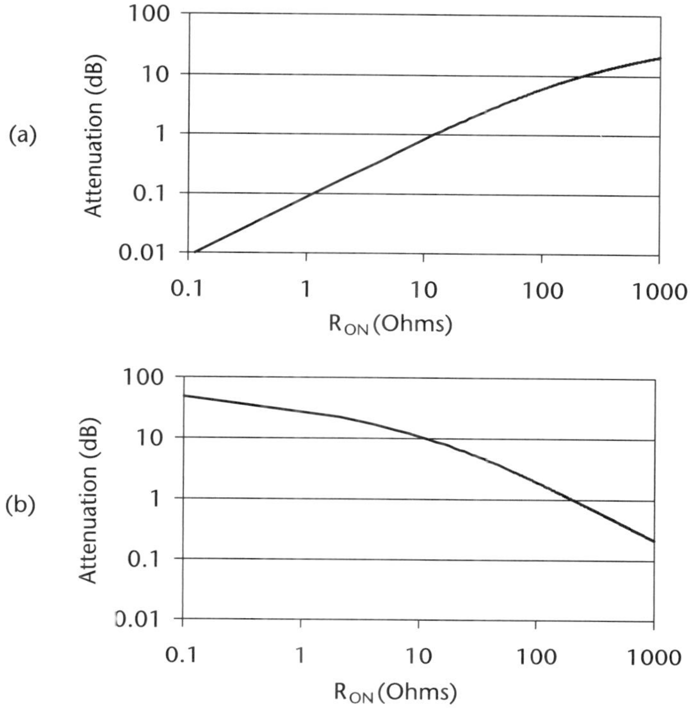

Figure 1.9 (a) Attenuation ATT in a series-reflective attenuator as a function of resistance R_{ON}. (b) Attenuation ATT in a shunt-reflective attenuator as a function of resistance R_{ON}.

Figure 1.9 shows that to achieve an attenuation of 10 dB, the series-connected device should exhibit an on-state resistance R_{ON} of approximately 220Ω, whereas the shunt-connected device would need to exhibit a resistance R_{ON} of approximately 11Ω. These resistances are easily met with current semiconductor control devices.

1.3.2 Matched Attenuators

One of the key advantages of the series and shunt-connected attenuator circuits outlined in the previous section is their simplicity; a single series or shunt control device can effectively manage the level of RF or microwave energy delivered to the load. One of the major drawbacks of the reflective attenuator is that the management of this RF energy is governed by reflection of RF energy back toward the generator. Depending on the power levels involved, the reflective energy would have to be dissipated by the generator and could exceed safe operating limits and cause undesired circuit operation, such as excessive heating or generation of unwanted signals. This reflected energy is minimized if the generator looks into a matched impedance.

For matched attenuators, there are two degrees of freedom that must be addressed in the design: the desired level of attenuation and the conditions for impedance matching. These two degrees of freedom indicate that, from the generator perspective, an attenuator circuit with two control resistances would be the simplest form. Consider such a circuit in Figure 1.10, where resistances R_1 and R_2 are the control

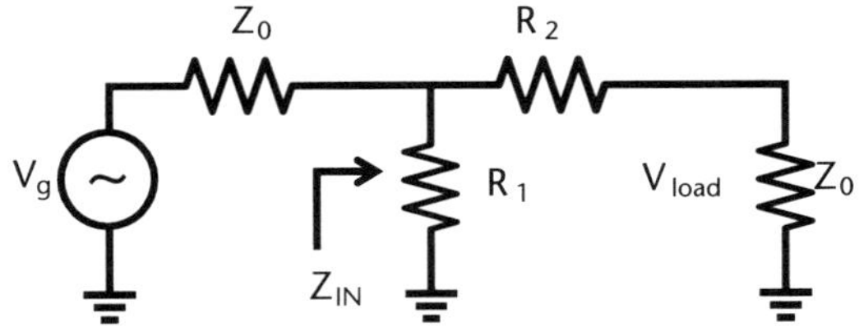

Figure 1.10 Example resistive attenuator circuit with an impedance match seen by the generator.

device resistances that will govern both the attenuation and the match condition in the Z_0 system (R_1 could also be placed in parallel with the load instead; the derivation of the relationships is similar). The attenuator input impedance seen by the generator of impedance Z_0 is the parallel combination of R_1 and $R_2 + Z_0$. For the generator to see a matched impedance of Z_0, this parallel combination Z_{IN} should be equal to Z_0 and provides a relationship between R_1, R_2, and Z_0:

$$Z_{IN} = Z_0 = \frac{R_1(R_2 + Z_0)}{R_1 + R_2 + Z_0} \Rightarrow R_1 = Z_0\left(1 + \frac{Z_0}{R_2}\right) \tag{1.29}$$

Since the input impedance of the matched attenuator Z_0 is the same as the generator impedance, half of the generator voltage $V_g/2$ is dropped across the attenuator input, and voltage division shows that the voltage drop across the Z_0 load is

$$V_{load} = \frac{V_g}{2}\frac{Z_0}{R_2 + Z_0} \tag{1.30}$$

The resulting attenuation ATT computed from the power loss ratio P_A/P_{load} in (1.13) can be derived as

$$ATT = 10\log_{10}\frac{P_A}{P_{load}} = 20\log_{10}\left(1 + \frac{R_2}{Z_0}\right)\text{ dB} \Rightarrow R_2 = Z_0\left(10^{ATT/20} - 1\right) \tag{1.31}$$

Therefore, for a given level of attenuation ATT, specific values of R_2 and R_1 using (1.31) and then (1.29) can be determined. Figure 1.11 shows the values of R_1 and R_2 based on (1.29) and (1.31) for attenuations over the range 3 to 25 dB. For example, a 6 dB matched attenuator would require R_2 of 50Ω with the resulting R_1 of 100Ω.

This simple two-resistor matched attenuator of the previous example is not a reciprocal network and presents a match only to the generator; the load sees a nonmatched condition that may not be desired in some applications. For example, the previous example of the 6 dB attenuator shows a reflection coefficient seen by the load of −6.6 dB. This simple structure, however, can be the basis for a reciprocal matched attenuator if the structure is repeated with a shunt resistor on the load side and a series resistor back toward the generator. If the two series resistors are then combined, the matched Π-connected attenuator structure is created (Figure

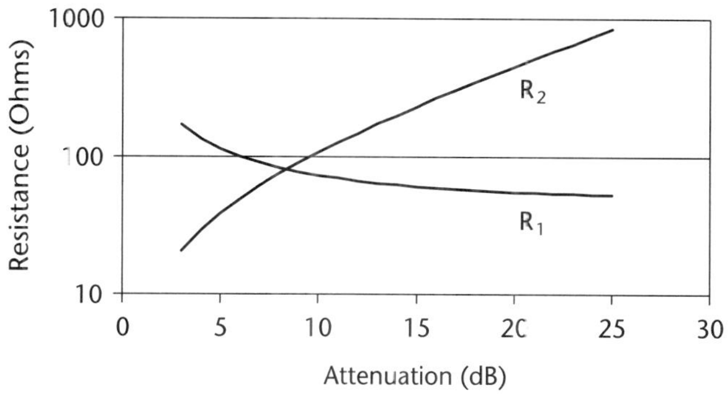

Figure 1.11 Simple two-resistor matched attenuator showing the two resistance values as a function of desired attenuation.

1.12(a)). Using a similar analysis as before, the two resistor values for a given level of attenuation and match for the Π attenuator are

$$ATT = 20\log_{10}\left(\frac{R_1 + Z_0}{R_1 + Z_0}\right)\text{dB} \Rightarrow R_1 = Z_0\frac{10^{\text{ATT}/20} + 1}{10^{\text{ATT}/20} - 1} \qquad (1.32)$$

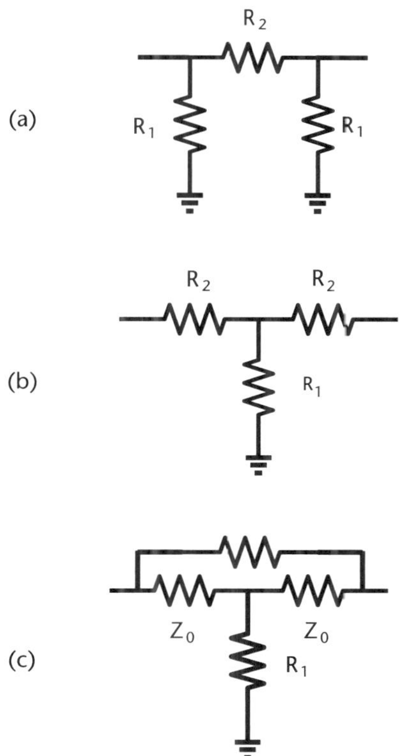

Figure 1.12 Matched attenuators: (a) Π-connected attenuator, (b) *T*-connected attenuator, (c) Bridged-*T* attenuator.

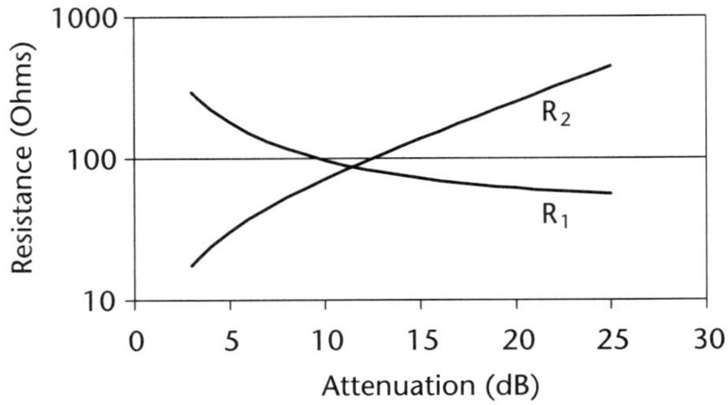

Figure 1.13 Matched Π-connected attenuator showing the two resistance values as a function of desired attenuation.

$$R_2 = \frac{2R_1Z_0^2}{R_1^2 - Z_0^2} \tag{1.33}$$

Figure 1.13 shows the range of R_1 and R_2 to provide matched attenuation over the range 0 to 30 dB. Other purely resistive matched attenuators include the T-type (Figure 1.12(b)) and the Bridged-T (Figure 1.12(c)), each with their own set of design equations. Note that the Bridged-T attenuator circuit only requires two variable resistances as compared with three in the other topologies; this reduction in the number of variable resistances will be useful in reducing component count when the variable resistances are eventually replaced with PIN diode or FET-based control elements. Table 1.1 summarizes the resistance equations for these three attenuators as a function of the desired attenuation. These and other matched attenuators will be covered in more detail in Chapters 7 and 8 on applications.

1.3.3 Phase Shifters

A number of RF and microwave applications require precise adjustment of signal phase for proper operation. A traditional example of this need for phase adjustment is in phased-array radar [26]. In a phased-array radar system, a spatially fixed number of antennas are carefully fed with well-defined phase shifts to form a tight

Table 1.1 Summary of Resistance Values for a Given Attenuation Level in Some Matched Attenuator Structures

Attenuator	R_1	R_2	*Attenuation (dB)*
Matched Π	$R_1 = Z_0\frac{10^{\text{ATT}/20}+1}{10^{\text{ATT}/20}-1}$	$R_2 = \frac{2R_1Z_0^2}{R_1^2 - Z_0^2}$	$20\log_{10}\left(\frac{R_1 + Z_0}{R_1 - Z_0}\right)$
Matched T	$R_1 = \frac{Z_0^2 - R_2^2}{2R_2}$	$R_2 = Z_0\frac{10^{\text{ATT}/20}-1}{10^{\text{ATT}/20}+1}$	$20\log_{10}\left(\frac{Z_0 + R_2}{Z_0 - R_2}\right)$
Matched Bridged-T	$R_1 = \frac{Z_0}{\left(10^{\text{ATT}/20}-1\right)}$	$R_2 = \frac{Z_0^2}{R_1}$	$20\log_{10}\left(1 + \frac{Z_0}{R_1}\right)$

beam that focuses the transmit energy into a small cone in a desired direction (RF/microwave beam forming). This directionality can be obtained without physical movement of the antenna array. The converse can also be performed under receive. Because the beam is required to sweep over a certain region of space, the phases of the signal feeds must be changed quickly and accurately, implying the need for a component that can not only be rapidly switched but also able to handle the high powers necessary in radar applications. This requirement is easily satisfied with semiconductor switching devices, with PIN diodes being more widely used because of the high powers (multi-kW) involved. Both PIN diodes and FET-based switches can be used in these systems for pulsed power levels up to several hundred Watts continuous (CW). There are two classic methods of phase shifting that will be covered in this section: loaded line or lumped element phase shifters and switched line phase shifters [26–29]. Other phase shift circuits are also possible [29–32], but the ones covered in this section are presented to provide the fundamental background needed to better understand more advanced phase shifting techniques.

1.3.3.1 Lumped or Loaded Line

Consider the shunt-connected control circuit shown previously in Figure 1.4 in which impedance Z_{CTL} is placed in shunt across a Z_0 load. Further, consider that

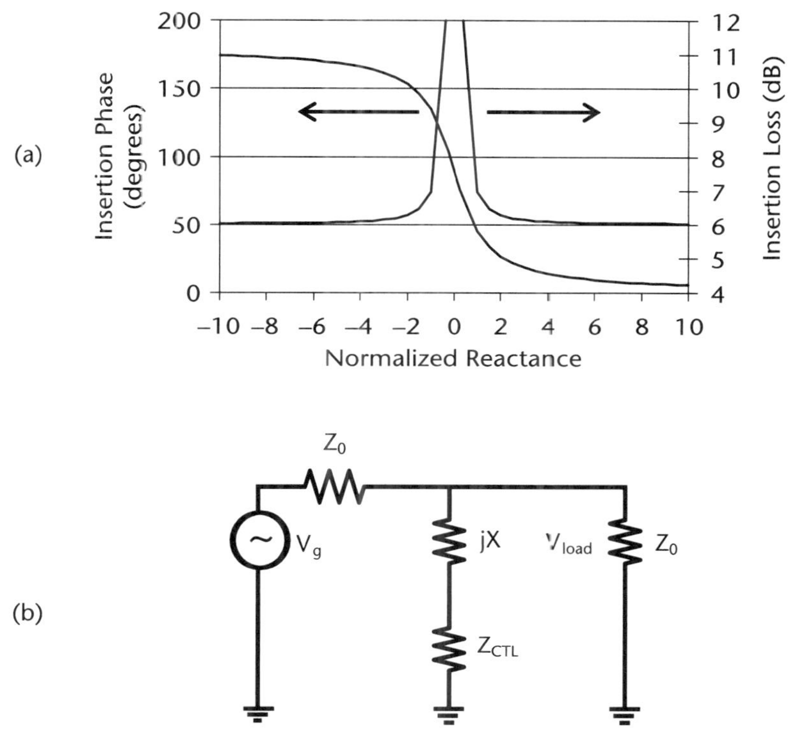

Figure 1.14 (a) Insertion loss and phase as a function of normalized reactance X/Z_0 in a shunt-connected circuit. (b) A control element Z_{CTL} in series with a shunt reactance can be used to switch X in and out of the circuit.

Z_{CTL} is a pure reactance jX. The ratio of the load to generator voltage can then be written as

$$\frac{V_{load}}{V_g} = \frac{Z_{CTL}}{2Z_{CTL} + Z_0} = \frac{jX}{2jX + Z_0} = \sqrt{\frac{X^2}{Z_0^2 + 4X^2}} e^{j\left(\frac{\pi}{2} - \tan^{-1}\left(\frac{2X}{Z_0}\right)\right)} \tag{1.34}$$

The voltage ratio in (1.34) shows both a reduction in amplitude and the introduction of phase, with a total phase shift (or insertion phase) of $\Delta\Theta = \pi/2 - \tan^{-1}(2X/Z_0)$. Figure 1.14(a) shows a plot of the insertion loss and insertion phase based on (1.34) as a function of normalized reactance X/Z_0. For small normalized reactance X/Z_0, a phase shift between the generator and load of 90° plus or minus a few degrees is observed but at the expense of a significantly attenuated signal at the load. For a large normalized reactance X/Z_0, insertion phases near zero or 180° are possible with minimal signal loss (the 6 dB insertion loss figure indicates $V_{load} = 1/2V_g$ under no loading). Placing a control element Z_{CTL} in series with the reactance (Figure 1.14(b)) suggests a method to switch the phase element jX in and out of the circuit. The phase element reactance jX can either be a lumped element or a length of transmission line; inductive or capacitance reactances can then be switched in and out of the circuit to adjust the phase shift desired. Note that insertion phases of less than 45° or between 135° and 180° have minimal impact on the insertion loss, and so an insertion phase of 45° is a typical upper limit for the use of this technique [28].

By imbedding the shunt combination of jX and Z_{CTL} in a section of Z_0 transmission line, the transmission coefficient of the line-load combination can be written as [28]

$$T = \sqrt{\frac{4X^2}{Z_0^2 + 4X^2}} e^{-j\tan^{-1}\left(\frac{Z_0}{2X}\right)}; \quad \Delta\Theta = \tan^{-1}\left(\frac{Z_0}{2X}\right) \tag{1.35}$$

with similar operating conclusions in terms of range of phase shifts associated with low levels of insertion loss.

The reflection from the shunt impedance causes the insertion loss associated with the phase shift element. If two identical shunt reactances of value jX are separated by a quarter wavelength section of Z_0 transmission line ($\lambda_g/4$), the 180° phase shift between the two reflections can partially cancel the overall circuit reflection [28]. The phase shift for this structure can be written as

$$\Delta\Theta \approx \frac{\pi}{2} + \frac{Z_0}{X} \tag{1.36}$$

for X/Z_0 small.

1.3.3.2 Switched Line

A length of transmission line, even if ideal and with zero loss, exhibits a phase difference in the voltage and current between the input and the output ports. The phase difference between these two ports is based on the ratio of the physical line length L and the wavelength in the transmission line λ_g and is given by standard transmission line theory as [28]

$$\Delta\Theta = 2\pi\frac{L}{\lambda_g} \rightarrow 360°\frac{L}{\lambda_g} \tag{1.37}$$

The switched line phase shifter takes advantage of this natural phase difference in a length of transmission line by the use of a switch matrix to switch different line length sections to create the desired input/output phase difference. Figure 1.15 shows such a switch matrix with three transmission lines of electrical lengths Θ_1, Θ_2, and Θ_3. Since the line lengths are usually a fraction of a wavelength, the transmission line loss is relatively low and the major component impacting the insertion loss is the loss introduced by the switch matrix. The phase difference then depends on which lines are selected. The input and the output switch are switched simultaneously (the so-called gang switch). The drawback of the switched line phase shifter is the relatively narrow bandwidth of the desired phase shift because of the changing electrical length of the lines as a function of frequency.

These and other types of phase shifters will be covered in more detail in Chapters 7 and 8 on applications

1.3.3.3 Other Important Circuits Used in Control

This section introduces basic circuits often seen in the RF and microwave control area. The circuits covered in this section are the quadrature hybrid, used in a number of control circuits, and several types of multiple device transmit-receive switches.

The Quadrature Hybrid

The quadrature or 3 dB hybrid (quad hybrid) circuit is used extensively in the RF and microwave field as part of more complex control circuits (Figure 1.16). The ideal matched quad hybrid is a four-port network, with two of the ports having equal amplitude through and coupled outputs (Ports 2 and 3) but a 90° phase shift between them. The fourth port (Port 4) is ideally isolated. The matched ideal quad hybrid has an S-parameter matrix given by [28]:

$$S = -\frac{1}{\sqrt{2}}\begin{bmatrix} 0 & j & 1 & 0 \\ j & 0 & 0 & 1 \\ 1 & 0 & 0 & j \\ 0 & 1 & j & 0 \end{bmatrix} \tag{1.38}$$

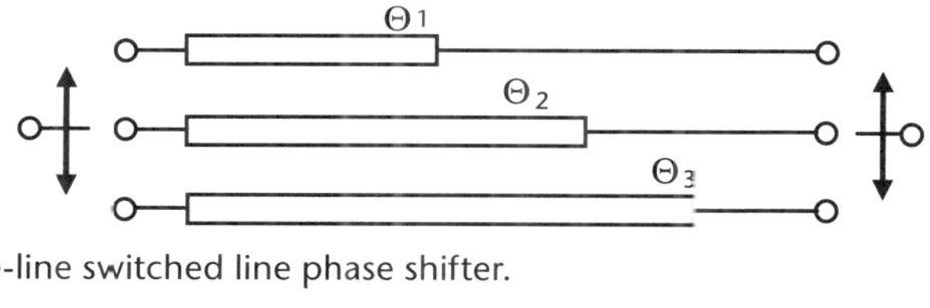

Figure 1.15 Three-line switched line phase shifter.

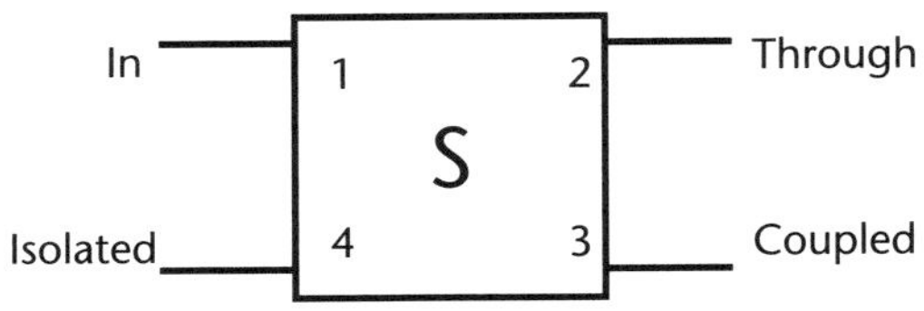

Figure 1.16 General port definitions for the four-port quadrature hybrid.

The utility of the quad hybrid in RF and microwave control can be seen if the two equi-amplitude ports (2 and 3) are terminated in impedances Z_{CTL} and the other two ports are the input and output ports (Figure 1.17). Using standard S-parameter circuit analysis on the four-port network, the output of Port 4 with respect to the input at Port 1 can be written as

$$\frac{V_4^-}{V_1^+} = j\Gamma = j\frac{Z_{CTL} - Z_0}{Z_{CTL} + Z_0} \tag{1.39}$$

where Γ is the reflection coefficient exhibited by the quad hybrid's two termination impedances. The beauty of the ideal quad hybrid-circuit is that regardless of the mismatch presented by Z_{CTL}, there is no reflected energy returning from the input back to the generator, so the circuit is matched.

The main drawback for the quad hybrid is that in actual quad hybrids, the S-parameters are frequency-sensitive and so the ideal S-matrix in (1.38) occurs only over a relatively narrow 10% bandwidth (Figure 1.19(a)). The reason for this relatively narrow bandwidth can be seen by looking at three different types of quad hybrids: distributed network, lumped element, and a combination lumped-distributed network. The classic distributed element quadrature hybrid is the branch line coupler where four $\lambda_g/4$ transmission lines of characteristic impedances Z_0 and $Z_0/\sqrt{2}$ are connected in a square configuration and each of the four ports are at the line intersections (Figure 1.18(a)) [28]. The lumped element version of the quad hybrid can be developed from the layout in Figure 1.18(a) with the knowledge that a $\lambda_g/4$ of Z_0 characteristic impedance transmission line can be replaced by a π-configured

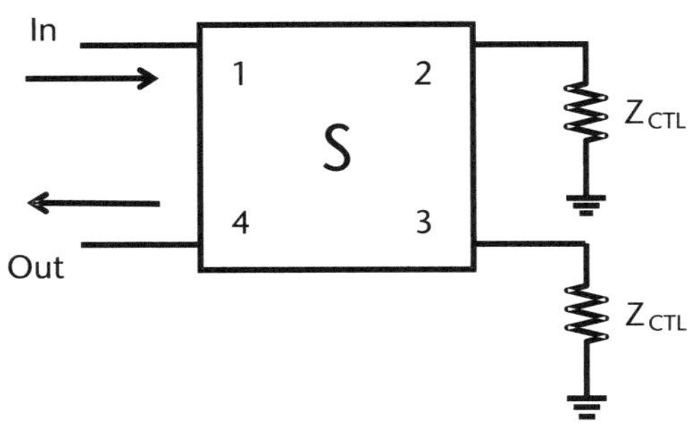

Figure 1.17 Quadrature hybrid terminated in impedance Z_{CTL}.

LC network (Figure 1.18(b)) where the inductance (L) and capacitance (C) values are functions of the center frequency F_0 and Z_0 [30–32]:

$$L_{QH} = \frac{Z_0}{2\pi F_0}; \quad C_{QH} = \frac{1}{2\pi F_0 Z_0} \tag{1.40}$$

The lumped element quad hybrid then is composed of four inductors (two each of identical value) and four capacitors (shunt capacitances at each of the four ports can be combined). This lumped element equivalent is valid for any $\lambda_g/4$ transmission line of characteristic impedance Z_0. The third circuit topology for the quad

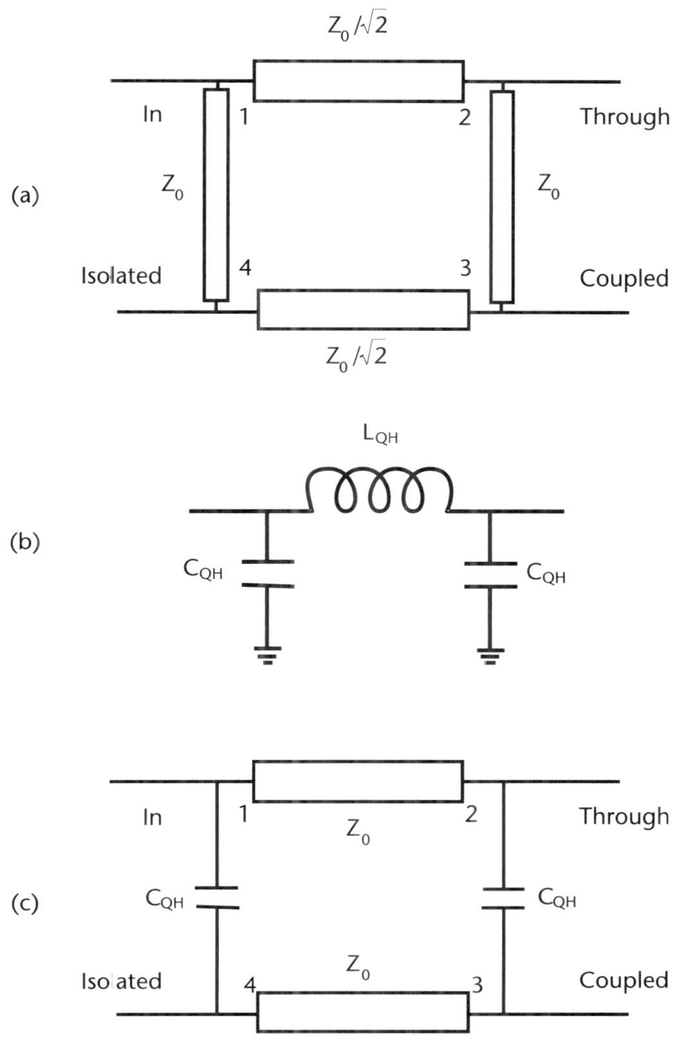

Figure 1.18 Three types of quadrature hybrids: (a) branch line coupler, (b) lumped element equivalent of the $\lambda_g/4$ line in (a), and (c) combination lumped and distributed element.

hybrid (Figure 1.18(c)) uses two $\lambda_g/8$ transmission lines of characteristic impedance Z_0 coupled by two capacitors of value $1/2\pi F_0 Z_0$. The magnitude of the frequency response of all three quad hybrids is shown in Figure 1.19(b) and assumes a center design frequency of 1000 MHz. Note that the responses for the three different circuit topologies meet at $|S_{21}| = -3$ dB at the design frequency of 1000 MHz but change away from the design frequency. The remaining S-parameters seen in Figure 1.19(b) have not been shown for the sake of clarity.

A matched attenuator can be created using a quad hybrid by varying the control element termination impedance Z_{CTL}. If Z_{CTL} is a pure resistance, R_{CTL}, then the attenuation ATT can be written as

$$ATT = 20\log_{20}\left|\frac{R_{CTL} + Z_0}{R_{CTL} - Z_0}\right| \text{ dB} \tag{1.41}$$

and an S-parameter analysis on Port 1 shows no reflected energy as long as the two loading resistances are the same. Since the two resistances have to track, the same control voltage can be used to control identical semiconductor control devices. Alternatively, different devices can be used for R_{CTL} with independent control signals, although this complicates the circuit design.

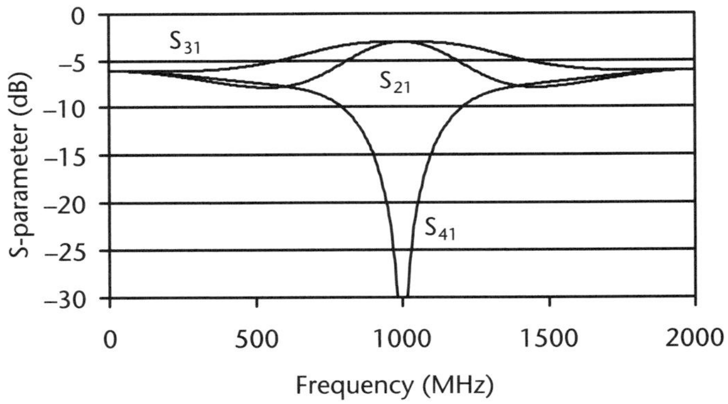

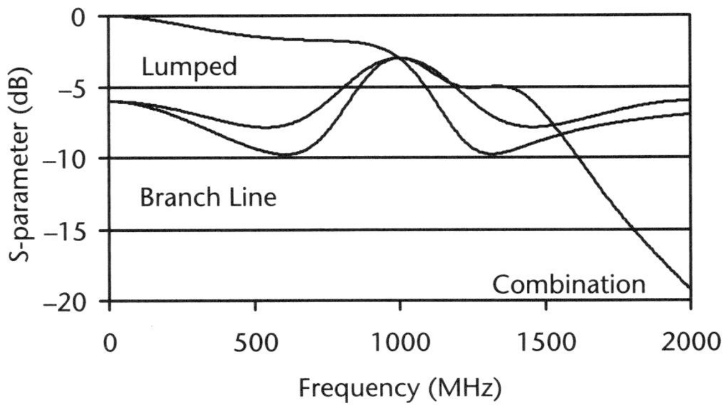

Figure 1.19 (a) S-parameters for an ideal distributed element branch line coupler design for a center frequency of 1000 MHz. (b) $|S_{21}|$ in dB for the branch line coupler and lumped and combination element equivalents.

A matched phase shifting circuit can be created using a quad hybrid by placing pure reactances jX_{CTL} at the two outputs. The reflection coefficient in this case can be written as

$$\frac{V_4^-}{V_1^+} = j\Gamma = j\frac{jX_{CTL} - Z_0}{jX_{CTL} + Z_0} = e^{j\left(\frac{\pi}{2} - 2\tan^{-1}\frac{X_{CTL}}{Z_0}\right)} \tag{1.42}$$

Placing a control component to switch the two reactances jX_{CTL} in and out of the circuit then creates a differential phase shift between the two switch states of $\Delta\Theta = 2\tan^{-1}X_{CTL}/Z_0$. Again, no reflected energy is seen at Port 1 if the two reactances track with each other. The reactance jX_{CTL} can be from the semiconductor control element itself or the control device can be used to switch a lumped or distributed element reactance in and out of the circuit.

Single-Pole, Double-Throw (SPDT) Transmit-Receive (TR) Switch

Many communications systems share an antenna between the transmitter and the receiver. The receiver is designed to take very low signals and amplify them with a minimum amount of internally generated noise. The transmitter, on the other hand, generates large signals in order to ensure the signal is sufficiently large enough at the receive end that a viable communications channel results. The typical transmitter output power is usually well above the level that can be safely handled by the receiver, and so the transmitter must be well isolated from the receiver to ensure the receiver is well protected. In addition, the transmitter should see a low insertion loss path to the antenna so that the majority of the signal is radiated and a minimum amount of energy is reflected by or dissipated in the control device. During receive, the receiver should see a low insertion loss path to the antenna as well and be well isolated from the transmitter circuit so that any noise or loading by the transmitter circuitry is minimized. The simplest switching structure, the single-pole double-throw (SPDT) switch, is composed of two series-connected low insertion loss, high isolation control components: one to make the transmitter-antenna connection, the other to make the receiver-antenna connection (Figure 1.20(a)). An alternative circuit structure that often reduces extra circuitry needed for biasing the control elements into their appropriate control state is the shunt-connection switch, as shown in Figure 1.20(b). The addition of the $\lambda_g/4$ lines are to ensure that the low impedance state of Z_{CTL} is transformed in the complementary switch arm to present a high impedance at the antenna connection node.

The isolation in the simple SPDT TR switches of Figure 1.20 can be further improved in the series-connected case by placing two more control devices in shunt across the output terminals and controlling them complementary to the series elements. This complementary switching is indicated in Figure 1.21 by $Z_{CTL\text{-}i}$, where the subscript "i" denotes the same state on transmit (T) or receive (R).

In Figure 1.21, when the transmitter is connected to the antenna, devices $Z_{CTL\text{-}T}$ are in a low impedance state and devices $Z_{CTL\text{-}R}$ are in a high impedance state. The receiver is then isolated by the series high impedance device, but any leakage energy is reflected by the low impedance shunt device, providing further receiver protection. The reverse is true during the receive state.

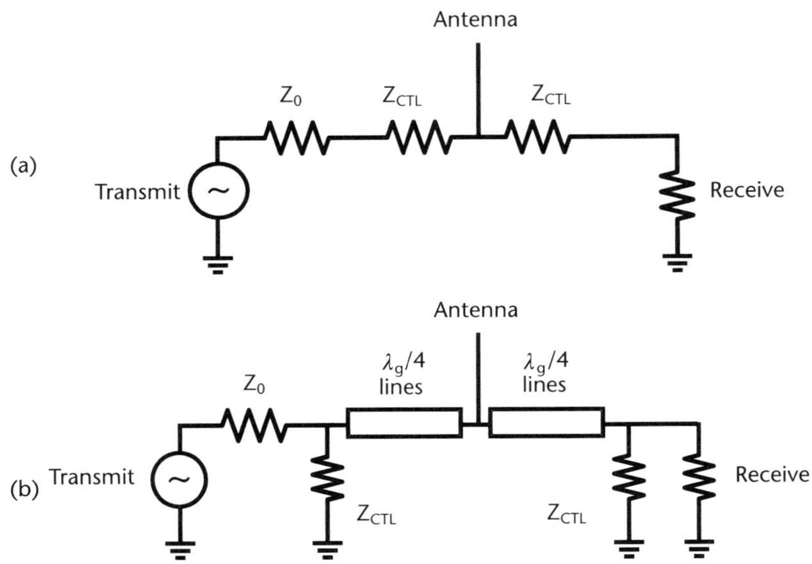

Figure 1.20 (a) Series and (b) shunt connected SPDT TR switches using two control elements.

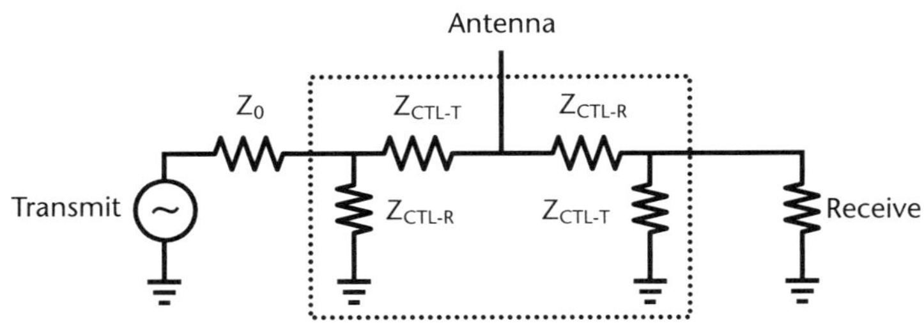

Figure 1.21 SPDT TR switch with improved isolation.

1.4 Noise

From the circuits shown in the previous sections, the RF and microwave control element will always be in the RF path, and any dissipative (resistive) losses can introduce noise into the system. For the RF and microwave circuit, the on-state switch is the more important of the two states because a dissipative network is placed directly inline with the system. In attenuators and in the off-state, noise introduced by the control element is usually not the main concern, although determining the control device's noise contributions is similar to those in the on-state if the noise in these module states needs to be considered. This section reviews some of the fundamental properties of noise in systems, how the control element contributes to overall system noise figure, and the computation of overall noise properties of a cascade system.

1.4.1 Resistive Noise Model

The random motion of charge carriers in any circuit causes minute instantaneous fluctuations in the current flow in any current-carrying structure, even in dc supplies.

In a resistor of value R, these current fluctuations Δi occur around the mean current I. The current fluctuations, in turn, create voltage fluctuations across R and give rise to so-called thermal resistive noise or just resistive noise. The noise power has been shown to be linearly dependent on the physical temperature T of the resistor [28]:

$$P = \alpha T = \Delta\overline{i}^2 R; \quad \Delta\overline{i}^2 = \alpha T / R = \alpha G T \tag{1.43}$$

where Δi is the rms value of the current fluctuations, and α is a function of the noise measurement. The mean square noise current can then be written in its traditional form as:

$$\Delta\overline{i}^2 = 4kTGB = 4kTB/R \tag{1.44}$$

where B is the bandwidth and k is the Boltzmann constant. The noise current form $\Delta\overline{i}^2$ is useful in circuit and system noise analysis because it provides an easy method of injecting a noise current into a circuit and calculating the circuit response. Traditional circuit theory can be used if $\Delta\overline{i} = \sqrt{\Delta\overline{i}^2}$ is assumed to be an ideal current source with a noiseless source resistance R (Figure 1.22(a)). Assuming that the noise source is driving a matched load of noiseless resistor R (Figure 1.22(b)), the available noise power (in Watts) delivered to noiseless load R is

$$P_N = \left(\frac{\Delta\overline{i}}{2}\right)^2 R = kTB \tag{1.45}$$

(a)

(b)

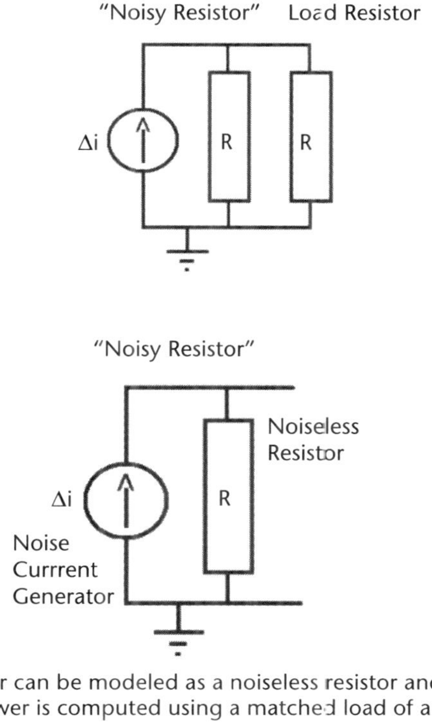

Figure 1.22 (a) A noisy resistor can be modeled as a noiseless resistor and a noise current generator. (b) The available power is computed using a matched load of a noiseless resistor of value R (from [33]).

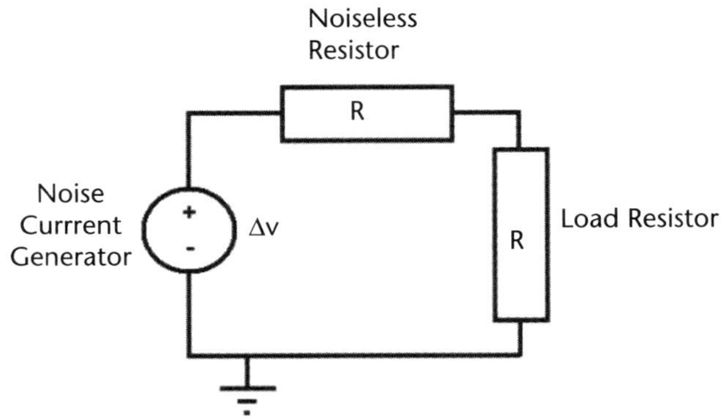

Figure 1.23 Thévenin equivalent noise circuit (from [33]).

A Thévenin equivalent circuit for the noise source can also be drawn as shown in Figure 1.23, where $\Delta\overline{v}^2 = 4kTRB$ and the available noise power $P_N = kTB$ is the same as before.

The noise power is independent of the value of resistance and so can be used to describe any element or system. In the general case, the noise temperature T is termed an equivalent noise temperature T_e of the element or system, and the IEEE standard noise reference temperature of 290K is defined with the symbol T_{ref}:

$$P_N = kT_eB; \;\; P_{\text{ref}} = kT_{\text{ref}}B \tag{1.46}$$

1.4.2 Noise Figure Model

Any electronic circuit, active or passive, will generate its own noise, and so a signal with noise at the input of the circuit will find its way to the output, but with additional noise due to that generated by the circuit itself. This effect is illustrated in Figure 1.24, where the desired signal P_{sig} and input circuit noise $P_{N\text{-in}}$ are applied to a gain stage (such as a low noise amplifier, LNA, at the front end of a receiver circuit) with gain G. These two signals, the desired signal and the noise, are both amplified, so the outputs of the amplifier are the amplified signal and noise, GP_{sig} and $GP_{N\text{-in}}$, respectively. The amplifier also has its own self-generated noise, P_{amp}. The input noise $P_{N\text{-in}}$ appears at the input, so for ease of calculation, the amplifier noise P_{amp} is assumed to be at the input as well, regardless of the actual origin of the noise in the amplifier; this is called the input-referred noise P_{amp} of the gain stage. With this assumption, the total noise power at the output can be calculated as the sum of the two noise powers at the input multiplied by the system gain: $G[P_{N\text{-in}} + P_{\text{amp}}]$.

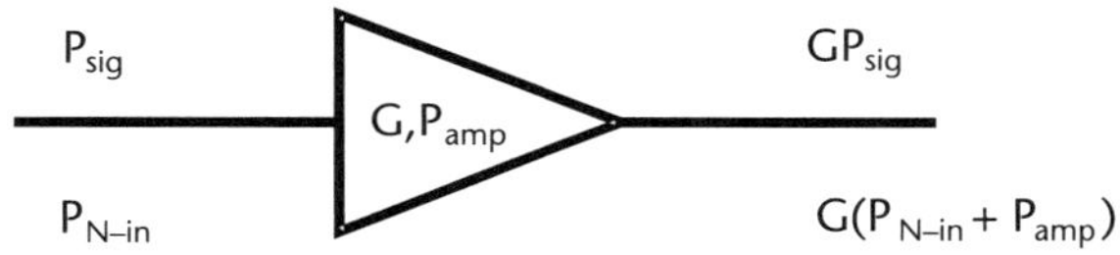

Figure 1.24 The amplifier multiplies both the signal and the noise by the gain and also adds self-generated noise (from [33]).

With the addition of the amplifier noise contribution, two signal-to-noise (SNR) values can be computed, one at the input (SNR_{in}) and the other at the output (SNR_{out}):

$$SNR_{in} = \frac{P_{sig}}{P_{N-in}};\quad SNR_{out} = \frac{GP_{sig}}{G\left[P_{N-in} + P_{amp}\right]} = \frac{P_{sig}}{\left[P_{N-in} + P_{amp}\right]} \tag{1.47}$$

From (1.47), SNR_{out} can be written in terms of SNR_{in} by rearranging terms and introducing a new term NF, called the noise figure:

$$SNR_{out} = \frac{P_{sig}}{\left[P_{N-in} + P_{amp}\right]} = \frac{P_{sig}/P_{N-in}}{1 + P_{amp}/P_{N-in}} = \frac{SNR_{in}}{NF};\quad NF = 1 + \frac{T_{amp}}{T_{N-in}} \tag{1.48}$$

where the noise powers have been replaced by their equivalent noise temperatures, T_{amp} and $T_{N\text{-}in}$, the amplifier equivalent noise temperature and input noise temperature, respectively. Further assuming that the input noise conforms to the IEEE standard allows NF to be written as

$$NF = 1 + \frac{T_{amp}}{T_{ref}} \tag{1.49}$$

where T_{ref} is the input or standard noise temperature. NF is always greater than unity and shows that the output SNR is always degraded be the noise figure of the system, irrespective of the gain G. The noise figure NF may also be placed in dB terms, which shows the degradation of SNR in terms of dB:

$$NF_{dB} = 10\log(NF);\quad SNR_{out\text{-}dB} = SNR_{in\text{-}db} - NF_{dB} \tag{1.50}$$

The control circuits of interest here are of course different than amplifiers and have losses associated with them, so instead of a gain G in the previous amplifier example, the circuit will exhibit a loss $L = 1/G$. The output noise power for a lossy element can be written for the system at temperature T as:

$$P_{out} = G\left(kTB + P_{N-loss}\right) \tag{1.51}$$

where $P_{N\text{-loss}}$ is the input-referred noise power of the lossy element. P_{out}, however, is kTB, and so $P_{N\text{-loss}}$ and its associated equivalent noise temperature can be written as

$$P_{N-loss} = \frac{1-G}{G}kTB = (L-1)kTB \tag{1.52a}$$

or

$$T_e = \frac{1-G}{G}T = (L-1)T \tag{1.52b}$$

with a corresponding noise figure NF of [28]:

$$NF = 1 + (L - 1)\frac{T}{T_{\text{ref}}} \tag{1.52c}$$

If the system is at temperature T_{ref}, then $NF = L$, and the noise figure NF is identical to the circuit loss L.

1.4.3 Cascade System Noise

Modern systems are not limited to a single stage or circuit, but rather have a number of stages, each providing gain (or loss) and their own noise contributions. In each stage, the stage block is defined with a power gain G_i (which could be a loss in the case of a control module) and an equivalent input referred noise temperature T_i (Figure 1.25).

A thermal noise source is placed at the input, set to the reference noise temperature T_{ref}. By tracking the input noise contributions at each input and multiplying them by the gain of each amplifier stage, the noise power output at the last state can be written as

$$P_{N-3} = kBG_1G_2G_3\left(T_1 + T_{\text{ref}}\right) + kBG_2G_3T_2 + kBG_3T_3 \tag{1.53}$$

The output noise power P_{N-3} can be written in terms of an equivalent noise temperature T_e at the input of a single stage amplifier with gain $G_1G_2G_3$, so that P_{N-3} in this single amplifier-equivalent can be written as

$$P_{N-3} = kBG_1G_2G_3\left(T_e + T_{\text{ref}}\right) \tag{1.54}$$

Equating (1.53) and (1.54) yields the equivalent noise temperature and noise figure NF for the entire system [34]:

$$T_e = T_1 + \frac{T_2}{G_1} + \frac{T_3}{G_1G_2}; \quad NF = 1 + \frac{T_e}{T_{\text{ref}}} = 1 + \frac{T_1}{T_{\text{ref}}} + \frac{T_2}{T_{\text{ref}}}\frac{1}{G_1} + \frac{T_3}{T_{\text{ref}}}\frac{1}{G_1G_2} \tag{1.55}$$

Equation 1.55 can be extended to any number of stages in a similar manner. Regardless of the total number of system stages, it is the first stage and its noise properties that have the greatest impact on the overall system noise. Because the RF

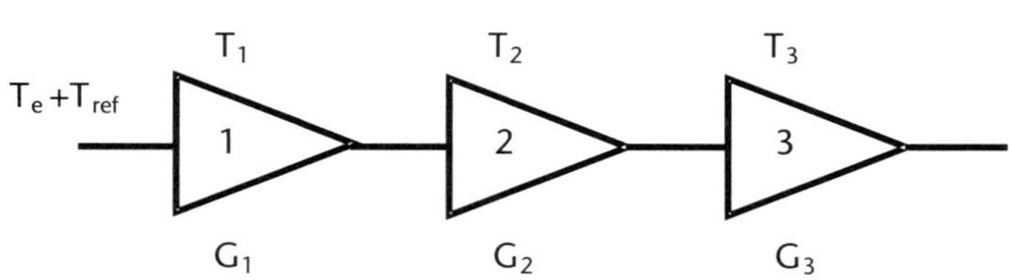

Figure 1.25 System cascade with gains and equivalent noise temperatures for cascade noise figure computation (from [33]).

and microwave control element is often placed immediately after the antenna and before the active components of the system, its loss (with noise properties described by its equivalent noise temperature) will add to the *NF* of the overall system. Using (1.52c) and (1.55) for a two-stage system with the control module as Stage 1 provides the information needed to compute the equivalent noise temperature and NF:

$$T_e = (L_1 - 1)T + L_1T_2; \quad NF = 1 + \frac{(L_1 - 1)T + L_1T_2}{T_{\text{ref}}} = 1 + \frac{(L_1 - 1)T}{T_{\text{ref}}} + L_1\frac{T_2}{T_{\text{ref}}} \tag{1.56}$$

where L is the loss of the control element. The impact of a control element on the overall system noise figure can be better seen with an example. Assume that the LNA at the front end of a system has a gain of 10 dB and an equivalent noise temperature of 50K (NF = 1.17 or 0.69 dB). An RF control module with 0.5 dB insertion loss (L = 1.122) held at temperature T_{ref} would exhibit a system equivalent noise temperature nearly double that of the amplifier of 91.48K and a noise figure of 1.19 dB:

$$T_e = (L_1 - 1)T + L_1T_2 = (1.122 - 1)290 + (1.122)50 = 91.48\text{K}$$

$$NF = 1 + \frac{(L_1 - 1)T + L_1T_2}{T_{\text{ref}}} = 1 + \frac{(1.122 - 1)290}{290} + 1.122\frac{50}{290} = 1.315 \Rightarrow 1.19\text{dB}$$

with a corresponding overall gain of the two stages of 9.5 dB.

1.5 Control Elements

In Sections 1.2 and 1.3, the RF and microwave control element was represented by a general impedance value, Z_{CTL}. This impedance value was assumed to be controlled with some circuit or system that toggles the control element between its high impedance and low impedance states. In a particular control device state, the ideal equivalent circuit in Figure 1.1 is used. This discussion was completely general, and Z_{CTL} could be an electromagnetic relay, a ferrite switch, a mechanical switch, or a semiconductor device. The focus of this book is the modeling and use of PIN diode and FET-based elements for use in RF and microwave and control applications. Micro-electrical-mechanical systems (MEMS), a concept introduced in the early 1960s [35–36] but with major technology investments introduced in the mid-1980s, are specifically not covered in this book as they operate on different physical and electrical principles than PIN diodes and FET-based control elements. Excellent references exist for the study of MEMS technology, and the reader interested in this technology is strongly encouraged to seek out these (and many other) references for more detailed information (for example, see [37–40]).

1.5.1 PIN Diode Control Elements

The PIN diode, first proposed for RF and microwave use by Uhlir [10], is a two-terminal device that exhibits the properties of dc current-controlled RF resistance when forward biased and small fixed capacitance when dc reverse biased. The term PIN comes from the layered structure of the diode: a thin intrinsic or highly resistive semiconductor layer (I-region) sandwiched between heavily doped p-type (P) and n-type (N) semiconductors. The equivalent circuit in Figure 1.1 is used to model this control behavior with a dc forward bias or on-state resistance R_{ON} and a reverse bias or off-state capacitance C_{OFF}. A common assumption used to define the switch cutoff frequency F_C is that $R_{ON} = R_{OFF}$ and so $F_C = 1/2\pi R_{ON}C_{OFF}$. This assumption is not entirely accurate because the on-state resistance is governed by the change in the I-region conduction properties by the dc forward current and in the off-state, the resistance is governed by contact and residual resistance in the diode. However, this definition is widely used and so is mentioned in the interest of better understanding the definitions used by control circuit designers.

A significant issue that will not be addressed in detail in the book but is also worth mentioning is dc biasing of the PIN diode. Being a two-terminal device, dc bias has to be introduced at terminals also in the RF path. The dc bias is usually introduced from an external circuit and so must be isolated from the RF path to prevent undesired RF loading by the dc source. One or more inductors and capacitors are used to decouple the dc bias source for controlling the PIN diode's state from the main RF and microwave path. These LC networks provide excellent power supply decoupling, but because they are reactive elements, the bandwidth of the decoupling circuit and therefore of the entire module can be limited. Large values of inductance and capacitance can also increase the time it takes to toggle between the two control device states, negatively impacting the control module's response time. Fortunately, a number of standard dc bias circuits and PIN diode biasing modules are available to best optimize the module response time.

1.5.2 FET-Based Control Elements

FET operation was first postulated and patented almost 100 years ago by Lilienfeld [41–42] as a three-terminal device, where current flow between two terminals was controlled by a voltage on a third terminal placed between the other two. At the time, the technology did not exist to fully implement these concepts; practical implementation was delayed until decades later. For the RF and microwave engineer, the two most important FETs are the metal-semiconductor FET (MESFET) and the metal-oxide-semiconductor FET (MOSFET). Although the physical operation of the two devices is different, they both rely on conduction between the drain (D) and source (S) nodes through a channel region that is controlled by a voltage on the gate (G) terminal. When a conducting channel exists between drain and source, the RF current flows through a low resistance path. When the conducting channel is suppressed, the only RF path that exists is through the FET capacitances. Based on this high-level definition of FET operation, the equivalent circuit in Figure 1.1 is again used to model this control behavior. The conducting device state is described in the equivalent circuit by the on-state resistance R_{ON} and the suppressed channel operating region by the off-state capacitance C_{OFF}. As in the case of the PIN diode, a common assumption used defining the switch cutoff frequency F_C is that $R_{ON} = R_{OFF}$ and so $F_C = 1/2\pi R_{ON}C_{OFF}$. This assumption is not entirely accurate for FET-based control devices either since the on-state resistance is governed by the channel conducting conduction properties, and in the off-state the resistance is governed by contact and residual resistance in the FET. However, as in the case of the PIN diode, this definition has been historically used and still is widely used by control circuit designers.

The third terminal in FET-based control devices simplifies the dc control decoupling circuit dramatically compared to that in PIN diode circuits. The third terminal (gate) is not directly in the RF path, and so the sometimes elaborate decoupling networks required in PIN diode circuits are often replaced by a single resistor of a few kΩ between the gate terminal and the control signal. Better isolation can be achieved with larger resistance values but at the expense of longer response times due to large RC values that can limit overall switching speed. Both the MESFET and the MOSFET have been successfully integrated with entire systems to create SoC microwave integrated circuits, but they are also available in discrete control module form. Of special interest is the MOSFET control device because of its ability to be easily integrated with other analog and digital circuits using standard low-cost CMOS digital fabrication technologies.

1.6 Additional Information

One of the key factors of a successful RF and microwave control circuit design (or any design for that matter) is the ability to accurately model the entire circuit's operation. Over the last several decades, a wide variety of computer-aided design (CAD) tools have been developed that have made the design engineer's predictions of actual circuit behavior increasingly accurate. There are many different CAD tools available for the design engineer. At the time of this book's writing, commercial CAD tools such as Keysight Technologies Advanced Design System (ADS), National Instruments Microwave Office, and Sonnett Software's High Frequency Electromagnetics Software, to name just a few, are widely used throughout the RF and microwave industry. For strict circuit simulation, freely available CAD tools based on the industry-standard SPICE (Simulation Program with Integrated Circuit Emphasis [43–44]) circuit simulator, such as Linear Technology's LT-SPICE, also find favor with design engineers. For the RF and microwave engineer charged with design at the integrated circuit level, SPICE simulations are crucial for verifying circuit operation prior to committing the financial resources required to fabricate a prototype. For mathematical modeling, Matlab from Mathworks and PTC's Mathcad are often employed.

Part of the RF and microwave engineer's development of understanding device and circuit operation is to employ these CAD tools in testing different circuit and device parameters and to observe their impact on circuit and module performance. Keeping with this need for RF and microwave engineers to "play computer games" as part of the circuit and systems development process, the book contains references to various Matlab scripts and SPICE netlists that can be used by the reader to develop further insight into device and circuit operation. These resources will have the usual "filename.m" structure for Matlab scripts and "filename.net" for SPICE circuit netlists. The SPICE version used for these simulations is LT-SPICE, which, at the time of this writing, is available at the Linear Technology website. Placing these software resources in a companion compact disk (CD) with the book results in a static "snapshot" of the tools at the time of publication. The author, with the gracious agreement of Artech House, has placed these resources on the web at a SourceForge repository:

http://sourceforge.net/projects/pindiodemodel/files/

The author welcomes reports of errors found in these resources, and with this dynamic resource repository, changes can be made quickly.

References

[1] Sengupta, D. L., T. K. Sarkar, and D. Sen, "Centennial of the Semiconductor Diode Detector," *IEEE Proc.*, Vol. 86, No. 1, January 1998, pp. 235–243.

[2] Ramsay, J. F., "Microwave Antenna and Waveguide Techniques Before 1900," *IRE Proc.*, Vol. 46, February 1958, pp. 405–415.

[3] Wiltse, J. C., "History of Millimeter and Sub-Millimeter Waves," *IRE Trans. on Microwave Theory and Techniques*, Vol. MTT-32, September 1984, pp. 1118–1127.

[4] Bose, J. C., "Detector for Electrical Disturbances," U.S. Patent 755 840, March 29, 1904.

[5] Coale, F. S., "A Switch-Detector Circuit," *IRE Trans. on Microwave Theory and Techniques*, Vol. 3, No. 6, December 1955, pp. 59–62.

[6] Armistead, M. A., E. G. Spencer, and R. D. Hatcher, "Microwave Semiconductor Switch," *IRE Proc.*, Vol. 44, December 1956, p. 1875.

[7] Garver, R. V., J. A. Rosado, and E. F. Turner, "Theory of the Germanium Diode Microwave Switch," *IRE Trans. on Microwave Theory and Techniques*, January 1960, pp. 108–111.

[8] Garver, R. V., "High-Speed Microwave Switching of Semiconductors—II," *IRE Trans. on Microwave Theory and Techniques*, Vol. 7, No. 2, April 1959, pp. 272–276.

[9] Garver, R. V., "Microwave Semiconductor Control Devices," *IEEE Trans. on Microwave Theory and Techniques*, Vol. 27, No. 5, May 1979, pp. 523–529.

[10] Uhlir, Jr., A. "The Potential of Semiconductor Diodes in High Frequency Communications," *IRE Proc.*, June 1958, pp. 1099–1115.

[11] Hall, R. N., "Power Rectifiers and Transistors," *IRE Proc.*, vol. 40, November 1952, pp. 1512–1518 [see also M. B. Prince, "Diffused p-n Junction Silicon Rectifiers," *Bell Sys. Tech. J.*, Vol. 35, May 1956, pp. 661–684.].

[12] Watson, H. A., ed., *Microwave Semiconductor Devices and Their Circuit Applications*, New York: McGraw-Hill 1969.

[13] White, J. F., *Semiconductor Control*, Dedham, MA: Artech House, 1977 [see also by the same author, *Microwave Semiconductor Engineering*, Van Nostrand Reinhold, New York, 1982].

[14] Pucel, R. A., "Looking Back at Monolithic Microwave Integrated Circuits," *IEEE Microwave Mag.*, Vol. 13, No. 4, May–June 2012, pp. 62–76.

[15] Liechti, C. A., "Microwave Field-Effect Transistors—1976," *IEEE Trans. on Microwave Theory and Techniques*, Vol. 24, No. 6, June 1976, pp. 279, 300.

[16] Ayasli, Y., R. A. Pucel, J. A. Vorhaus, and W. Fabian, "A Monolithic X-Band Single-Pole, Double-Throw Bidirectional GaAs FET Switch," *Proc. IEEE GaAs IC Symp.*, 1980.

[17] Ayasli, Y., "Microwave Switching with GaAs FETs," *Microwave J.*, Vol. 25, November 1982, pp. 61–74.

[18] Kurokawa, K., and W. O. Schlosser, "Quality Factor of Switching Diodes for Digital Modulators," *IEEE Proc.*, Vol. 38, January 1970, pp. 180–181.

[19] Hines, M. E., "Fundamental Limitations in RF Switching and Phase Shifting Using Semiconductor Diodes," *IEEE Proc.*, Vol. 52, June 1964, pp. 697–708.

[20] Atwater, H. A., and R. W. Sudbury, "Use of Switching Q in the Design of FET Microwave Switches," *1981 IEEE MTT-S Int. Microwave Symp. Dig.*, Vol. 1, June 1981, pp. 370–372.

[21] Gopinath, A., and J. B. Rankin, "GaAs FET RF Switches," *IEEE Trans. on Electron Devices*, Vol. 32, No. 7, July 1985, pp. 1272–1278.

[22] "IRE Standards on Antennas and Waveguides: Definitions of Terms, 1953," *IRE Proc.*, Vol. 41, No. 12, December 1953, pp. 1721–1728.

[23] R. W. Beatty, "Insertion Loss Concepts," *IEEE Proc.*, Vol. 52, No. 6, June 1964, pp. 663, 671.

[24] http://dictionary.reference.com/browse/isolation.

[25] http://dictionary.reference.com/browse/attenuate.

[26] Mailloux, R. J., "Phased Array Theory and Technology," *IEEE Proc.*, Vol. 70, No. 3, March 1982, pp. 246, 291.

[27] Ulaby, F. T., R. K. Moore, and A. K. Fung, *Microwave Remote Sensing: Active and Passive, Volume I: Fundamentals and Radiometry*, Addison-Wesley, Reading, MA: Addison-Wesley, 1981.

[28] Pozar, D., *Microwave Engineering*, 2nd ed., New York: John Wiley, 1998.

[29] Skolnik, M. I., *RADAR Handbook*, New York: McGraw-Hill, 1970.

[30] Application Note 922: Applications of PIN Diodes, Hewlett-Packard, 1997.

[31] Application Note: Design with PIN Diodes, Skyworks Solutions Inc., 2009.

[32] *The PIN Diode Designer's Handbook*, Microsemi Corp., 2007.

[33] R. Caverly, *CMOS RFIC Design Principles*, Norwood, MA: Artech House, 2006.

[34] Friis, H. T., "Noise Figures of Radio Receivers," *IRE Proc.*, Vol. 32, No. 7, July 1944, pp. 419–422.

[35] Kabell, L. J., Semiconductor Beam Strain Gauge, U.S. Patent 3161844 A, Filing date December 5, 1961, Publication date December 15, 1964.

[36] Mason, W. P., Stress Sensing Semiconductive Devices, U.S. Patent 3277698 A, Filing date November 15, 1963, Publication date October 11, 1966.

[37] De Los Santos, H. J., *RF MEMS Circuit Design for Wireless Applications*, Norwood, MA: Artech House, 2001.

[38] Maluf, N., and K. Williams, *An Introduction to Microelectromechanical Systems Engineering*, 2nd ed., Norwood, MA: Artech House, 2003.

[39] Nguyen, C. T.-C., "MEMS-Based RF Channel Selection for True Software-Defined Cognitive Radio and Low-Power Sensor Communications," *IEEE Communications Magazine*, Vol. 51, No. 4, April 2013, pp. 110–119.

[40] Rebeiz, G. M., C. D. Patel, S. K. Han, K. M. J. Chih-Hsiang Ko, "The Search for a Reliable MEMS Switch," *IEEE Microwave Magazine*, Vol. 14, No. 1, January–February 2013, pp. 57–67.

[41] Lilienfeld, J. E., Method and Apparatus for Controlling Electric Current, U.S. Patent 1745175 A, Filed Date October 8, 1926, Publication Date January 28, 1930.

[42] Lilienfeld, J. E., Device for Controlling Electric Current, U.S. Patent 1900018 A, Filed Date March 29, 1928, Publication Date March 7, 1933.

[43] Nagel, L. W, and D. O. Pederson, SPICE (Simulation Program with Integrated Circuit Emphasis), *Memorandum No. ERL-M382*, University of California, Berkeley, April 1973.

[44] Nagel, L. W., SPICE2: A Computer Program to Simulate Semiconductor Circuits, *Memorandum No. ERL-M520*, University of California, Berkeley, May 1975.

CHAPTER 2

Nonideal Device Behavior in Control Circuits

Chapter 1 introduced the characteristics of RF control circuits using idealized impedances for the circuit elements. The expressions supplied in Chapter 1 provide a good first-pass attempt at design or analysis of RF and microwave control circuits, but they do not tell the entire story of semiconductor control device behavior and its influence on the overall control circuit and attached system. Chapters 3 through 6 of the book provide a detailed look at modeling some of these nonideal behaviors in PIN diodes and FET-based RF and microwave control circuits, but to fully understand these device behaviors, a review of several important types of nonideal operation will help in better understanding the material in these later chapters. The chapter begins with a discussion of the origin of nonideal or parasitic behavior of semiconductor control modules when packaged or used in fully integrated SoC applications. The origin of these parasitics is based on the physical characteristics of the connections and package (for packaged devices) and can be modeled using either lumped or distributed elements. Thermal and nonlinear behaviors are also covered with examples to help illustrate these concepts. Thermal resistance and thermal time constant are discussed, and an example on computing temperature rise is provided. Nonlinear behavior is analyzed mathematically, with the results of the analysis being used to define the distortion intercept points, a specification often seen on manufacturers' datasheets.

2.1 Control Device Parasitics

The overall frequency response of the microwave and RF switching and control module is governed not only by the semiconductor device itself but also the package, since the package is part of the RF path. Depending on the package or semiconductor device, the package itself can ultimately influence the final module bandwidth, and so being able to model the package is an important part of the overall modeling picture. Even if the RF switching device is not packaged but is fully integrated as part of a SoC system, on-chip interconnections between the intrinsic control circuit and the remaining on-chip system components need to be modeled as well. The first part of this section will look at the parasitics of packaged devices, and the second part of this section will focus on these on-chip interconnections.

2.1.1 Device Packages

There are many different types of packages used to house and protect the various microwave and RF control components, with a wide variety of abbreviations and layouts for attaching to circuits and transmission lines. For surface mounting, there are small outline integrated circuit (SOIC) packages such as SOP (small outline package) and SOT (small outline transistor), for example, and a variety of other definitions used by various manufacturers that create an acronym-rich array of packages styles and types, each with their own advantages in terms of RF and thermal performance. Plastic, glass axial (for two-terminal control components such as the PIN diode) and ceramic packages are available, with the plastic package frequently chosen for low-cost and high-volume RF and microwave control modules; ceramic packages are used for very high frequencies because of their lower loss. However, all these package types, regardless of their abbreviations, share a number of characteristics that will be reviewed in this section.

At the center of the RF and microwave control module package is a platform or paddle where the semiconductor control component die is attached. This platform is usually metallic and can be extended through the bottom of the package, helping to dissipate any heat produced by the control component as well as providing a sturdy mechanical base. Depending on how much heat is to be dissipated, the platform may have a connection to the exterior of the package for external heat sinking. The die is attached to the paddle with an adhesive that promotes good thermal conductivity between the semiconductor die and the paddle. This adhesive may also be electrically conductive to improve module grounding. Close to the platform are package pads or leads that go through the package wall and provide the pin connections to the package exterior. Bond wires are thermally and/or ultrasonically attached between the internal connections to the package pads and the two heavily doped end regions (anode and cathode) of the PIN diode or to the gate/drain/source nodes of FETs, and are the interface between the "internal world" of the RF control component and the "external world" of the circuit or system input and output. Once the semiconductor die has been attached to the platform and the bond wires connected to the package, the interior of the package may be filled with an encapsulant material that seals the package and freezes the bond wires in place. Figure 2.1 shows several examples of PIN diode packages, with Figure 2.1(b) showing a package style with a metallic heat dissipation tab spanning the underside of the package. The central tab can be soldered or thermally attached with heat sinking compound to a small metallic tab on the board to better dissipate the heat generated by the semiconductor device.

These connections introduce a number of parasitic passive elements in series and shunt with the intrinsic RF control device and can cause a degradation of RF and microwave performance of the entire control module. This degradation can go all the way down to dc due to series resistance (R_{pack}) inherent in the metal-semiconductor interfaces as well as the bond wires, the inductance of the bonding wires themselves, and various highly conducting regions separated by short distances that give rise to shunt and coupling capacitance. This section will discuss these parasitic components.

(a)

(b)

Figure 2.1 Examples of PIN diode packages: (a) 0402 PIN diode package showing anode and cathode connections. (b) PIN diode package showing anode and cathode connections (ends) as well as bottom-side metal pad for heat conduction and dissipation away from the internal semiconductor die. (*Scurce:* Cobham Metelics, 2015. Used with permission.)

2.1.1.1 Bond Wire Inductance

Bond wires are used to connect the package pins or platform to the semiconductor die (Figure 2.2). The bond wires are gold or gold-plated to reduce the series resistance R_{pack} as much as possible [1]. A general rule of thumb in computing the inductance of the bond wire is 1 nH per millimeter of bonding wire length [2]. There is a weak dependence of inductance with bond wire diameter, with the inductance slightly decreasing as the bond wire diameter increases. Multiple parallel bond wires are often used in packages because the number of parallel wires used reduces the overall inductance of the bond; another positive effect of the additional wires is an increase in the current-carrying capability (Figure 2.2(b)). Bond wires are required for all connections, and so there may be several in the RF control path (Figure 2.3 shows the mask level layout of a SPDT MOSFET integrated switch with RF

(a)

(b)

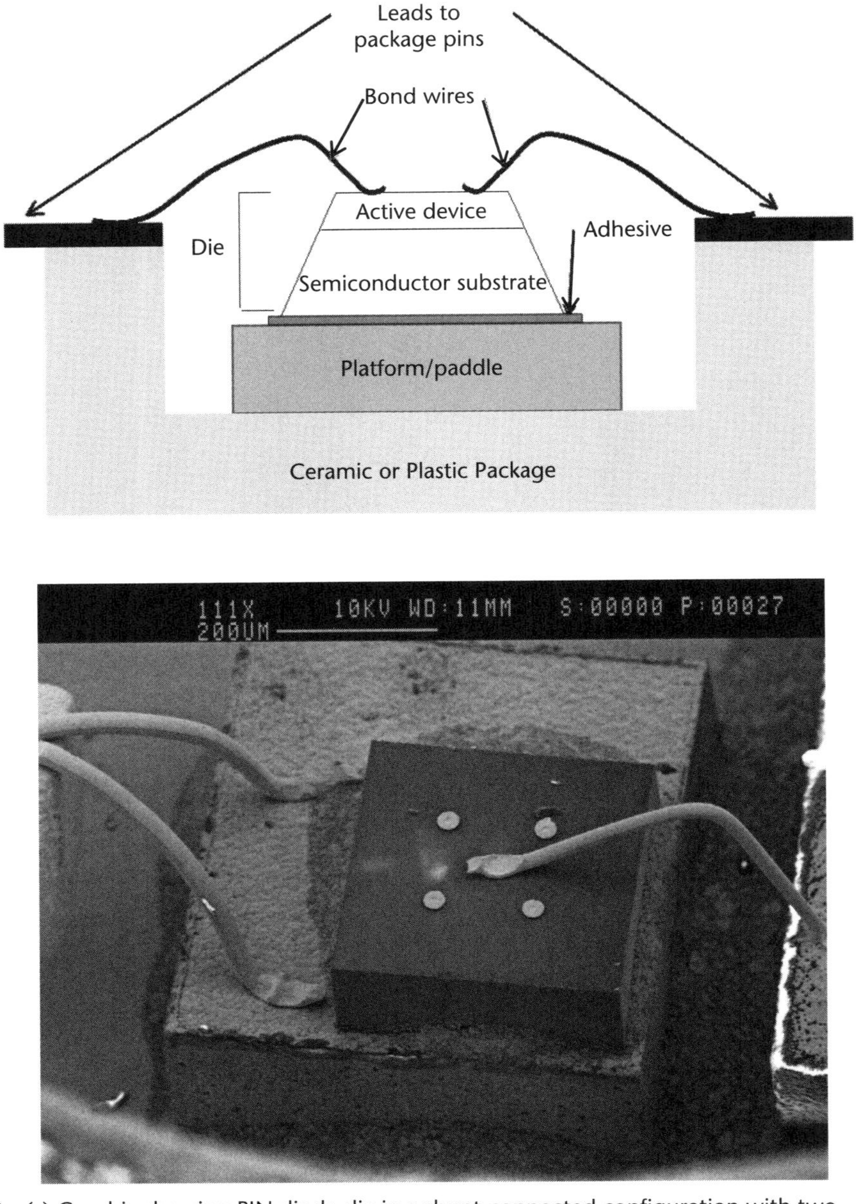

Figure 2.2 (a) Graphic showing PIN diode die in a shunt-connected configuration with two bond wires connecting the diode die between leads. (b) Scanning electron microscope image of bond wire connections between a PIN diode die and its package. Note the two bond wires to the paddle for reduced overall inductance. The adhesive can also be seen to the left and top of the die. (*Source:* Courtesy of Cobham Metelics, 2015. Used with permission.)

inputs and RF common indicated). There are limitations to the number of bond wires that can be used in lower power control component packages because only a limited number (usually no more than three) may be able to fit on a particular pad or semiconductor die. For higher power devices, many bond wires can be used for both inductance and current density reduction, but the pads and semiconductor

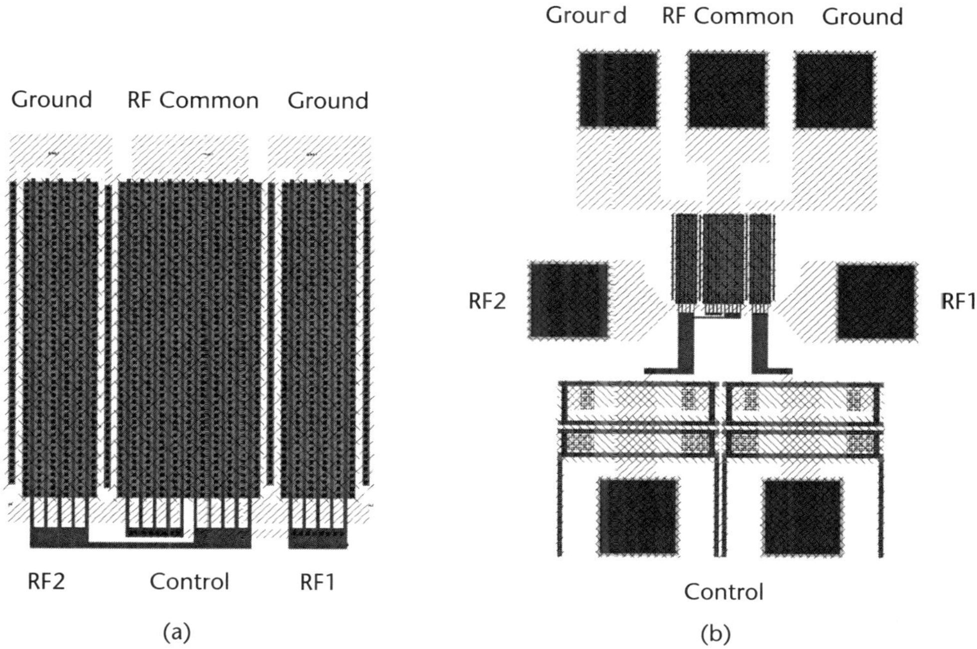

Figure 2.3 (a) Mask level layout of the core 0.8 micron SPDT MOS microwave and RF control element. (b) The control element with the input/output connection pads. The input control pads (with electrostatic discharge—ESD—protection) are along the bottom with the RF inputs (RF1, RF2) and RF common (RF Common) port indicated (after [8]).

die are then physically larger and will exhibit increased capacitance, reducing the high-frequency response.

The package end of the bond wire connection will be a short metallic section that connects to the external pin(s) or pad(s) of the package. This short metallic section can be modeled as a simple inductance or a short section of transmission line. For the transmission line option, knowledge of the dielectric constant of the encapsulant and the dimensions of the metallic line are required to estimate the characteristic impedance. For the low gigahertz range and below, a simple LC model is often sufficient. For higher frequencies, transmission line sections using standard analysis equations are often used for more accurate device and package modeling [3–7].

2.1.1.2 Series Resistance, R_{pack}

The bonds that make up the connections between the semiconductor die and the package are primarily mechanical in nature and are not electrically perfect and so can introduce additional series resistance R_{pack} in the connection. In addition to this resistance, mechanical contacts using bond wires are typically made to highly doped semiconductors, either n-type or p-type, which are called ohmic contacts. There is also the resistance of the mechanical connection with the bond wires to the metallic package pads. Each one of these contributions is a fraction of an Ohm, but the various resistive contributions are all in series and so add together. Even though gold is used as much as possible since it is a soft, ductile material and makes a good, low resistive contact, it does have some resistance. The overall package resistance associated with the various contributions may be as high as several tenths of an Ohm. In an RF control device that may itself have only a few Ohms resistance, the added series resistance R_{pack} becomes a relatively large fraction of the overall resistance, making it difficult to obtain extremely low insertion loss. This series resistance impacts the RF control device at high frequencies and may even have an impact all the way down to dc. For example, in some high-power PIN diodes, dc forward currents of 1.0A or higher may be needed, and since this resistance is in series with the "natural" PIN diode, dc voltage drops of several tenths of a volt and the resulting dc power dissipation of hundreds of milliwatts could occur.

2.1.1.3 Package Capacitance

The package and its connections are part of the input and output RF path and so must be modeled for simulation of the performance of the entire package. From the RF control component pad attachment point (which will be capacitive in nature), there is a bond wire (or wires) connecting to the package pad, putting an inductance in series with the RF path. The pin pad, having a relatively large area footprint, will exhibit a capacitance to RF ground, with a coupling capacitance to adjacent pins occurring as well. The leads (even so-called leadless chip carriers have metallic extensions from the internal connections to the package exterior) will also present an inductance and a small capacitive component. These effects suggest the general form for an RF package connection model (Figure 2.4) for a single connection from the port (that is, the connection to the circuit) to the control device. The values of the inductance typically range from 0.1 to 0.7 nH in small and medium signal power applications (although up to 1 or 2 nH in some high-power, low-frequency packages [1]) and package capacitance of 0.1 to 0.4 pF (although higher power packages could be up to 1 pF [1]).

2.1.1.4 Grounding

A key component to successful circuit operation is proper grounding. Component and device parasitics and undesired circuit coupling or crosstalk can all be controlled or minimized to varying degrees by application of various techniques to ensure proper RF grounding. At the package level, the control module will eventually be placed in some test fixture or into the system itself, and the two grounds (that

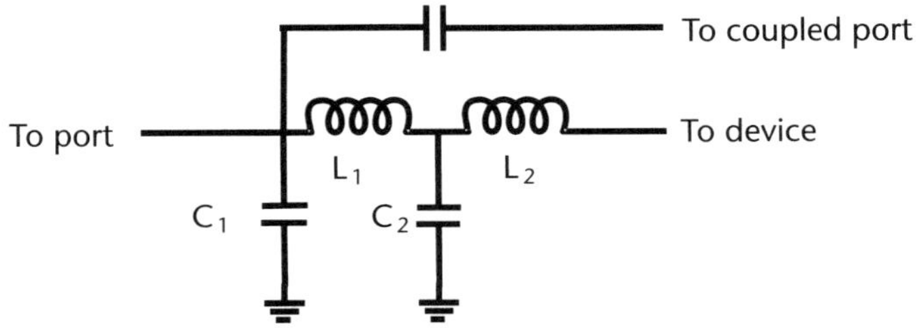

Figure 2.4 RF equivalent circuit for single connection of an RF package, showing the pin and bond wire inductance and pad and package capacitance.

is, packaged module and the system grounds) have to be connected. The package ground and the semiconductor die ground can be connected with an inductive bond wire, solder bump, or other internal inductive connection. The ground-connection inductance tends to "lift" the device RF ground from the overall RF system ground as the frequency increases due to the increasing inductive reactance of the ground bonding wire(s). Multiple ground connections from the die to the package platform internally and with multiple ground connections at the exterior of the package can reduce (but never completely eliminate) the ground inductance due to the parallel nature of the connections. In microstrip and stripline-based circuits, multiple RF grounds on the packaged RF control device will need to be connected to the ground plane. Multiple ground vias (top-side to bottom ground plane connections) help to improve grounding as well. In active device applications such as amplifiers, this RF grounding issue could result in unwanted feedback paths that cause instability and turn the amplifier into an oscillator. However, in RF and microwave control applications, the modules are inherently stable and so oscillations are not an issue, although degradations in insertion loss and isolation can occur.

2.1.1.5 Modeling Example

A common and inexpensive package type is the surface mount SOT package; a simple sketch of a two-terminal version is shown in Figure 2.5(a). This is a popular package for use at frequencies below approximately 2.0 GHz (the SOT-23 and SOT-323 being the most common [9]) but was not initially made for use at this frequency [9]. For this modeling example, a simple two-terminal package suitable for a series-connected PIN diode will be assumed. This simple example shows the important packaging considerations that the RF control designer must consider when modeling the semiconductor device in the package. From Figure 2.5(a), there are two inductances needed to model the connection to each pin: the bond wire inductance from the paddle to the pin on the top (point A in Fig. 2.5a) and the bond wire inductance from the die to the pin on the bottom (B). The inductance from paddle to top pin will be approximately half the inductance of the die to pin inductance because of the two parallel bond wires. A second inductor for each pin models the effect of the finite pin length and starts at the bond point for the bond wire on each pin. The dotted ovals in Figure 2.5(a) indicate the locations of the parasitic capacitance associated with the package. There is a capacitance associated with the bond point on each pin (C), which is relatively small since the bond wire is much smaller in diameter compared with the lead dimensions. The second capacitance region, modeled at the pin end (D), is larger than the bond point capacitance because it encompasses a larger area relative to the bond wire point. Using this model, extensive measurements yield the equivalent parasitic circuit for this package, as shown in Figure 2.5(b) [9]. Note that there is no coupling capacitance between the two pins; in this case, the pins are sufficiently far apart that very low coupling can be assumed. However, if the package had more pins in close proximity, coupling between the pins would have to be considered.

The impact of the package parasitics using this model can be illustrated by a simple example. In this example, a 5.0Ω resistance is used as a substitute for a semiconductor-based control element. A series-connected RF control circuit is assumed,

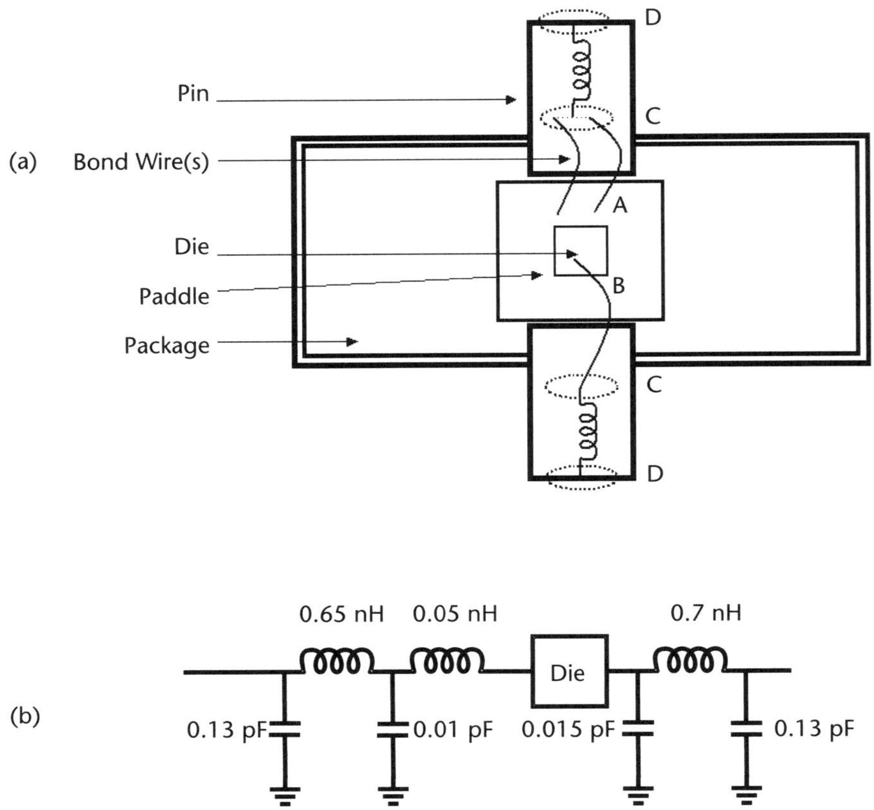

Figure 2.5 (a) Example of internal and external connections for an SOT-23 package (dotted ovals indicate locations where parasitic capacitances occur). (b) SOT-23 equivalent circuit parameters as measured in [9].

and the package parasitics shown in Figure 2.5(b) are used in the simulation. The results of the circuit simulation, shown in Figure 2.6, show that, below approximately 1.0 GHz, the insertion loss does not vary by more than approximately 0.10 dB from its 0.4 dB low-frequency value. However, the insertion loss deteriorates rapidly above this frequency, reaching 2.0 dB insertion loss at 4.3 GHz and 3.0 dB at 5.5 GHz. This behavior is to be expected, since the inductances at both input and output exhibit a reactance on the order of the resistance value at 1.0 GHz, with the corresponding voltage drops contributing to the increased insertion loss. Since the inductive reactance increases with frequency, the insertion loss deteriorates with increasing frequency. At 1.0 GHz, the capacitive reactances are still relatively large and therefore of minor impact, leaving the parasitic inductance as the main factor in the degraded insertion loss. The capacitance reactance decreases with frequency, however, and will manifest itself with increased insertion loss degradation at higher frequencies. It is for this reason that bond wire inductance is often the main focus in discussions of improved packaging.

Another approach for determining the package parasitics is the use of CAD to electromagnetically model the package and derive the equivalent circuit for the

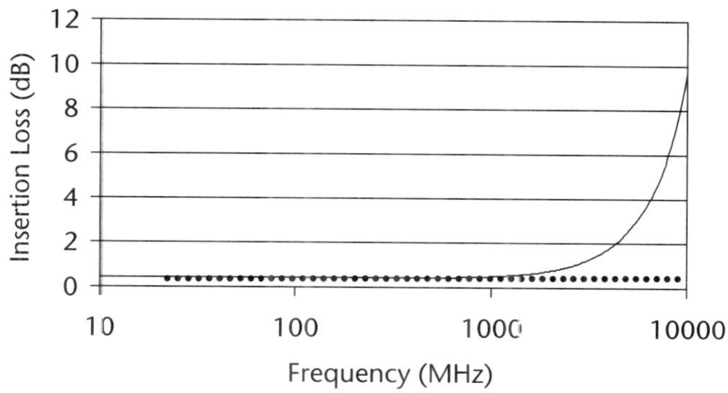

Figure 2.6 Insertion loss of a 5.0Ω resistor in an SOT-23 package as a function of frequency. Degraded insertion loss occurs at frequencies above 1.0 GHz. The dotted line indicates the insertion loss of the 5.0Ω resistor without the impact of the package parasitics.

die to pin connections. This approach can be quite time consuming, however, as it requires the creation of the package layout geometry and the computation time, but is worth the effort to ensure an accurate understanding of the complete module RF and microwave performance.

2.1.2 Interconnections (On-Chip)

The increased emphasis on SoC applications requires that all system components, including the RF and microwave control components, be fully integrated on a single semiconductor die. The connections between the RF and microwave control circuit and the various SoC subsystems can impact the frequency response, time delay, and phase shift between the circuit blocks [7]. Modeling these interconnects is crucial for obtaining an overall system level of performance estimate. In general, the longer the interconnection, the greater the impact on SoC performance. Two approaches are often used for estimates of the interconnect behavior: (a) models based on simple interconnect resistance and capacitance (RC) and (b) models based on distributed or transmission lines. Both of these approaches to interconnect modeling will be discussed in this section. These modeling approaches can also be used for control modules in other nonintegrated applications.

2.1.2.1 Simple RC Models

The simplest RC model describes the interconnect in terms of the interconnect layer resistance and interconnect layer capacitance. The interconnect in the RC model case is assumed to be a conducting layer of finite thickness t_{layer} over an insulating layer of thickness t_{ins} as shown in Figure 2.7. The model elements are usually computed assuming a simple slab resistance and parallel plate capacitance structure using the following equations:

$$R_{IC} = \frac{L}{W}\frac{\rho_{\text{layer}}}{t_{\text{layer}}} = \frac{L}{W}R_{SH}; \quad C_{IC} = \frac{\varepsilon_{\text{ins}}}{t_{\text{ins}}}WL = WLC_A \tag{2.1}$$

where L is the length of the interconnect, W is the width of the interconnect, R_{SH} is the sheet resistance of the interconnect layer (ρ_{layer}/t_{layer}), and ε_{ins} is the dielectric permittivity of the insulating material under the interconnect (alternatively, C_A, the capacitance per unit area, can be used). To keep the resistance associated with the interconnect to a minimum, metal interconnection layers are employed. In addition, the highest level of metal on an integrated circuit yields the lowest interconnect capacitance. The simple RC model routinely over-estimates the actual delay associated with an interconnect because the interconnect should ideally be modeled as a distributed RC line. From time delay calculations [10] it has been shown that the two-element RC interconnect model yields time delays that are about 30% too high, resulting in a twofold increase in RC time constant for the structure. Better estimates of the interconnect resistance and capacitance are given by (2.2):

$$R_{IC} = \frac{1}{\sqrt{2}}\frac{L}{W}R_{SH}; \quad C_{IC} = \frac{1}{\sqrt{2}}WLC_A \tag{2.2}$$

As an example to show the calculation of these two model elements, consider the following structure: a $W = 5.0\ \mu$m wide interconnect trace over a 3.0 μm thick insulating layer ($\varepsilon_r = 4$) with an $L = 50\ \mu$m interconnect length and a metallic layer with $R_{SH} = 0.1\ \Omega$/square. Using (2.1) shows the resistance and capacitance elements to be

$$R_{IC} = \frac{L}{W}R_{SH} = \frac{50}{5}0.1 = 1.0\Omega$$

$$C_{IC} = \frac{\varepsilon_{ins}}{t_{ins}}WL = \frac{4 \cdot 8.85 \cdot 10^{-12}}{3 \cdot 10^{-6}}50 \cdot 10^{-6} \cdot 5 \cdot 10^{-6} = 3\ \text{fF}$$

The RC time constant for the interconnect is approximately 3 femto-seconds. Note that the $R_{IC}C_{IC}$ product is independent of line width W but proportional to L^2, implying that the time constant quadruples for every doubling of the interconnect line length.

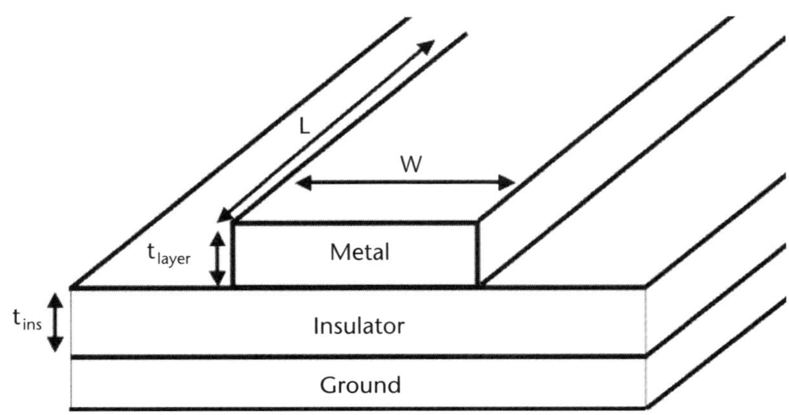

Figure 2.7 Structure for modeling an integrated circuit interconnection using an RC modeling approach.

2.1.2.2 Transmission Line Models

Transmission line models more accurately describe the distributed nature of the interconnect and require looking at four parameters associated with the interconnect: the resistance, inductance, capacitance, and conductance per unit length of the connection. The complex layout of the integrated circuit virtually ensures that there will be inductance effects due to the current loops created by the ground return path(s) [2]. While electromagnetic (EM) numerical modeling will yield the most accurate results, this method is also very time consuming and not easily integrated with equivalent control device circuit models. EM numerical modeling requires detailed knowledge about the dimensions and material parameters that the designer may not know, making it difficult to build an accurate model of the interconnect.

Partial element equivalent circuit methods and the concept of the effective distance to ground yield closed form estimates for the resistance and inductance per unit length can be made. A so-called complex effective inductance per unit length L^* can be derived using a complex image approximation using the closed form microstrip inductance formula derived by Wheeler [11] with the substrate height being related to the effective ground location, t_{eff}, assuming the bottom-side of the semiconductor substrate is at ground potential [12]:

$$L^*(\omega) = \frac{\mu_0}{4\pi}\ln\left[1 + 32\left(\frac{W}{t_{eff}(\omega)}\right)^{-2}\left(1 + \sqrt{1 + \left(\frac{\pi W}{8t_{eff}(\omega)}\right)^2}\right)\right]$$

$$t_{eff}(\omega) = t_{ins} + \frac{1-j}{2}\delta_S \tanh\frac{(1+j)t_{sub}}{\delta_S} \qquad (2.3)$$

$$\delta_S = \sqrt{\frac{2\rho}{\omega\mu_0}}$$

where W is the trace width, t_{ins} is the insulator thickness, t_{sub} is the substrate thickness, and δ_S is the skin depth in the semiconductor with ρ substrate resistivity [2, 12]. The resistance and inductance per unit length are functions of L^*:

$$R(\omega) = -\omega IM(L^*) \quad L(\omega) = RE(L^*) \qquad (2.4)$$

The capacitance per unit length can be found using the simple form for C in (2.1):

$$C_L = \frac{C_{IC}}{L} = \frac{\varepsilon_{ins}}{t_{ins}}W = WC_A \qquad (2.5)$$

A characteristic line impedance for the on-chip interconnect can be defined based on the inductance and capacitance per unit length in (2.4) and (2.1) as [13]

$$Z_0 = \sqrt{\frac{L(\omega)}{C_L}} \qquad (2.6)$$

assuming the low-loss definition for the transmission line.

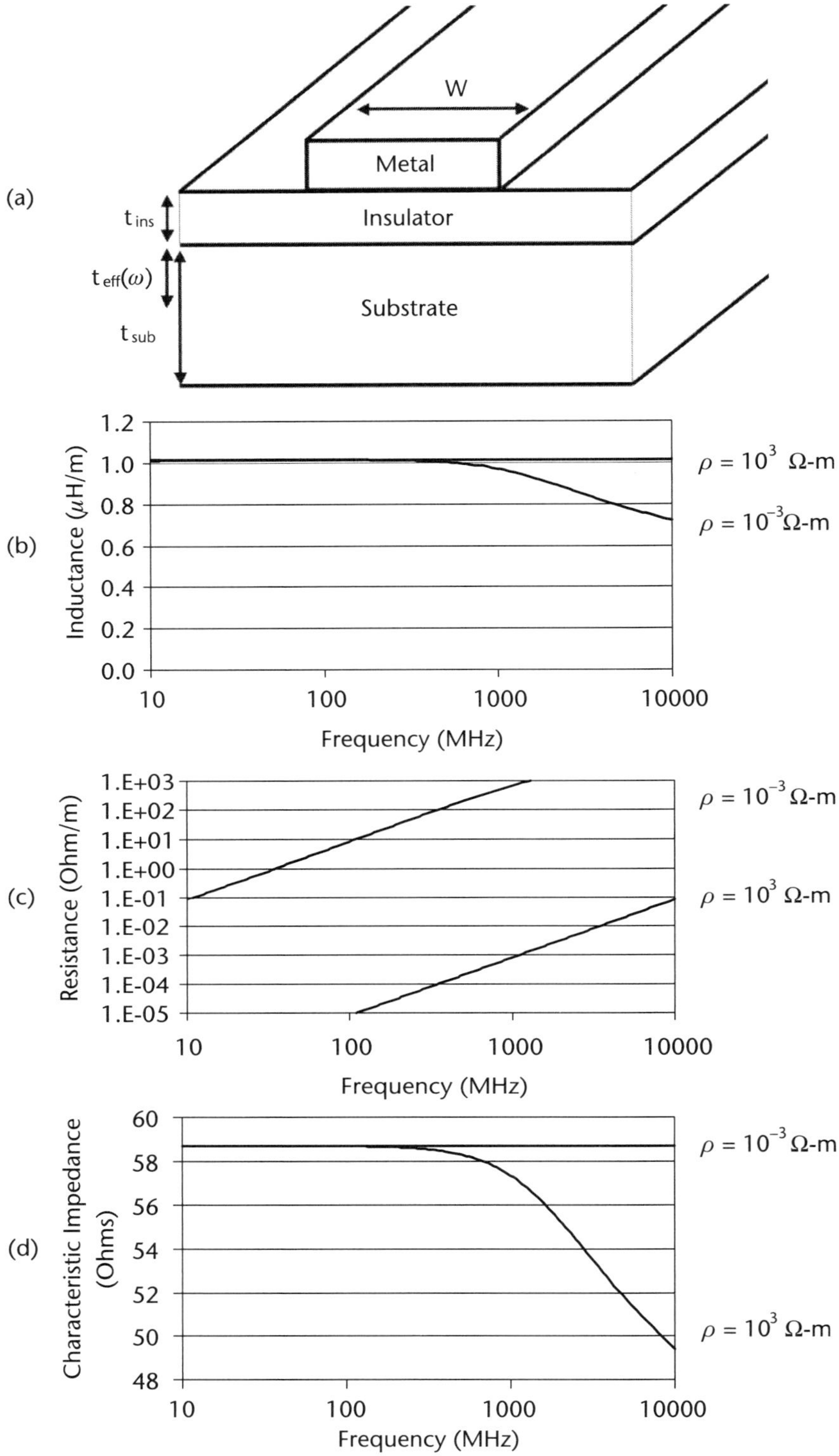

Figure 2.8 A transmission line interconnect modeling example for on-chip connection of RF control circuits: (a) geometry definitions for interconnect calculations, (b) inductance per unit length, (c) resistance per unit length, and (d) characteristic impedance.

As an example to show the frequency dependence of these two model elements, consider the following structure: a $W = 25$ μm-wide interconnect trace over a 3.0 μm-thick insulating layer ($\varepsilon_r = 4$) and a 500 μm-thick substrate with two different substrate resistivity (ρ) values: 10^{-3} and 10^3 Ω-m, representing a conductive (silicon CMOS, for example) and insulating (GaAs, for example) substrate (Figure 2.8(a)). Application of (2.4) and (2.6) show that the inductance per unit area and characteristic impedance do not vary significantly with frequency for insulating substrates and are only weakly dependent on frequency for conducting substrates below approximately 1.0 GHz using the parameters in this example. However, above approximately 1.0 GHz, the inductance per unit length (Figure 2.8(b)) and characteristic impedance (Figure 2.8(d) show a strong frequency dependence for the conducting substrate case. The loss term (Figure 2.8(c)) shows increasing loss with frequency for all substrate types, although the loss is very small for the insulating substrate. The interconnect line length is then all that is needed to accurately model the impact of the connection. These results were calculated using the "effective_height.m" Matlab script, part of the software tools that accompany this book.

2.2 Modeling Thermal Behavior

In RF and microwave control circuits, the circuit elements all exhibit some degree of loss. In the presence of RF and microwave signals, especially high-power ones, this loss converts the lost RF energy into heat that must be dissipated for successful device operation. Semiconductor die temperatures as low as 150°C can degrade performance temporarily or catastrophically, so power dissipated by the RF control circuit must be removed for proper operation. As an example, consider a relatively low insertion loss RF SPST series-connected switch. A 5Ω control device resistance will exhibit an insertion loss IL of approximately 0.4 dB. If this device is controlling a 50W continuous wave (CW) RF signal in a 50Ω system, 10% or 5.0W of heat needs to be dissipated by the RF control device. If the RF signal is pulsed, the device dissipates less average power, depending on factors such as the pulse width, duty cycle, and the thermal time constant of the semiconductor. Although knowledge of the actual dissipated power is certainly an important measure, the dissipated power is not a complete measure for determining the temperature rise in the control component. The next section describes a method to help relate the power dissipated in an RF control device to the associated temperature rise in the device.

2.2.1 Thermal Resistance

A measure of a package's ability to remove heat from the semiconductor die, then dissipate it to the ambient and estimate the associated die temperature rise, is needed for full understanding of the operation of semiconductor microwave and RF control devices. Such a parameter, termed the thermal resistance, is used to estimate this temperature rise and is important enough that it is frequently provided by manufacturers, usually given the symbol Θ (with dimensions °C/W). The thermal resistance is governed by the control device geometry and is a function of the area (A)

of the active RF circuit on the semiconductor die, the semiconductor die thickness (F), and the thermal conductivity of the semiconductor material (K_{TH}). The thermal resistance is estimated as $\Theta = F/A\ K_{TH}$ [14]. The temperature rise above the ambient temperature of the RF control device is based on this thermal resistance and the difference between the input powers (RF plus any dc) and the RF output power [15]:

$$T = T_{\text{ambient}} + \Theta\left[\left(P_{dc} + P_{\text{RF-in}}\right) - P_{\text{RF-out}}\right];\ \Delta T_{\text{rise}} = \Theta\left[\left(P_{dc} + P_{\text{RF-in}}\right) - P_{\text{RF-out}}\right] \quad (2.7)$$

Equation 2.7 shows the effective or total temperature rise as a function of a single thermal resistance number. This thermal resistance number has various contributions from the heat flow at various locations in the semiconductor. Consider a single semiconductor circuit element with thermal resistance describing the circuit to air heat path Θ_{AD} of 100°C/W. In this scenario, heat is generated by the circuit and flows from the circuit to the ambient (air). In real semiconductor control devices, the active circuit component is fabricated on a semiconductor substrate, and so two thermal paths now exist: the original circuit-air path but now also the circuit-substrate-ambient path with thermal resistance Θ_D. These two paths both conduct heat, and so the equivalent thermal circuit for this scenario is shown in Figure 2.9. Since the substrate is physically much larger than the electronic circuit, it has a smaller thermal resistance; a typical value for this thermal resistance is 30°C/W. Thermal resistances are modeled as an electrical resistance analog, and so the overall thermal resistance of the circuit is computed as the "parallel combination" of the two thermal resistances or 23.1°C/W.

The circuit-substrate-ambient path, described by thermal resistance Θ_D, is actually made up of various subcomponents as part of the overall packaging scheme. These heat paths are "in series" and are the semiconductor to the paddle (Θ_{SP}), paddle to package/package heat sink (Θ_{PP},) and package to air (Θ_{PA}) heat paths. A modified thermal resistance "circuit" that includes this more-detailed heat flow path description is shown in Figure 2.10. Typical values for these thermal resistances are shown in Table 2.1.

As an example of the calculation procedure, consider a problem where the RF control circuit designer is asked to determine the maximum RF power (assuming no

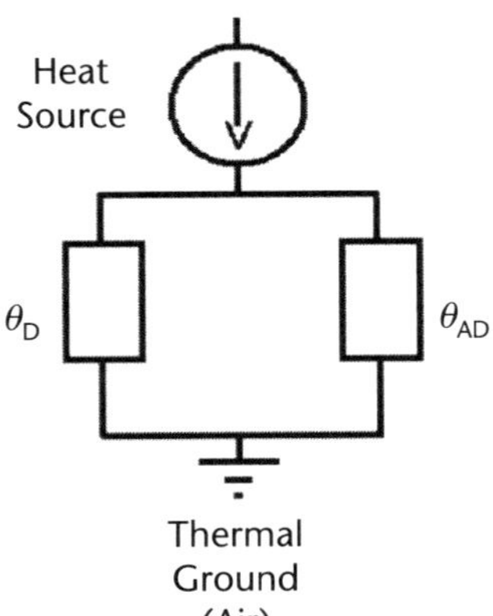

Figure 2.9 Simple thermal model of heat flow from a heat source to a heat sink (from [2]).

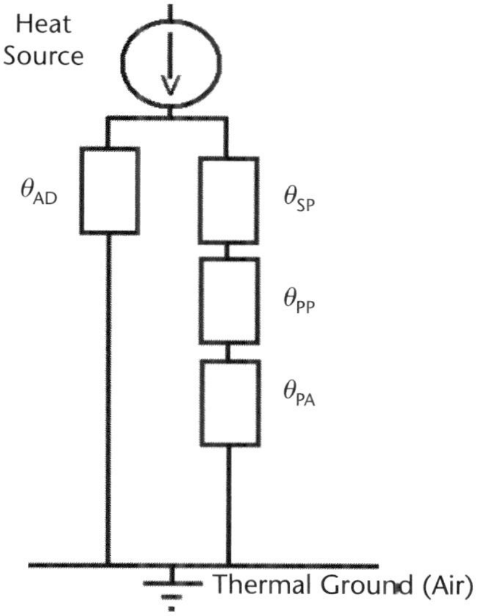

Figure 2.10 Expanded thermal model of heat flow from a heat source to a heat sink through two paths, reducing the overall thermal resistance to heat flow (from [2]).

dc power) that the control circuit can safely handle in a packaged and unpackaged environment, assuming a maximum safe circuit temperature of 125°C and an ambient temperature of 25°C. The total allowed RF power for the packaged (Θ_{AD}, Θ_D) and unpacked (die only, Θ_{AD}) environment can be computed using (2.7), assuming a thermal resistance of 100°C/W for the unpackaged environment:

$$P_{\text{TOT}} = \frac{T - T_{\text{ambient}}}{\Theta_{\text{TOT}}} = \frac{125 - 25}{100} = 1\text{W}$$

and 23.1 °C/W for the packaged environment:

$$P_{\text{TOT}} = \frac{T - T_{\text{ambient}}}{\Theta_{\text{TOT}}} = \frac{125 - 25}{23.1} = 4.32\text{W}$$

These results indicate that it takes more than 4.0W of power to raise the temperature of the die 100°C above the ambient compared with 1.0W for the unpackaged

Table 2.1 Table of Typical Thermal Resistance Values

Θ	*Thermal Resistance °C/W*	*Description*
Θ_{AD}	100	circuit-air heat path
Θ_{SP}	15	semiconductor-paddle path
Θ_{PP}	10	paddle-package
Θ_{PA}	5	package-air
Θ_D	30	$\Theta_{SP} + \Theta_{PP} + \Theta_{PA}$

device, showing how providing better thermal paths to ambient can increase the power-handling capability of the circuit.

2.2.2 Thermal Time Constant

The thermal resistance Θ provides an estimate of the temperature rise associated with RF power being dissipated as heat in the control device but does not provide any information about how long it takes for the final temperature to be achieved after RF energy has been applied (or how long it takes to cool off after the RF energy has been removed). The semiconductor die takes time to respond to the applied RF energy, the response being described by the thermal time constant τ_{TH}. Similar to other time constants in engineering, τ_{TH} describes the length of time it takes for the temperature to rise to 63% of its final value when RF power is applied or to drop to 37% of its initial temperature should RF power be removed from the circuit (Figure 2.11). The thermal time constant τ_{TH} is a function of the semiconductor die thickness and density (d_{semi} and ξ_{semi}), K_{TH} and the specific heat of the semiconductor, χ_{semi} [14]:

$$\tau_{TH} = \left(\frac{2d_{\text{semi}}}{\pi}\right)^2 \left(\frac{\xi_{\text{semi}}\chi_{\text{semi}}}{K_{TH}}\right) \tag{2.8}$$

Equation 2.8 shows that thicker semiconductor dies (d_{semi}) exhibit longer thermal time constants and so take longer for the die temperature to rise to its final value with application of RF power (or take longer to cool off with removal of RF power).

Using the RC circuit electrical analogy and applying it to thermal behavior, the thermal time constant can be written as the product of the thermal resistance and thermal capacity C_{TH}: $\tau_{TH} = \Theta C_{TH}$. This analogy allows an equivalent circuit model to be developed for the thermal circuit with current excitation acting as the heat source, with the current (heat) flowing through the circuit to the circuit ground

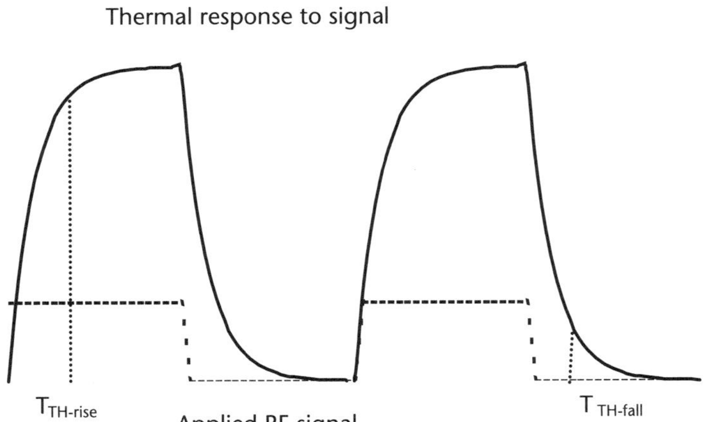

Figure 2.11 Temperatures rise and fall in an RF control element with the application and removal of RF energy. The rise and fall times are described by the thermal time constant τ_{TH}.

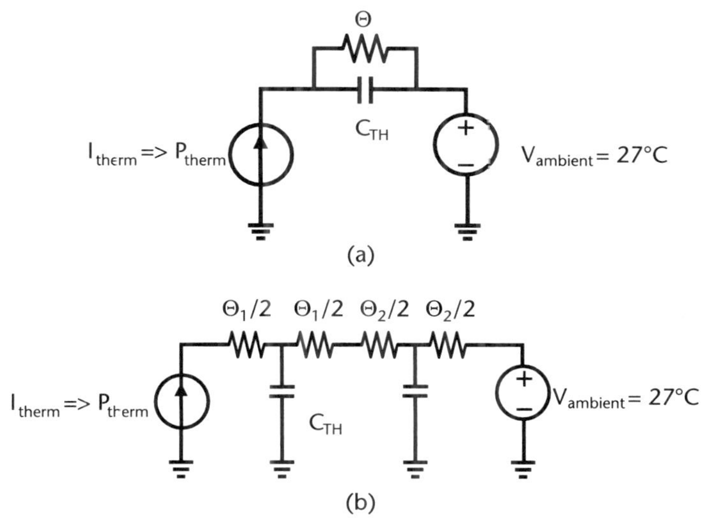

Figure 2.12 (a) Foster form for the thermal equivalent circuit using a single thermal time constant model. (b) Cauer form for thermal equivalent circuit using a double thermal time constant model.

(ambient temperature, T_j). There are two possibilities for the layout of this thermal equivalent circuit: the Cauer (or continued fraction circuit) and the Foster (or partial fraction circuit) equivalent circuit forms [16], as shown in Figure 2.12. Of the two, the Cauer is the more physically intuitive circuit, as the elements and element connections better correspond to the real, physical configuration [16]. The Foster and Cauer thermal model forms are useful in circuit and device simulation because they relate the power dissipated by the device (P_{therm}) to the temperature variation in the device. The dissipated power can be determined from simulator computations of the device voltage drop and current draw.

2.3 Device Nonlinearity

In the ideal system described in the frequency domain, the sum of two or more signals of the same or different frequency at the input yields an output that is composed of the same sum times the system response. This relationship is expressed mathematically in the frequency domain as

$$\begin{aligned} Y(f) &= Y_1(f) + Y_2(f) + \ldots = H(f)\left[X_1(f) + X_2(f) + \ldots\right] \\ &= H(f)X_1(f) + H(f)X_2(f) + \ldots \end{aligned} \tag{2.9}$$

where $X_i(f)$ is the input stimuli, $Y(f)$ is the output response, and $H(f)$ is the system transfer function that does not change with time. The time-domain equivalent of (2.9) involves a convolution of the two responses. This system is often termed a linear time invariant system (LTI), in which superposition principles can be used to construct the final response from the individual stimuli. In real systems, however, the transfer function H(f) does change with time because the elements that make up the various circuits and subsystems often respond to the applied signal by changing their key component values or parameters as a function of time. Whenever a component changes value as a function of time, a time-variance is said to occur, in which case the component changes cause the system to behave in a nonlinear fashion, and so the superposition principle no longer applies. A very simple example of this property is to look at a system with a quadratic response. For $X_1(f)$ as the input, the output becomes $Y(f) = X_1^2(f)$. If the input is now the sum of two excitations $X_1(f) + X_2(f)$, the output response is then $Y(f) = [X_1(f) + X_2(f)]^2 = X_1^2(f) + X_2^2(f) + 2X_1(f) X_2(f)$. In this response, the first two terms are the two quadratic response terms as expected, but there is another response that appears, and that is the product of the two input stimuli. The quadratic response is a nonlinear response but clearly shows that superposition cannot be applied to nonlinear systems. The study of nonlinear systems is quite complex mathematically, but there are simplifications to the mathematics that can provide a link to nonlinear properties that the RF and microwave engineer might encounter in the laboratory or when studying component data sheets. These are the topics that appear in the section.

2.3.1 Origin of Nonlinearity

Any active device, including the active devices used in microwave and RF control circuitry, can exhibit nonideal behavior in the form of a nonlinear response to the applied signal.[1] This nonlinear behavior will generate new or spurious signals in response to the input as shown in the previous example using a quadratic system response. These newly generated spurious signals (sometimes referred to as spurs) will occur at frequencies both harmonically and nonharmonically related to the stimulus. The easiest way to see how this nonlinear behavior can generate new signals is to consider one of the most basic nonlinear semiconductor devices, the PN junction diode. The current-voltage characteristic for the PN diode is given by the Shockley diode equation [17]:

$$I_d(V) = I_0\left(e^{V/V_T} - 1\right) \tag{2.10}$$

where $V_T = kT/q$ and the diode ideality factor is assumed to be unity. It is immediately apparent that the voltage and current in (2.10) are not linearly related but are instead related by an exponential. Attempting to describe the nonlinear behavior using this exponential form for the current-voltage relations does not lend itself to a straightforward mathematical analysis of nonlinear behavior. However, by writing (2.10) in terms of a Taylor series expansion, the current in terms of powers of the voltage V are more easily observed:

1 There is a source of distortion termed "passive intermodulation distortion" (PIM) that theoretically originates from purely passive components such as resistors or connectors. However, PIM is usually caused by such factors as rectifying contacts due to corrosion or metallurgical differences and so is not truly from purely "passive" components. More information about PIM can be found in the literature [18, 19].

$$I_d(V) = \sum_{i=1}^{\infty} \frac{I_0}{i!\left(V_T^i\right)} V^i = \sum_{i=1}^{\infty} g_i V^i = g_1 V + g_2 V^2 + g_3 V^3 + \ldots \tag{2.11}$$

where the expansion coefficients g_i are the associated derivatives of (2.11). The term $i = 1$ is linear in voltage, with g_1 being the linear conductance. For i greater than unity, the order of the nonlinearity is simply the power of the relevant voltage term: V^2 is the second order or quadratic response; V^3 is the third order or cubic response, and so on. All terms should be considered for the most accurate mathematical representation of the signal. The next section covers the impact of the nonlinearity order on the current-voltage response.

2.3.2 Order of Nonlinearity

The impact of the nonlinear current-voltage relationship in (2.11) can be seen if one first considers a single frequency sinusoid at fundamental frequency ω_a for the voltage drop across the diode:

$$V = V_A \cos\left(\omega_a t\right) \tag{2.12}$$

The resulting form for the current using the Taylor series expansion in (2.11) is

$$\begin{aligned} I_d(t) &= \sum_{i=1}^{\infty} g_i V^i = g_1 V_A \cos\left(\omega_a t\right) + g_2 \left[V_A \cos\left(\omega_a t\right)\right]^2 + g_3 \left[V_A \cos\left(\omega_a t\right)\right]^3 + \ldots \\ &= \left[g_1 V_A + \frac{3}{4} g_3 V_A^3\right] \cos\left(\omega_a t\right) + \frac{1}{2} g_2 V_A^2 + \frac{1}{2} g_2 V_A^2 \cos\left(2\omega_a t\right) + \frac{1}{4} g_3 V_A^3 \cos\left(3\omega_a t\right) + \ldots \end{aligned} \tag{2.13}$$

Equation 2.13 shows that the second order nonlinearity has generated a dc term[2] and a term at twice the frequency of the fundamental (second harmonic), $2\omega_a$, which are both proportional to V_A^2 and the g_2 term. The third order nonlinearity has generated a signal at three times the fundamental (third harmonic), $3\omega_a$, but also a signal at the fundamental, ω_a, both of which are proportional to V_A^3 and the g_3 term. Depending on the phase relationship between g_1 and g_3, a reduction or compression of the fundamental can occur, especially as the amplitude of V_A increases. Note that for this so-called single tone test, all the nonlinear responses are harmonically related to the fundamental frequency, ω_a.

The result of (2.13) is interesting because it shows that an infinite spectrum of signals at integer multiples of the fundamental frequency can be generated. In a circuit with a nonlinear device, the RF engineer is usually interested in the fundamental frequency circuit behavior and so these harmonically related signals need

2 The dc current generated by the diode with an applied RF signal is related to the square of the voltage, a measure of the power in the signal. This nonlinear response is the origin for the use of diodes in power or square law detectors.

to be significantly reduced. The reduction of these spurs in the laboratory can be achieved through filtering (either a band pass around the fundamental or a low pass with cutoff between the fundamental and second harmonic) or operating at a low RF drive level (V_A) so that the spurs are at low levels compared with the fundamental. The reduction in these spurs at the circuit level, however, requires detailed knowledge of the voltage and current behavior of the individual RF control devices in the RF control circuits. These reductions in spurs generated by the RF control devices used in the control circuits will be discussed in detail in later chapters.

Most RF engineering problems do not involve single frequency sinusoids but rather multiple signals or signals over some bandwidth. Here, multiple discrete signals or a continuum of signals over the bandwidth can create spurs at noninteger frequencies related to the fundamental and so are much more problematic to deal with. An estimate of the spurious signals originating from these more complex excitation signals requires the application of a sum of two signals of different amplitude and frequency (the so-called two-tone test), which can be described mathematically as

$$V(t) = V_A \cos(\omega_a t) + V_B \cos(\omega_b t) \tag{2.14a}$$

or, in terms of the current

$$I_d(V) = \sum_{i=1}^{\infty} g_i \left[V_A \cos(\omega_a t) + V_B \cos(\omega_b t) \right]^i \tag{2.14b}$$

In the following discussion, only orders up to $i = 3$ will be studied, although the technique can be applied to any order or number of applied fundamental signals. The resulting derivation is quite lengthy, requiring a number of trigonometric substitutions (and sheets of paper) but can be simplified by looking at each order separately. The first order expression ($i = 1$) is simply the linear response and is conductance g_1 times the expression in (2.14a). The second order response is related to the square of the voltage ($i = 2$) and can be written as

$$I_{d-2} = g_2 \left\{ \begin{array}{l} \dfrac{V_A^2}{2} + \dfrac{V_B^2}{2} + \dfrac{V_A^2}{2}\cos(2\omega_a t) + \dfrac{V_B^2}{2}\cos(2\omega_b t) \\ \quad + V_A V_B \cos\left[(\omega_a + \omega_b)t\right] + V_A V_B \cos\left[(\omega_a - \omega_b)t\right] \end{array} \right\} \tag{2.15}$$

The first two terms are dc components and represent a change in the dc level that is a function of the squares of the RF amplitudes. The second two terms represent second order harmonics (H) or signals that occur at twice the fundamentals. The third set of terms represents the second order intermodulation (IM), or the so-called sum and difference or beat frequencies. Note that for closely spaced frequencies, the IM products are between $2\omega_a$ and $2\omega_b$ at $\omega_a + \omega_b$, and $\omega_a - \omega_b$ is at a low frequency (but not quite dc). During the RF two-tone test, the fundamental amplitudes V_A and V_B are usually set equal to one another, with the result being that the second order IM products are always 6 dB above the level of the harmonics (twice the level of the harmonics).[3] Since the second order products are proportional to the voltage amplitude squared, for every 1 dB increase in the fundamental amplitude, the second order products will increase by 2 dB.

The third order response ($i = 3$) requires more mathematics and further applications of trigonometric identities (higher orders are even more complex) but is actually of more interest to the RF and microwave control device designer. In its full mathematical form, the third order current can be written as

$$I_{d-3} = g_3 \left\{ \begin{array}{l} \left[\left(\frac{3V_A^{\,3}}{4} + \frac{3}{2}V_A V_B^{\,2}\right)\right]\cos(\omega_a t) + \left[\left(\frac{3V_B^{\,3}}{4} + \frac{3}{2}V_B V_A^{\,2}\right)\right]\cos(\omega_b t) \\ + \frac{V_A^{\,3}}{4}\cos(3\omega_a t) + \frac{V_B^{\,3}}{4}\cos(3\omega_b t) \\ + \frac{3}{4}V_A^{\,2}V_B\cos\left[(2\omega_a + \omega_b)t\right] + \frac{3}{4}V_A V_B^{\,2}\cos\left[(2\omega_b + \omega_a)t\right] \\ + \frac{3}{4}V_A^{\,2}V_B\cos\left[(2\omega_a - \omega_b)t\right] + \frac{3}{4}V_A V_B^{\,2}\cos\left[(2\omega_b - \omega_a)t\right] \end{array} \right\} \tag{2.16}$$

Equation 2.16 shows that the third order response during a two-tone test creates quite a rich spectrum of signals. The first two terms are spurious components that have been converted (or beat) down to the level of the fundamentals and will influence the fundamental signal. Depending on the magnitude and phase of g_3 with respect to g_1, the fundamental signal component can vary due to this third order component, with an overall reduction or compression of the fundamental signal. The second group of terms shows third order harmonic signals. The last set of signals is especially important to the RF and microwave control circuit designer because the two IM difference terms $(2\omega_a - \omega_b)$ and $(2\omega_b - \omega_a)$ actually beat down to frequencies very close to the two fundamentals, ω_a and ω_b. During the two-tone test, equal drive signals are usually applied, and so the third order IM terms will be 9.54 dB higher (factor of three) than the third order harmonic terms.[3] All third order terms increase at three times the rate of the fundamentals (in decibel terms), so that for every 1 dB increase in fundamental signal power increase, the third order IM products increase by 3 dB. The spurious third order IM signals (and indeed all odd order IM signals) tend to be the most problematic to deal with from a circuit perspective. All the other high frequency signals (second order H, IM, and third order H and the sum IM products) can be significantly reduced through filtering. The difference IM products, however, residing so close to the fundamentals, are difficult if not impossible to adequately filter without serious degradation of the desired signals. Any improvement in in-band spurs needs to occur through reduction in the generation of these signals by the RF control components themselves, the

3. This assumes operation of the device well below its compression point and therefore in the small-signal regime. This small regime provides what has been called "behaving" distortion because the spurious signals generated "behave" according to the trigonometric theory [Gerald Hiller, personal communication, sometime during the 1980s]. The 6 dB difference between intermodulation and second harmonic signals and the corresponding 9.54 dB between intermodulation and third harmonic signals when measured in the laboratory indicate that "behaving" measurements are being performed.

major reason why details on RF control component nonlinearities will be discussed in later chapters.

Figure 2.13 shows an example frequency spectrum with the two fundamentals and the second and third order H and IM products for a two-tone test indicated. During a two-tone test, as the fundamental drive power is increased, the second and third order H and IM powers increase at twice and three times the rate (in decibels) as the fundamentals, respectively. If the fundamental, H and IM products are plotted on a graph of output power versus input fundamental power, the difference in slopes of the various output components can be seen (Figure 2.14). By extrapolating the second and third order power curves up to the level of the fundamental, the curves intersect the fundamental power curve at distinct power levels. The power levels at these intersections (or intercepts) are termed the second and third order intercept points, or IP2 and IP3, respectively. IP2 and IP3 provide a convenient metric for comparison of device nonlinearities. In addition, IP2 and IP3 also provide a means to compute the actual nonlinear power level (or distortion) for any reference fundamental power level:

$$P_{2D\text{-dBm}} = 2P_{F\text{-dBm}} - \mathrm{IP2}_{\mathrm{dBm}} \quad P_{3D\text{-dBm}} = 3P_{F\text{-dBm}} - 2 \cdot \mathrm{IP3}_{\mathrm{dBm}} \tag{2.17a}$$

where $P_{XD\text{-dB}}$ is the X order distortion power (2 or 3 for this discussion) and $P_{F\text{-dBm}}$ is the fundamental power in dBm (IP2 and IP3 are usually specified in terms of dBm). In absolute terms, the second and third order distortion powers can be computed using the following relationships:

$$P_{2D} = \frac{P_F^2}{\mathrm{IP2}}; \; P_{3D} = \frac{P_F^3}{\mathrm{IP3}^2} \tag{2.17b}$$

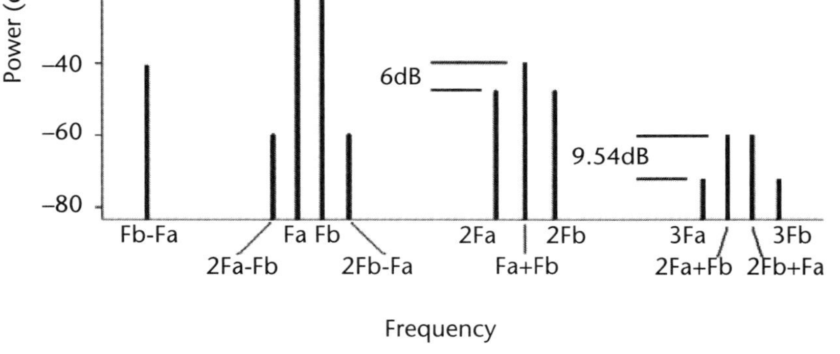

Figure 2.13 Spectrum plot showing fundamental and second and third order harmonic and intermodulation products (from [2]).

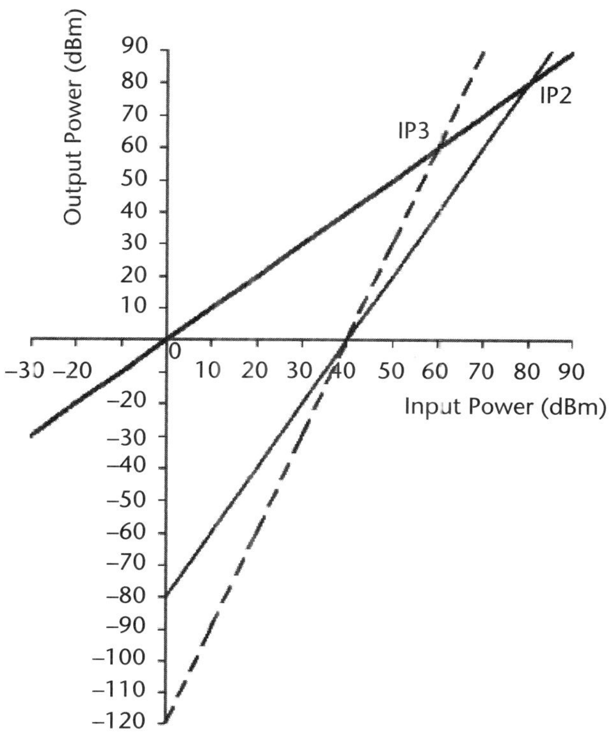

Figure 2.14 The intercept point can be computed given measured fundamental and distortion powers (from [2]).

These two expressions indicate that higher IP levels are desired to keep the generated nonlinear distortion power as low as possible.

If the RF two-tone test powers (fundamental and all the higher orders) are all measured at the output of the system block or circuit, the resulting nonlinear response is referred to as output IP2 or output IP3, or OIP2 and OIP3, respectively. An alternate measure is to refer the intercept points to the input of the system block or circuit, in which case they are referred to as input referred intercept points, IIP2 and IIP3. The relationship between input-referenced and output-referenced distortion intercept points is simply the gain of the system or circuit:

$$\mathrm{IIP2} = \mathrm{OIP2} - G \qquad \mathrm{IIP3} = \mathrm{OIP3} - G \tag{2.18}$$

where the gain G is in terms of dB (or, in the case of an RF or microwave control circuit, the insertion loss or isolation, which both have a negative gain in dB terms).

The first step in determining the IP values is to place the RF control circuit in its desired state (low insertion loss or high isolation) and measure the power levels at the fundamental and the second and third order IM components, usually at the system output. As an example, consider a two-tone test measurement with the two fundamentals adjusted to provide +20 dBm at the output of the RF control circuit. A measurement on a spectrum analyzer shows the second IM term at $\omega_a - \omega_b$ of −30 dBm and the third order, in-band IM term at $2\omega_a - \omega_b$ of −70 dBm. Using (2.17a) and rearranging to find the IP terms provides the following values of OIP2 and OIP3:

$$\begin{aligned} OIP2_{dBm} &= 2P_{F\text{-}dBm} - P_{2D\text{-}dBm} = 2(20) - (-30) = 70 \text{ dBm} \\ OIP3_{dBm} &= 1.5P_{F\text{-}dBm} - 0.5P_{3D\text{-}dBm} = 1.5(20) - 0.5(-70) = 65 \text{ dBm} \end{aligned}$$

The input-referenced distortion IP can be calculated simply by subtracting the insertion loss (isolation) in the on-state (off-state) from the output-referenced IP terms in the above example.

References

[1] Philpot, K. R., "A Guide to Microwave Diode Package Styles and Their Performance," *High Frequency Electronics*, February 2005, pp. 1–4.

[2] Caverly, R., *CMOS RFIC Design Principles*, Dedham, MA: Artech House, 2006.

[3] Edwards, T. C., *Foundations for Microstrip Circuit Design*, 2nd ed., West Sussex, UK: John Wiley, 1991.

[4] Edwards, T. C., and M. B. Steer, *Foundations of Interconnect and Microstrip Design*, 3rd ed., West Sussex, UK: John Wiley, 2000.

[5] Sutono, A., N. G. Cafaro, J. Laskar, and M. M. Tentzeris, "Experimental Modeling, Repeatability Investigation and Optimization of Microwave Bond Wire Interconnects," *IEEE Trans. on Advanced Packaging*, Vol. 24, No. 4, November 2001, pp. 595–603.

[6] Grabinski, H., B. Konrad, and P. Nordholtz, "Simple Formulas to Calculate the Line Parameters of Interconnects on Conducting Substrates," *Proc. Dig. IEEE 7th Topical Meeting Electrical Performance of Electronic Packaging*, 1998, pp. 223–226.

[7] Deutsch, A., P. W. Coteus, G. V. Kopcsay, H. Smith, C. W. Surovic, B. L. Krauter, D. C. Edelstein, and P. J. Restle, "On Chip Wiring Design Challenges for GHz Operation," *IEEE Proc.*, Vol. 98, April 2001, pp. 529–555.

[8] Caverly, R., "Linear and Nonlinear Characteristics of the Silicon CMOS Monolithic 50 Ohm Microwave and RF Control Element," *J. Solid-State Circuits*, Vol. 34(1), January 1999, pp. 124–126.

[9] Brown, B., and Hiller, G., "Circuit Models for Plastic Packaged Microwave Diodes," *Microwave Symposium Digest, 1996, IEEE MTT-S International*, Vol. 3, June 1996, pp. 1779–1782.

[10] Weste, N., and K. Eshraghian, *Principles of CMOS VLSI Design: A Systems Perspective*, 2nd ed., Reading, MA: Addison-Wesley, 1992.

[11] Wheeler, H. A., "Transmission-Line Properties of a Strip on a Dielectric Sheet on a Plane," *IEEE Trans. on Microwave Theory and Techniques*, Vol. MTT-25, August 1977, pp. 631–647.

[12] Weisshaar, A., H. Lan, and A. Luoh, "Accurate Closed-Form Expressions for the Frequency-Dependent Line Parameters of On-Chip Interconnects on Lossy Silicon Substrate," *IEEE Trans. Advanced Packaging*, Vol. 25(2), May 2002, pp. 288–296 [see also A. Weisshaar, Workshop Slides, *IEEE Radio Wireless Conf. Workshop*, January 2006].

[13] Pozar, D. M., *Microwave Engineering*, 4th ed., New York: John Wiley, 2011.

[14] Hewlett Packard Applications Note, "High Frequency Transistor Primer, Part III: Thermal Properties," 1998.

[15] "Determining the Junction Temperature from Device Thermal Resistance for Plastic Semiconductor Devices," *Thermal App. Note SMI-200*, Sirenza Microdevices.

[16] Schutze, T., "Thermal Equivalent Circuit Models," *Infineon App. Note AN2008-03*, 2008.

[17] Sze, S., *Physics of Semiconductor Devices*, New York: Prentice Hall, 1982.

[18] Shitvov, A., A. G. Schuchinsky, M. B. Steer, J. M. Wetherington, "Characterisation of Nonlinear Distortion and Intermodulation in Passive Devices and Antennas," *2014 8th European Conference on Antennas and Propagation (EuCAP)*, April 2014, pp. 1454–1458.

[19] Sun, T., F. Xu, J. Lu, Y. Xu, and K. Liu, "Novel and Simplified Passive Intermodulation Distortion Measurement Using Vector Network Analyzer," *2014 IEEE 9th Conference on Industrial Electronics and Applications (ICIEA)*, June 2014, pp. 1688–1690, 9–11.

CHAPTER 3

Modeling PIN diodes—Linear Behavior

3.1 Introduction

The PIN diode has been used for microwave and RF control for more than half a century. One of the first detailed descriptions of the microwave operation of the PIN diode was by Uhlir [1] in 1958, although control of microwave energy with diodes had been discussed by early microwave pioneers such Coale, Garver, and Mortensen [2–7]. A classic book by White [8] helped bring great understanding of both the operation of the PIN diode at microwave frequencies as well as practical considerations of their use in microwave high-power design and applications. Since that time, the PIN diode has been used in various applications, ranging from millimeter waves all the way down to the megahertz range (and even below that). Microwave and RF design engineers should always have a good understanding of the devices they are using, not only in the range of the desired frequency of operation but also outside of it. This understanding can take many forms: many hours at the laboratory bench making detailed measurements, many hours spent using a circuit simulator, or many hours understanding the physics and associated mathematics underlying the PIN diode's operation. The next two chapters in the book will look at the physical operation and use various levels of mathematics to describe the PIN diode's operation.

3.2 PIN Diode Modeling—Simple

Detailed analyses of PIN diodes by many workers over the years have shown interesting phenomenon at low frequencies as well as provided an accurate description of just what is meant by low frequencies. This chapter presents several levels of physical understanding of PIN diode behavior, starting with the classical charge control PIN diode model that yields the standard forward bias resistance and reverse bias capacitance descriptions. The presentation then extends to a detailed mathematical analysis of the forward bias behavior, including high injection and surface influences on the device behavior. Reverse bias behavior is then discussed.

3.2.1 Simple Lumped Element Modeling

3.2.1.1 Open Circuit PIN Diode

The standard PIN diode is often described as an intrinsic layer of material sandwiched between two heavily doped regions, with n-type material (N^+) on one side and p-type material (P^+) on the other (Figure 3.1(a)). The primary semiconductor element in widespread use is silicon, although GaAs, SiC, and GaN-based PIN diodes are also in use. Regardless of semiconductor material used, the nice cylindrical shape as shown in the figure is not the actual case but does provide a convenient way to view the general construction of the device. The typical discrete PIN diode is fabricated in a multiple-step process (Figure 3.1(b)). First, a semiconductor substrate is heavily doped with typically N^+ doping. The second step involves an epitaxial layer of semiconductor grown to the desired thickness, usually with a mesa-style construction. Finally, the P^+ doping is introduced on the top (or cap) of the mesa; metallization on both this cap and the substrate side constitutes the two PIN diode leads. With this construction and that shown in Figure 3.1, the term used for the diode is PIN. However, reversing the two heavily doped regions (P^+ substrate, N^+ cap) creates what is termed an NIP diode, the choice from a designer's and modeler's perspective being the direction of the dc biasing. Regardless of PIN or NIP, the anode will always refer to the P^+ side of the diode with the cathode on the N^+

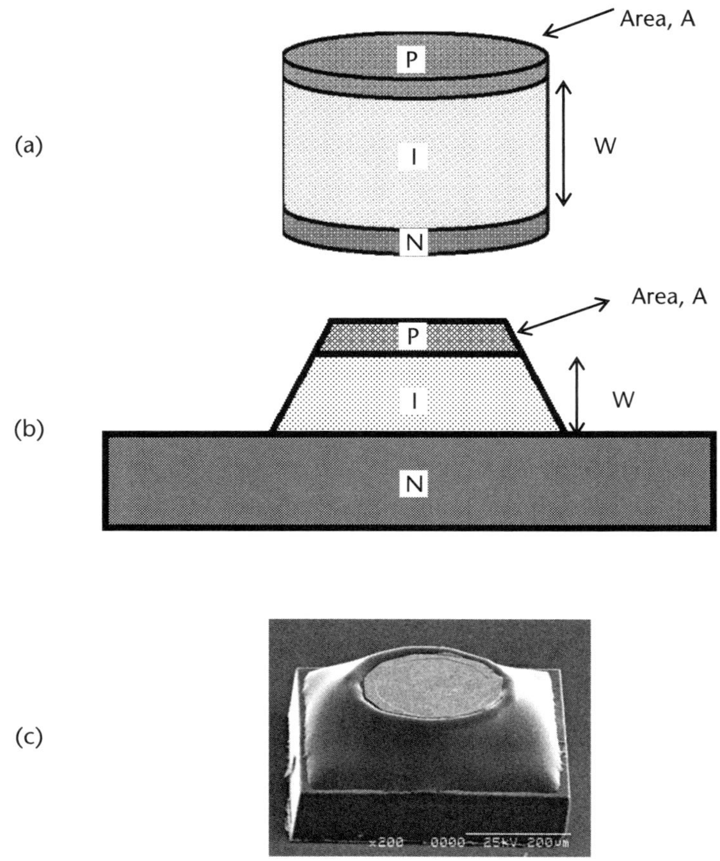

Figure 3.1 (a) Ideal PIN diode showing I-region thickness W and cross-section A. (b) Typical fabricated PIN diode showing mesa construction with P^+ material cap. (c) Scanning electron microscope photograph of a PIN diode with 150 μm mesa diameter. (Source: Courtesy of Cobham Metelics, 2015. Used with permission.)

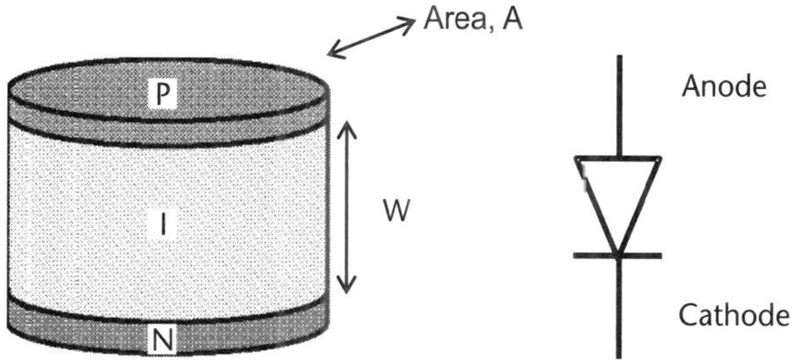

Figure 3.2 Circuit symbol for the PIN diode, with the Anode corresponding to the P+ contact.

end (see Figure 3.2). In the remainder of this book, PIN and NIP diodes will be used interchangeably.

In actual fabrication, a true intrinsic region is not really possible; however, current fabrication techniques can eliminate most background impurities to about two orders of magnitude above the intrinsic carrier limit ($n_i = 1.5 \times 10^{10}$ cm^{-3} in silicon at 300K), yielding a high resistivity I-region. Using mobility values in Table 3.1 for silicon, the theoretical limit for I-region resistivity (ρ) at 300 K is about 3,000 Ω-m (300,000 Ω-cm) in silicon, although the practical resistivity using current fabrication techniques is in the range of 3,000 to 10,000 Ω-cm. A dielectric relaxation frequency F_D for the I-region can be defined by using this resistivity value and the semiconductor permittivity value as

$$F_D = \frac{1}{2\pi\rho\varepsilon_r\varepsilon_0} \tag{3.1}$$

Knowledge of F_D is important because it is above this frequency that the dielectric permittivity can be considered constant. Using the free space permittivity value ε_0 = 8.85 pF/m and a silicon I-region resistivity of 10,000 Ω-cm (100 Ω-m, a value achievable commercially [10]) yields an F_D of approximately 15 MHz. For most microwave and RF applications of PIN diodes, operation should always be above F_D. For example, 20 to 50 Ω-cm I-region resistivity is usually sufficient for operation above 1,000 MHz, whereas I-region resistivities of 10,000 Ω-cm are necessary for operation in the megahertz range [10]. Operation near and below F_D also requires careful design to accommodate the varying permittivity [10]. The dielectric relaxation properties of the heavily doped end regions are of much less influence than that of the I-region and so are not usually considered.

Table 3.1 Table of Some Semiconductor Parameters [9]

Parameter	*Symbol*	*Silicon (Si)*	*Gallium Arsenide (GaAs)*
Intrinsic carrier concentration ((cm^{-3}) at 300K)	n_i	10^{10}	10^6
Electron mobility (m^2/V-s)	μ_n	0.15	0.85
Hole mobility (m^2/V-s)	μ_p	0.045	0.05
Relative permittivity (relative dielectric constant)	ε_r	11.9	16
Ambipolar Mobility (m^2/V-s)	μ_a	0.064	0.094

An alternative representation of (3.1) that uses lumped elements can be derived if the I-region is considered an unbiased (and open-circuited) cylinder of a lossy material (resistivity ρ and permittivity ε_r) with width W and cross-section A. Using the simple relationships for resistance and capacitance of a lossy cylinder, expressions for the open circuit resistance R_{oc} and open circuit capacitance C_{oc} are:

$$R_{oc} = \rho \frac{W}{A}; \quad C_{oc} = \frac{\varepsilon_r \varepsilon_0 A}{W} \tag{3.2}$$

The dielectric relaxation frequency F_D can be rewritten using these pure lumped elements as models for the open-circuited PIN diode (Figure 3.3):

$$F_D = \frac{1}{2\pi R_{oc} C_{oc}} \tag{3.3}$$

The open circuit values of R_{oc} and C_{oc} show that the PIN diode is not a true RF open circuit if placed across a transmission line but causes lossy and reactive loading. An order of magnitude approximation of this transmission line loading can be seen using the following example. Consider a nearly ideal silicon PIN diode with an I-region width $W = 50\ \mu$m and diameter $100\ \mu$m ($A = 7.85 \times 10^{-9}$ m^2). Using (3.2), the parallel equivalent circuit parameters of this open-circuited PIN diode assuming a 100 Ω-m I-region resistivity are

$$R_{oc} = \rho \frac{W}{A} = 10^2 \frac{50 \cdot 10^{-6}}{7.85 \cdot 10^{-9}} = 637\,\text{K}\Omega$$
$$C = \frac{\varepsilon_r \varepsilon_0 A}{W} = \frac{11.9 \cdot 8.85 \cdot 10^{-12} \cdot 7.85 \cdot 10^{-9}}{50 \cdot 10^{-6}} = 0.016\ \text{pF}$$

with a resulting dielectric relaxation frequency of approximately 15 MHz, as before.

Silicon-based PIN diodes are the most widely used device type in microwave and RF control applications. Other semiconductor technologies such as GaAs, GaN, and SiC have been used in the fabrication of PIN diodes. These wider band gap semiconductors exhibit higher I-region resistivities and better thermal properties than their silicon counterparts [38] and also are more easily integrated into SoC applications based on the same semiconductor technologies [39]. They are, however, more expensive to produce, still leaving the silicon-based PIN diode as the predominant diode-based microwave and RF control technology.

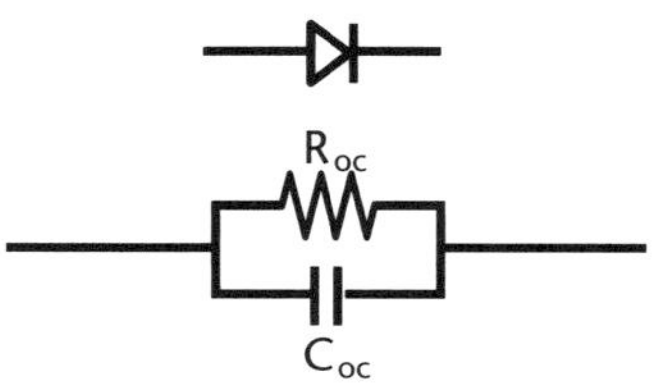

Figure 3.3 Open-circuited PIN diode lumped element equivalent circuit.

3.2.2 Forward Bias Operation

The PIN diode exhibits two PN-like junctions—the PI and IN interfaces—and behaves at dc in a similar manner as the traditional PN junction diode: if the anode is placed at a higher potential than the cathode, the energy barrier lowers, current flows, and forward bias conditions apply. The distinct difference from the PN junction case is that, under forward bias, holes injected from the heavily doped P^+ region and electrons injected from the heavily doped N^+ region both enter the highly resistive I-region, dramatically decreasing the effective resistivity of the region in a process called conductivity modulation. I-region resistivity drops many orders of magnitude from its open-circuited case as described above due to the increased amount of free or stored charge Q located in the I-region. Since higher dc currents cause more charges to be injected into the I-region, the resistivity lowers and is inversely proportional to this dc current. The injected holes and electrons do not immediately recombine in the I-region but rather exist for some average time termed the ambipolar (because both carrier types are present) I-region carrier lifetime, τ, or simply carrier lifetime (or sometimes even shortened to lifetime). This carrier lifetime has no relation to the failure time of the device but has an indirect relationship to the switching time (a topic for Section 3.5.2). The relationship between the stored charge Q and carrier lifetime τ is the dc forward current I_{dc}:

$$Q = I_{dc}\tau \tag{3.4}$$

The carrier lifetime can vary by several orders of magnitude, from less than 1 nanosecond (nS) to greater than 10 microseconds (μS).

This dc-injected stored charge Q reduces the I-region resistivity to very low values and is governed by a diffusion process, similar to the traditional PN junction device. Diffusion processes in semiconductors are relatively slow, and when the PIN diode is used at RF and microwave frequencies, junction effects have little influence and so the stored charge Q is unable to change with the varying signal and remains constant throughout the RF cycle. The constant stored charge Q keeps the resistivity of the I-region low and hence dramatically lowers the corresponding resistance R_S of the diode. The I-region RF resistance can be calculated with the aid of the resistance expression in (3.2). If we assume there are N carriers per unit volume (equal numbers of holes and electrons for an electrically neutral I-region) injected from their respective end regions of the PIN diode, the total stored charge Q in the I-region is the product of the electronic charge q, the carrier density per unit volume N, and the I-region volume AW, so the stored charge may be written as $Q = qNAW$. Since the resistivity is defined as $\rho = 1/\mathrm{N}q(\mu_n + \mu_p)$, an approximate expression for R_S can be written as

$$R_S = \rho\frac{W}{A} = \frac{W}{A}\frac{1}{qN\left(\mu_n + \mu_p\right)} = \frac{W}{A}\frac{1}{2qN\mu_a} = \frac{W}{A}\frac{AW}{2\mu_a Q} = \frac{W^2}{2\mu_a Q} = \frac{W^2}{2\mu_a I_{dc}\tau} \tag{3.5}$$

where the last term is the standard approximation of the PIN diode's RF resistance. In (3.5), the mobility sum $(\mu_n + \mu_p)$ has been replaced by the ambipolar mobility $\mu_a = 2\mu_n\mu_p/(\mu_n + \mu_p)$, which provides a more accurate representation of behavior under the high-injection condition of dc forward bias [8,9]. From a modeling perspective, this resistance R_S can be treated just like a traditional lumped element resistance at microwave and RF frequencies. From a designer's perspective, the designer has direct control of the dc bias current I_{dc}, indirect control of I-region thickness W and carrier lifetime τ, and no control over the mobility for a specific semiconductor type. The indirect control of W and τ is based on the PIN diode device type selected from the manufacturer's data sheets for a given device or for a particular process, in the case of custom fabrication.

At very low frequencies, the PIN diode RF resistance is primarily governed by the two junction regions and is equal to the slope of the dc I-V characteristic. An approximation to R_S in the low-frequency limit is

$$R_S = \frac{2V_T}{I_{dc}} \tag{3.6}$$

where V_T is called the thermal voltage, kT/q. At room temperature, V_T is approximately 26 mV and so the low-frequency limit of R_S is approximately 52Ω at 1 mA dc forward current.

The PIN diode has to somehow be connected to the outside world, and this usually requires physical wires or interconnects from the end cap metallizations on the heavily doped N^+ and P^+ to the package or other parts of the circuit. These interconnections will be inductive in nature, and so a more accurate model for the forward-biased PIN diode should include this inductance, usually termed the bond wire or contact inductance, L_{bond}, in series with intrinsic PIN diode (Figure 3.4).

(a)

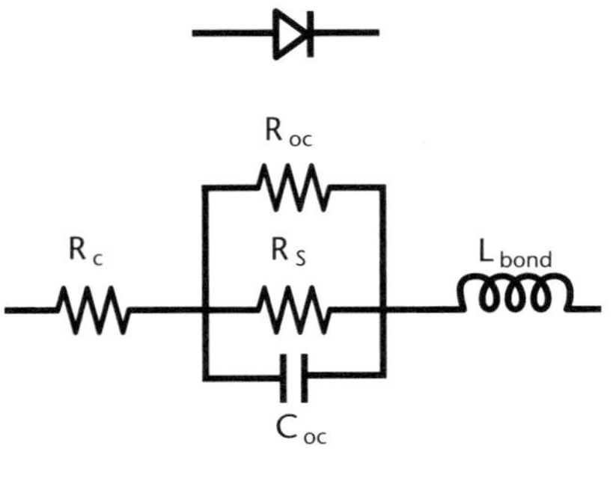

(b)

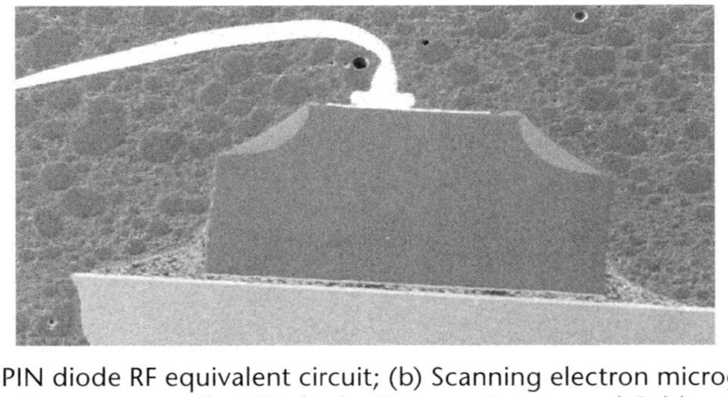

Figure 3.4 (a) Simple PIN diode RF equivalent circuit; (b) Scanning electron micrograph of a bond wire attached to the mesa cap of a PIN diode. (Source: Courtesy of Cobham Metelics, 2015. Used with permission.)

Finally, the resistance of the metalized contacts (so-called contact resistance) can be modeled as a single lumped resistance R_c.

Equation (3.5) indicates that PIN diode resistance R_S decreases monotonically from its open circuit value (which can be placed in parallel for more accurate modeling) with increasing dc forward current I_{dc}. Using the same characteristics as the previous example, the RF resistance as a function of dc forward current I_{dc} for a range of carrier lifetimes between 0.1 μS and 10.0 μS is plotted in Figure 3.5 assuming $R_c = 0.1\Omega$. Note that at low dc forward currents, the open-circuited resistance R_{oc}'s influence is observed, especially for the 0.1 μS device, as the resistance does not strictly follow a $1/I_{dc}$ relationship. For large forward currents, resistance values of less than 1 Ohm can be obtained, including contact resistance. An experimental way to determine R_c is to plot the RF resistance as a function of $1/I_{dc}$ (not I_{dc}) and noting the zero intercept ($1/I_{dc} = 0$) on the resistance axis.

3.2.3 Reverse Bias Operation

Reversing the dc bias polarity on the PIN diode (making the anode more negative with respect to the cathode) causes a similar effect as in a PN junction. However, instead of just sweeping charge out of the depletion region as in the PN junction

(a)
(b)

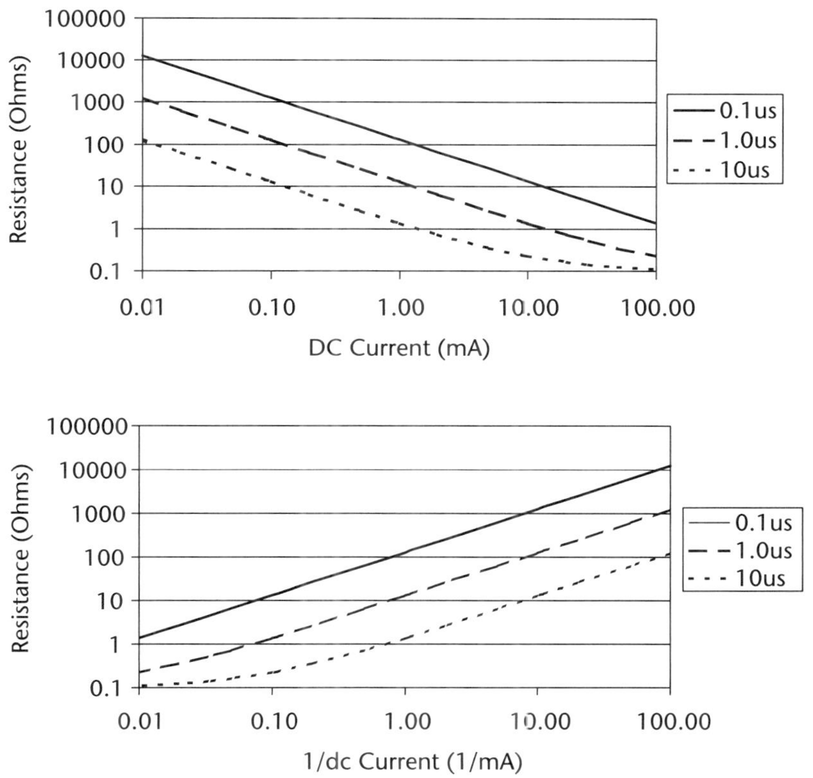

Figure 3.5 (a) RF resistance of the example PIN diode as a function of dc forward current using the carrier lifetime as a parameter. (b) RF resistance plotted versus $1/I_{dc}$ to determine the contact resistance.

device, charge is swept out of both the two junction regions as well as the I-region, causing the I-region to be fully depleted of charge and therefore acting as an insulating material. Since the two heavily doped end regions are highly conductive, a simple parallel plate capacitance structure is obtained and the reverse bias capacitance is the same as the open circuit value in (3.2). In practical PIN diodes, however, there is always a small residual resistivity in the I-region that can be caused by either the I-region itself or the passivation (covering around the PIN diode for protection) and this reverse bias element is usually placed in shunt with the capacitance and is called the shunting resistance or R_P. R_P values are usually in the 100 kΩ to 1 MΩ range.

This simple description of the PIN diode reverse bias capacitance is actually too simplistic. There is variation of the capacitance under reverse bias since at zero dc bias, the assumption of a fully depleted I-region (and hence perfectly insulating) is not completely valid [8]. Under PIN diode reverse bias, two distinct regions in the I-region of the device emerge: one completely free of mobile charge carriers (the so-called "swept" region), the other ("unswept") containing some charge carriers (Figure 3.6(a)) due to incomplete penetration of the depletion boundary [8]. Standard PN junction theory provides a means to measure this depletion boundary penetration W_D of the I-region as

$$W_D \cong \sqrt{\frac{2\varepsilon_r\varepsilon_0}{qn_I}\left(V_{bi} - V\right)}; \quad W_D \le W$$
$$V_{bi} = \frac{kT}{q}\ln\frac{Nn_I}{n_i^2} \tag{3.7}$$

where V is the applied voltage, n_I is the I-region impurity concentration, n_i is the intrinsic carrier concentration, and N is the doping on one of the heavily doped sides. W_D is about 2.5 μm for an I-region impurity concentration of 10^{12} cm^{-3} and V_{bi} = 0.7V.

At low frequency (less than the dielectric relaxation frequency of the I-region), the reverse shunt resistance is large and dependent on the unswept region's resistance. C_{RB} is primarily governed by the swept region capacitance, C_s. As the frequency increases, the reverse capacitance becomes the series combination of C_u and C_s, and since C_s is typically smaller than C_u, C_{RB} is the effective reverse capacitance. The shunt resistance becomes constant at high frequencies for a given reverse bias voltage. As the reverse voltage V increases in the negative direction, the depletion boundary W_D (dotted line in Figure 3.6(a)) moves so that at the so-called punchthrough voltage V_{PT}, the entire intrinsic region is finally "swept" clean of mobile charges, at which point a constant capacitance described by (3.2) is obtained. This capacitance value is termed the punchthrough capacitance, C_{PT}. Once punchthrough has been achieved, in addition to the constant capacitance C_{PT}, there is also a shunt equivalent resistance (of typically high value, R_P). Using the same simple physical models as in (3.2) yields the following equivalent circuit model expressions for $W_D < W$:

$$R_U = \rho_I\frac{W - W_D(V)}{A}; \quad C_U = \varepsilon_r\varepsilon_0\frac{A}{W - W_D(V)}; \quad C_S = \varepsilon_r\varepsilon_0\frac{A}{W_D(V)} \tag{3.8}$$

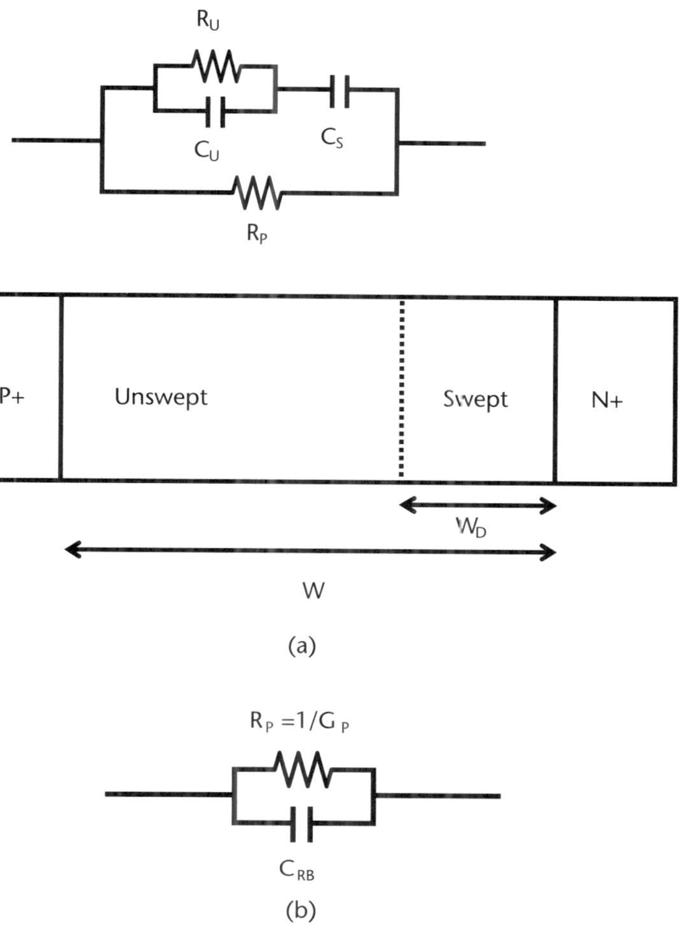

Figure 3.6 (a) Reverse bias lumped element equivalent circuit for PIN diode showing the physical origin of the various components. (b) Equivalent circuit for the reverse bias PIN diode impedance.

or, with explicit voltage dependence:

$$R_U = \rho_I \frac{W - \sqrt{\frac{2\varepsilon_r\varepsilon_0}{qn_I}\left(V_{bi} - V\right)}}{A}; \quad C_U = \varepsilon_r\varepsilon_0 \frac{A}{W - \sqrt{\frac{2\varepsilon_r\varepsilon_0}{qn_I}\left(V_{bi} - V\right)}};$$

$$C_S = \varepsilon_r\varepsilon_0 \frac{A}{\sqrt{\frac{2\varepsilon_r\varepsilon_0}{qn_I}\left(V_{bi} - V\right)}} \tag{3.9}$$

Note that when $W_D = W$, the I-region is completely swept clean of carriers and the unswept parallel equivalent circuit is now gone, leaving only $C_S = \varepsilon_r\varepsilon_0 A/W$,

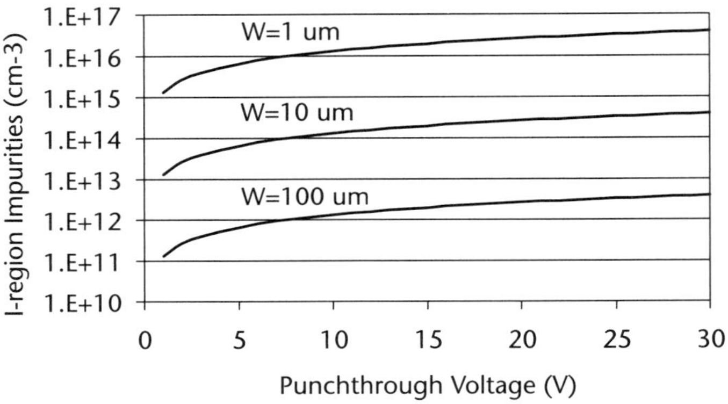

Figure 3.7 Punchthrough voltage V_{PT} as a function of background I-region impurity concentration for three different PIN diode I-region widths.

the value of the punchthrough capacitance C_{PT}. The voltage at which the I-region capacitance becomes constant is the voltage where $W = W_D$:

$$V_{PT} = \frac{qn_I W^2}{2\varepsilon_r \varepsilon_0} - V_{bi} = \frac{qn_I W^2}{2\varepsilon_r \varepsilon_0} - \frac{kT}{q}\ln\frac{Nn_I}{n_i^2} \approx \frac{qn_I W^2}{2\varepsilon_r \varepsilon_0} \tag{3.10}$$

In many analyses [8], the V_{bi} term is just left off of the equation for V_{PT}. For low I-region background carrier density n_I, the punchthrough voltage is nearly zero, in which case the PIN diode is referred to as a zero bias punchthrough device. Figure 3.7 shows a plot of the punchthrough voltage as a function of I-region thickness W and I-region background impurity n_I. Note that for an I-region background impurity concentration of 10^{12} cm^{-3}, for example, both the 1 μm and 10 μm diodes go to a V_{PT} of near zero (and therefore are zero bias punchthrough and fully depleted at zero bias), whereas the 100 μm PIN diode requires additional reverse bias of approximately 7V to have a fully depleted I-region. A background I-region impurity concentration of 10^{13} cm^{-3} will require more than 30V of reverse bias to be fully depleted at punchthrough. For this reason, the lowest I-region impurity concentration that can be obtained through the fabrication process is needed for these thick PIN diodes to keep the punchthrough voltage down to a reasonable level and not require higher reverse voltages to reach the constant capacitance state.

Parallel resistance R_P in Figure 3.6 models any residual I-region or passivation effects on the reverse bias impedance. These passivation effects are highly resistive (but not infinite resistivity) and are often paths along the surface of the PIN diode at the interface of an oxide or glass covering around the diode for protection. These passivations are typically very good, and so the shunting resistance R_P can easily be 1 MΩ and is usually much higher.

3.3 PIN Diode Equivalent Circuit Models

The previous discussion shows that simple lumped elements can be used to model the ideal behavior of the PIN diode at microwave and RF frequencies. Various workers, however, have come up with interesting variations on these lumped element models, and this section will outline those methods.

3.3.1 Lumped Element Model

The lumped elements outlined in the previous section can be used in creating a simple model useful in modeling PIN diode behavior in circuits. Figure 3.8 shows the models separately for the two device states: dc forward bias (or on) and dc reverse bias (or off). For the on-state, R_S is usually much smaller than the open circuit resistance and so R_{oc} can easily be removed from the circuit model. Likewise, the capacitive reactance of C is usually much higher than the resistive value of R_S and so can be removed as well, resulting in the equivalent circuit model in Figure 3.8(a). In the off-state, R_S and R_{oc} can be removed from consideration since they are ideally infinite, leaving the diode capacitance as the critical modeling element (Figure 3.8(b)). The extra component values L_{bond} and R_c, however, must be included in both state's models.

The utility of using these two models, depending on bias state, is that no connection to dc biasing networks has to be considered; simply place the models in the circuit simulator using lumped elements or use standard circuit analysis techniques. The downside, of course, is that the circuit modeler has to change the circuit elements, primarily R_S, for every change in I_{dc}, which also does not help in simulating dynamic behavior of the device.

3.3.2 Current and Voltage-Dependent Models

In the simple lumped element model, there is no means to include the PN junction as part of the simulation unless added by the modeler. A number of next-level lumped element-based models include a PN diode as part of the overall PIN diode model. Extensive research efforts over the decades have led to sophisticated SPICE (Simulated Program with Integrated Circuit Emphasis)-based models for the PIN diode [11, 12]. A number of these next-level models add one or two PN diodes (two diodes added to model the PI and IN junctions separately) and use traditional SPICE parameters for their description.

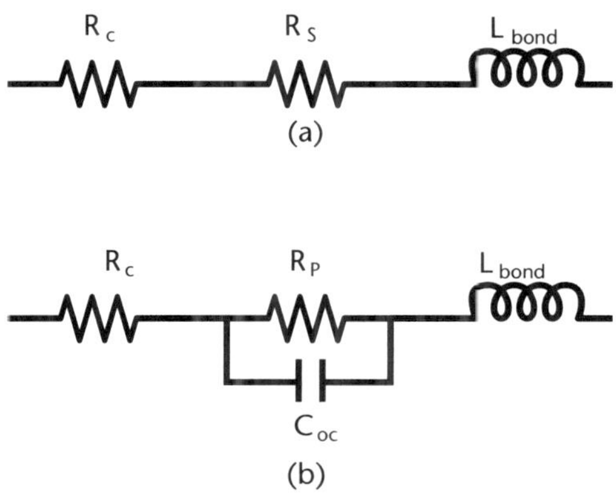

Figure 3.8 (a) On-state (forward bias) and (b) off-state (reverse bias) lumped element models for the PIN diode.

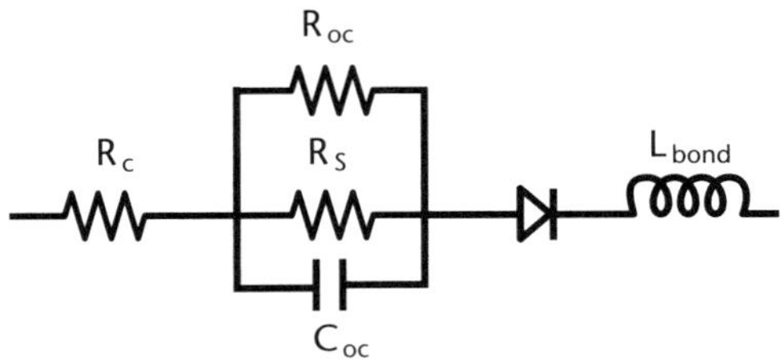

Figure 3.9 Walston PIN diode model suitable for SPICE simulation.

One of the more popular next-level PIN diode models was developed by Walston [11] and builds on the lumped equivalent circuit model in Figure 3.8 with an added series PN junction (Figure 3.9). In the Walston model, the series resistance R_S, being a function of the dc forward current, I_{dc}, is modeled as a current-dependent resistor based on the current flowing through the circuit, including the PN junction. Resistance R_S is modeled using the equation

$$R_S = \frac{K}{I_{dc}^N} \tag{3.11}$$

Inspection of (3.11) shows that the numerator value of K can be estimated as

$$K = \frac{W^2}{\left(\mu_n + \mu_p\right)\tau} \tag{3.12}$$

N can be easily varied to fit measured dc data but is usually left as unity. The other elements in the Walston model (R_c, L_{bond} and C_{oc}) have their same meaning as before. SPICE parameters can be used to model the PN junction for the Walston PIN diode model, and Table 3.2 shows some of the more important parameters and their definitions.

With the Walston model, the variation of PN junction capacitance is covered by the SPICE model equations; however, the variation of the I-region capacitance with voltage is not modeled. This can be a problem with using the Walston model at low frequencies, but this problem is minimized at frequencies significantly above the punchthrough voltage and the associated constant capacitance. Another limitation

Table 3.2 Abbreviated Table of SPICE PN Junction Modeling Parameters [13]

Parameter	*Definition*	*Unit*	*Default*
BV	Reverse breakdown knee voltage	Volt	infinite
CJO	zero-bias *p-n* capacitance	farad	0.0
IS	saturation current	amp	1E-14
TT	transit time	sec	0.0
VJ	*p-n* junction potential	volt	1.0

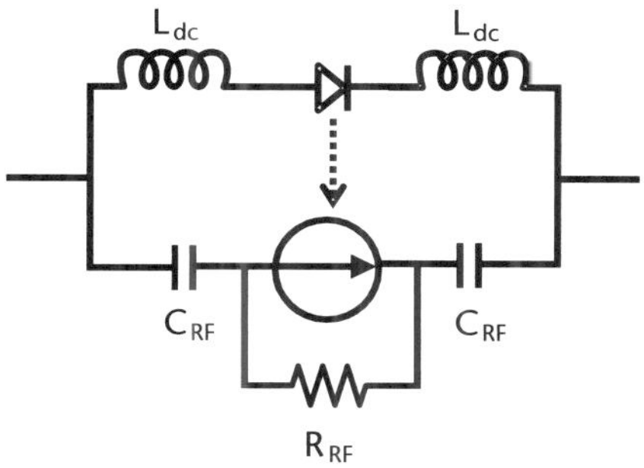

Figure 3.10 Jang model for the PIN diode showing the dc and RF paths separated by passive reactive routing (L_{dc}, C_{RF}).

of the Walston model is the I-region is modeled only as a pure resistance; later sections of this book will show that there are significant diffusion effects in the I-region that must be considered and will influence the device's lower frequency behavior.

A SPICE-based PIN diode model that shares similar aspects of the Walston model was developed by Jang and colleagues [12], who split the diode model into two different paths, one to model the dc behavior and the other to model the RF behavior (see Figure 3.10). The PN junction device parameters are set to model the dc *I-V* curve of the PIN diode but are removed from the RF path by large value inductors (and their respective high reactances) at both the anode and cathode. The I-region resistance is modeled using a current controlled current source in the RF path, with the controlling current the dc current in the PN junction diode. The controlled current source gain provides the resistance conversion. Large value capacitors at the input and output of the controlled current source remove the controlled source from the dc path but provide low reactance at RF frequencies. The Jang model also shares many of the limitations as the Walston model but does remove the RF resistance in series with the PN junction from influencing the dc response.

3.4 Integral-Based PIN Diode Model—Forward Bias

The previous section looked at purely lumped element-based models for the PIN diode or used lumped elements with the addition of traditional SPICE-based elements to model the PN junctions. A number of assumptions were made that dramatically simplified the forward bias analysis and subsequent models:

The frequency of operation was high enough so that only the I-region behavior was considered; PN junction effects were negligible.

A homogenous level of stored charge Q originating from carriers injected from the P^+ and N^+ end regions resided in the I-region and did not change with applied RF signal level.

The I-region stored charge does not vary with applied RF signal and is strictly governed by injected dc current.

When PIN diodes were first introduced, they were used primarily at VHF frequencies and higher, and so the first assumption was a valid one at the time and provided the physical basis for the subsequent development of the first PIN diode models. However, applications using PIN diodes at frequencies in the HF band and lower (less than a few megahertz) have uncovered limitations to these models. In addition, the first assumption suggests the question, "what is a high enough frequency?" One metric, the dielectric relaxation frequency, has already been discussed, but this metric is solely based on the intrinsic properties of the I-region (ρ and ε). A variety of studies have shown that another metric for the "high enough" frequency is related to the inverse of the carrier lifetime $1/\tau$. Since τ can range from nS to μS, the high-frequency metric can range from kilohertz to hundreds of megahertz. Standard PN junctions also show frequency dependent behavior in the forward direction that needs to be fully described to understand and accurately model the PIN diode.

The second assumption, homogenous charge in the I-region, is also too simple of an assumption. The dc charge injected from the two heavily doped regions does suffer some recombination in the I-region the further the carriers transit into the I-region. This recombination is also related by the I-region carrier lifetime, τ, through the recombination length parameter, L_a (not to be confused with the inductance).

The third assumption, no change in stored charge Q with applied RF signal, is too simplistic as well. The applied RF signal can modulate the I-region stored charge, but the level of charge modulation depends on a number of factors including the carrier lifetime, dc forward current, and level of applied RF current. White [14] provided a very simple explanation of this by looking at the time dependent charge equation, noting that once charge ceases to be injected, charge decays at a rate described by the simple time function $e^{-t/\tau}$. White then applied the same reasoning to the alternating cycles of an RF signal. A 1.0 GHz signal has a period of 1.0 nS or 0.5nS for each half cycle. For a 1.0 μS carrier lifetime PIN diode, the $e^{-t/\tau}$ equation indicates 99.95% of the charge will still be present at the end of the negative half cycle, with a corresponding 0.05% added during the positive half cycle. On the other hand, the same diode with a 1.0 MHz signal indicates approximately 60% of the charge remains, so clearly low-frequency signals (compared to the carrier lifetime) can cause significant changes to the level of I-region stored charge and can impact the RF resistance accordingly. Ryder aptly described this charge phenomenon that, for the PIN diode, the RF signal is actually the *small* signal and the dc forward current is the *large* signal [15].

These phenomena and the assumptions used in the simpler models require a very detailed look at the fundamental equations governing carrier transport in all regions of the PIN diode. In this section, a full mathematical treatment of the PIN diode for each of the three regions (PI junction, I-region and IN junction) will be performed to more accurately model the diode at all frequencies. This resulting expression can be used for any type of PIN diode, whether Si, GaAs, or other technology since the only factors that govern the difference between the technologies are the carrier mobilities, dielectric permittivities, and the intrinsic carrier concentrations. The derivation, outlined in detail in such references as [16–19], seeks to develop an expression for the voltage drop across all three major regions of the PIN diode. This derivation then provides the mathematical basis for a total PIN diode impedance that exhibits both resistive and reactive parts, $R_x + jX_x$ (where the subscript x indicates the PI or

Figure 3.11 Detailed forward bias model for the PIN diode with contributions from all three regions of the device and includes series inductance and contact resistance.

IN junctions and the I-region, PI, IN, I) for each region. The impedance calculated is valid from low frequencies where the impedance is described by the slope of the dc I-V characteristic ($2V_T/I_{dc}$) to its value at microwave and RF frequencies (3.6). The simple equivalent circuit in Figure 3.8 needs to be significantly expanded to include these new elements (Figure 3.11).

3.4.1 Linear Modeling—One Dimensional

The full mathematical treatment for the PIN diode starts with the ambipolar carrier transport equation in a single dimension, x, given by [18–20]

$$\frac{\partial n(x,t)}{\partial t} = -\frac{n(x,t)}{\tau} + D_a \frac{\partial^2 n(x,t)}{\partial x^2} \tag{3.13}$$

where $n(x,t)$ is the carrier density throughout the I-region (Figure 3.12), τ is the ambipolar carrier lifetime, and D_a is the ambipolar diffusivity given by

$$D_a = \frac{kT}{q}\frac{2\mu_n\mu_p}{\mu_n + \mu_p} \tag{3.14}$$

The analysis assumes $n(x,t)$ is a linear combination of the dc and RF charge densities, with the static dc carrier density term denoted $n_0(x)$ and the time-varying term $n_{RF}(x)e^{j\omega t}$:

$$n(x,t) = n_0(x) + RE\left[n_{RF}(x)e^{j\omega t}\right] \tag{3.15}$$

The carrier transport equation as shown in (3.13) assumes that quasi-charge neutrality holds in the I-region: $n \approx p$; $\partial n/\partial x \approx \partial p/\partial x$. The carrier transport equation in (3.13) can be split into two components, one for each of the carrier densities in (3.15):

$$\frac{\partial^2 n_0(x,t)}{\partial x^2} - \frac{n_0(x,t)}{L_a^2} = 0; \quad \frac{\partial^2 n_{RF}(x,t)}{\partial x^2} - \frac{n_{RF}(x,t)}{\lambda_a^2} = 0 \tag{3.16}$$

where $L_a^2 = D_a\tau$ and $\lambda_a^2 = D_a\tau/(1 + j\omega\tau)$ are the dc and RF recombination lengths in the I-region, respectively. The carrier density $n(x,t)$ is supported by both a dc and a single RF current flowing through the device:

$$I(t) = I_{dc} + RE\left(I_k e^{j\omega_k t}\right) = I_{dc} + I_{RF}\cos(\omega t) \tag{3.17}$$

where I_{dc} supports n_0 and I_{RF} supports n_{RF}.

At the PI boundary ($x = -W/2$), the boundary conditions require a hole-dominated diffusion current with the boundary condition written as

$$I(t) = I\left(-\frac{W}{2}\right) = -2qD_pA\left.\frac{\partial n(x,t)}{\partial x}\right|_{x=-W/2} \tag{3.18a}$$

Similarly, for electrons at the IN boundary ($x = +W/2$)

$$I(t) = I\left(\frac{W}{2}\right) = 2qD_nA\left.\frac{\partial n(x,t)}{\partial x}\right|_{x=+W/2} \tag{3.18b}$$

Combining the results of (3.16) through (3.18) yields both dc and RF current densities in the I-region:

$$n_0(x) = \frac{I_{dc}L_a}{2qD_nD_pA}\frac{D_p\cosh\left(\frac{W+2x}{2L_a}\right)+D_n\cosh\left(\frac{W-2x}{2L_a}\right)}{\sinh\frac{W}{L_a}} \tag{3.19a}$$

and

$$n_{RF}(x) = \frac{I_{RF}\lambda_a}{2qD_nD_pA}\frac{D_p\cosh\left(\frac{W+2x}{2\lambda_a}\right)+D_n\cosh\left(\frac{W-2x}{2\lambda_a}\right)}{\sinh\frac{W}{\lambda_a}} \tag{3.19b}$$

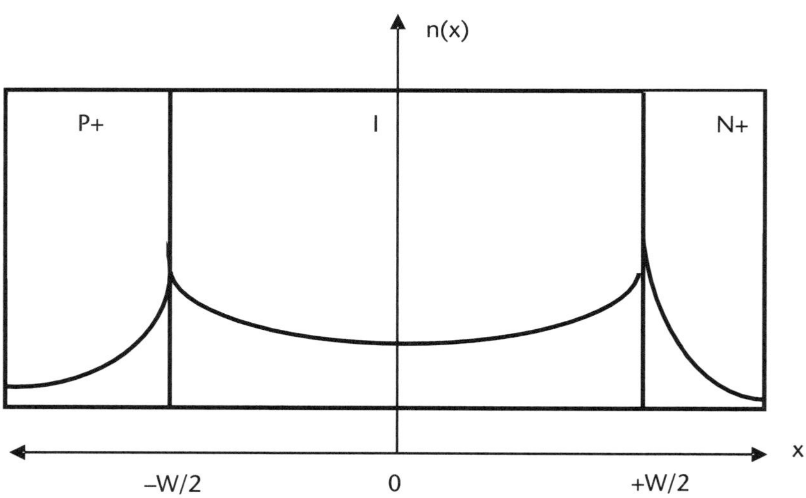

Figure 3.12 PIN diode carrier density as a function of position variable *x*.

Using the electromagnetic relationship between electric field E, current density $J(I/A)$, and conductivity σ, the I-region electric field $E(x,t)$ can be written as

$$E(x,t) = \frac{J(t)}{\sigma} = \frac{I(t)}{A}\frac{1}{qn(x,t)\left(\mu_n + \mu_p\right)} \tag{3.20}$$

where $E(x,t)$ is simply the I-region current density $J(t)$ divided by the I-region conductivity.

Earlier in this chapter, the I-region charge change due to the RF signal was shown to be relatively small, and using the approximation $1/(1 + x) \approx 1 - x$ for small x provides a way to linearize the $1/n(x,t)$ term in (3.20):

$$\frac{1}{n(x,t)} = \frac{1}{n_0(x) + n_{\mathrm{RF}}(x)} = \frac{1}{n_0(x)\left(1 + \dfrac{n_{\mathrm{RF}}(x)}{n_0(x)}\right)} \cong \frac{1}{n_0(x)}\left(1 - \frac{n_{\mathrm{RF}}(x)}{n_0(x)}\right) \tag{3.21}$$

Looking at the coefficients of the two charge density equations (3.19a) and (3.19b) shows that for the "small x" assumption to hold, the following expression must be less than unity:

$$\frac{\dfrac{I_{\mathrm{RF}}\lambda_a}{2qD_nD_pA}}{\dfrac{I_{dc}L_a}{2qD_nD_pA}} = \frac{I_{\mathrm{RF}}}{I_{dc}\sqrt{1 + j\omega\tau}} \tag{3.22}$$

which is approximately $I_{\mathrm{RF}}/(I_{\mathrm{dc}}\sqrt{\omega\tau})$ for large $\omega\tau$. The assumption is even valid at extremely low frequencies ($\omega\tau$ less than unity) as long as the RF current is less than the dc forward current. For example, with 10 mA dc forward bias current, the approximation is always valid up to 10 mA of RF current (approximately 4 dBm in a 50Ω system). For $\omega\tau = 100$ (frequency significantly higher than $1/\tau$), up to 100 mA of RF current (approximately 24 dBm in a 50Ω system) will still keep the assumption valid even though I_{dc} remains fixed at 10 mA.

The I-region voltage drop is the integral of $E(x,t)$ over the I-region length $-W/2$ to $W/2$, mathematically written as

$$V_I(t) = \frac{1}{qA\left(\mu_n + \mu_p\right)}\int_{-W/2}^{W/2}\left(I_{\mathrm{RF}}(t)\frac{1}{n_0(x)} - I_{\mathrm{dc}}\frac{n_{\mathrm{RF}}(x,t)}{n_0^2(x)}\right)dx \tag{3.23}$$

The I-region impedance is then written as the ratio of the I-region voltage drop divided by the RF current, $V_I(t)/I_{\mathrm{RF}}(t)$:

$$Z_I = R_I + jX_I = \frac{V_I(t)}{I_{\mathrm{RF}}(t)} = \frac{1}{qA\left(\mu_n + \mu_p\right)}\int_{-W/2}^{W/2}\frac{1}{n_0(x)}\left(1 - \frac{I_{\mathrm{dc}}}{I_{\mathrm{RF}}}\frac{n_{\mathrm{RF}}(x)}{n_0(x)}\right)dx \tag{3.24}$$

where the expressions for $n_0(x)$ and $n_{RF}(x)$ are described by (3.19).

The PI and IN junction voltage drops are computed from standard PN "law of the junction" [9, 21–24] theory, which states that the injected carrier density is the product of the background carrier density (n_i in the case of the I-region) and the PN junction voltage exponential driving function $e^{qv_J/kT}$:

$$n(X,t) = n_i e^{qV_J(X,t)/kT}$$
$$\Rightarrow V_J(X,t) = \frac{kT}{q}\ln\left(\frac{n(X,t)}{n_i}\right) = \frac{kT}{q}\ln\left(\frac{1}{n_i}n_0(X)\left(1+\frac{n_{RF}(X,t)}{n_0(X)}\right)\right) \tag{3.25}$$

where X is $-W/2$ for the PI junction and $W/2$ for the IN junction. Using the approximation $\ln(1 + x) \approx x$ for small x since the RF charge density will be smaller than the dc charge density (3.22), and keeping only those time-varying terms yields the junction voltage and corresponding impedance as V_J/I_{RF} at boundary X:

$$V_J(X,t) = \frac{kT}{q}\frac{n_{RF}(X,t)}{n_0(X)} \tag{3.26}$$

and

$$Z_J = R_J + jX_J = \frac{V_J(X)}{I_{RF}} = \frac{kT}{q}\frac{n_{RF}(X)}{n_0(X)}\frac{1}{I_{RF}} \tag{3.27}$$

where X is $-W/2$ for the PI junction (Z_{PI}) and $W/2$ for the IN junction (Z_{IN}). The total PIN diode impedance is the series combination of the three impedance terms in (3.24) and (3.27) (two junctions):

$$\begin{aligned} Z_{PIN} &= Z_{PI} + Z_I + Z_{IN} = RE\left(Z_{PI} + Z_I + Z_{IN}\right) + jIM\left(Z_{PI} + Z_I + Z_{IN}\right) \\ &= \left(R_{PI} + R_I + R_{IN}\right) + j\left(X_{PI} + X_I + X_{IN}\right) \end{aligned} \tag{3.28}$$

which shows that the PIN diode impedance contains both resistive and reactive components. These resistance and reactive terms are not the parasitics that were mentioned earlier (L_{bond} and R_c, for example) but are inherent to the PIN diode.

3.4.2 Recombination in the Heavily Doped Regions

Recombination in the heavily doped end regions, a phenomenon not treated in the previous analysis, has been studied to help explain the lower impedance at low frequencies [25]. The full mathematical treatment in [25] of this phenomenon shows an overestimation of the diode impedance at low frequencies and a slight underestimation at high frequencies using the previous method. The end region recombination modifies the boundary condition at the junctions (3.18) and can be written for the PI junction as [25]

$$I(t) = I\left(-\frac{W}{2}\right) = -2qD_pA\left[\frac{\partial n(x,t)}{\partial x} - \frac{h_p}{D_p}n^2(x,t)\bigg|_{x=-W/2}\right] \quad (3.29a)$$

and for the IN junction as

$$I(t) = I\left(\frac{W}{2}\right) = 2qD_nA\left[\frac{\partial n(x,t)}{\partial x} + \frac{h_n}{D_n}n^2(x,t)\bigg|_{x=W/2}\right] \quad (3.29b)$$

where h_n and h_p are emitter recombination parameters defined in [26]. Note that if h_n and h_p are zero, the boundary conditions are the same as the previous result and so the carrier densities will be as well. The modified dc and RF carrier densities are calculated as before and may be written as

$$n_0(x) = \frac{L_a}{\sinh\frac{W}{L_a}}\left[\begin{array}{l}\left(\frac{1}{2}\frac{I_{dc}}{qAD_n} - \frac{h_n}{D_n}n^2{}_0\left(\frac{W}{2}\right)\right)\cosh\left(\frac{W-2x}{2L_a}\right)\\ +\left(\frac{1}{2}\frac{I_{dc}}{qAD_p} + \frac{h_p}{D_p}n^2{}_0\left(-\frac{W}{2}\right)\right)\cosh\left(\frac{W-2x}{2L_a}\right)\end{array}\right] \quad (3.30a)$$

and

$$n_{RF}(x) = \frac{\lambda_a}{\sinh\frac{W}{\lambda_a}}\left[\begin{array}{l}\left(\frac{1}{2}\frac{I_{RF}}{qAD_n} - \frac{h_n}{D_n}2n_0\left(\frac{W}{2}\right)n_{RF}\left(\frac{W}{2}\right)\right)\cosh\left(\frac{W+2x}{2\lambda_a}\right)\\ +\left(\frac{1}{2}\frac{I_{RF}}{qAD_p} + \frac{h_p}{D_p}2n_0\left(-\frac{W}{2}\right)n_{RF}\left(-\frac{W}{2}\right)\right)\cosh\left(\frac{W-2x}{2\lambda_a}\right)\end{array}\right] \quad (3.30b)$$

Note that (3.30) and (3.19) are identical if the h-parameter is zero (in other words, no recombination). The calculation of PIN diode impedance using these modified carrier densities follows the same approach as above. This more accurate approach taking into account end region recombination is mostly observed under very high injection where n_0 is large. The end region recombination tends to suppress the PIN diode impedance at lower frequencies, as it is primarily a junction effect. In any case, the two models approach similar results at frequencies well above $1/\tau$.

3.4.3 I-Region Charge Density

Equation (3.19a) can be used to plot the dc I-region charge density that constitutes the stored charge Q. Figure 3.13 shows the shape of this dc charge density as a function of normalized position x/W with the recombination length ratio W/L_a used as a parameter for a Si PIN diode. The figure shows that for small W/L_a, the dc charge density does not vary much throughout the I-region, whereas for larger W/L_a, there is a deep minimum toward the center of the I-region. The figure also

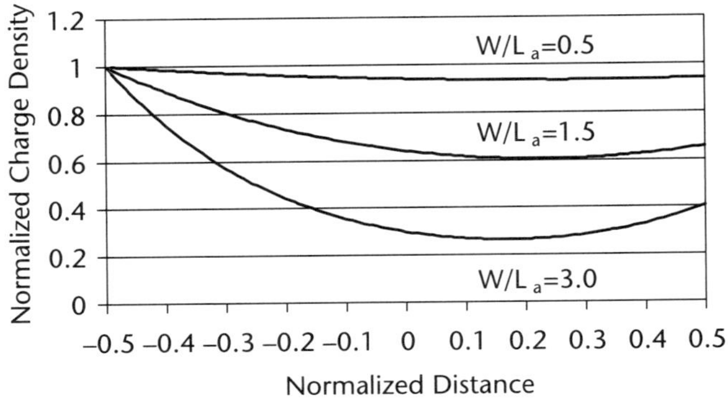

Figure 3.13 Normalized injected I-region charge density (normalized to the value at $X/W = -0.5$) as a function of normalized distance into the I-region (0 is the center of the I-region). Silicon hole and electron mobilities have been assumed.

shows there are a higher number of electrons injected from the P^+ region than holes from the N^+ region in all cases.

In Section 3.2.2, it was mentioned that there was little change in I-region stored charge from the injected RF current. Figure 3.14 shows a plot of the ratio $|n_{RF}(x)/n_0(x)|$ as defined by (3.19) as a function of position in the I-region for $\omega\tau$ values of 0.1, 1.0, and 10, $W/L_a = 0.5$ and $I_{RF}/I_{dc} = 0.1$. The figure shows that at low frequencies ($\omega\tau$ unity or less), the injected RF charge follows the stored charge, implying that the I-region carrier density closely tracks the RF signal, with significant amount of charge being injected or removed during each RF cycle. At higher frequencies ($\omega\tau$ greater than 10), the injected RF charge is significantly less than the dc stored charge, meaning that the dc stored charge is the dominant charge storage mechanism in the I-region. At these higher frequencies, the traditional microwave/RF resistance equation (3.5) holds. For frequencies much below $10/2\pi\tau$, a more detailed analysis of the RF impedance must be performed. That discussion is the focus in the next section.

Figure 3.15(a) shows the additional charge injected by the RF signal in terms of the dc charge density for $\omega\tau$ equal to 100. Note that the larger figure shows very little additional charge in the I-region from the RF signal; Figure 3.15(b) uses an

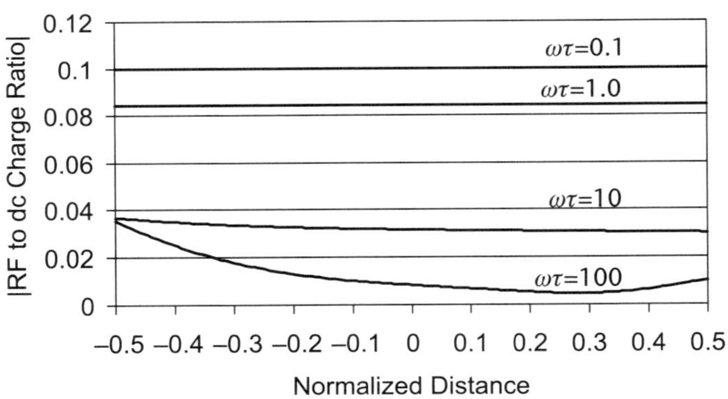

Figure 3.14 Ratio of $|n_{RF}/n_0|$ in the I-region as a function of normalized distance in the I-region for $I_{RF}/I_{dc} = 0.1$. For small $\omega\tau$, the ratio is 0.1 and decreases as $\omega\tau$ increases.

enhanced scale. The largest additional charge occurs at the PI junction and is only about 4% additional charge due to the RF signal, with only about 1% additional charge injected from the IN junction. The same conclusion can be reached as above—the I-region stored charge is essentially constant at high frequencies.

The variation of the PIN diode forward bias characteristics with RF current does not just impact the impedance-frequency characteristic. The RF current does move small amounts of charge in and out of the I-region as a function of time, making Z_{PIN} also a function of time. This time-varying impedance (not to be confused with the frequency varying impedance of the previous discussion) leads to nonlinear behavior of the PIN diode under forward bias, a phenomenon covered in detail in the next chapter.

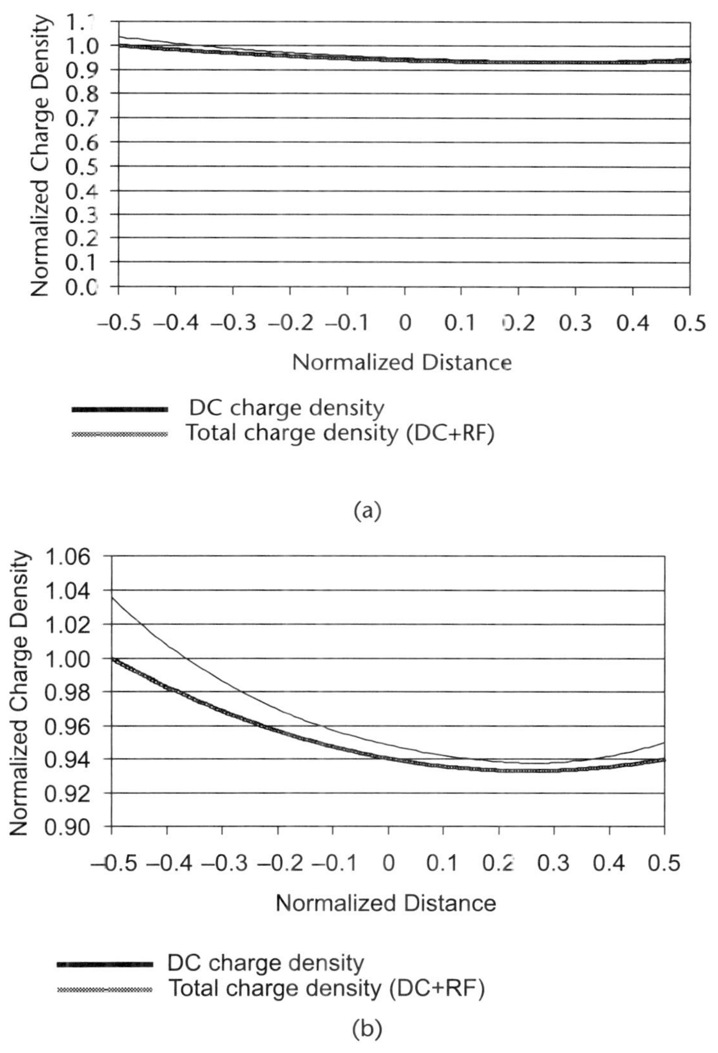

Figure 3.15 (a) Additional charge injected into the I-region as a function of normalized distance from an applied RF signal of 10% of the dc bias current. (b) Expanded view of the additional injected charge.

3.4.4 Linear Modeling—Multidimensional

The ideal three-dimensional model of the PIN diode assumes a cylindrical shape (Figure 3.16), whereas the previous discussion only looked at the single axial x-dimension of the PIN diode. In the earlier, simpler case, the carrier lifetime was assumed constant throughout the I-region, regardless of the shape of the diode itself. In practical PIN diodes, the passivation of the side wall surface impacts the overall carrier lifetime in the device. If the dc-injected carriers are completely constant throughout the cross section right out to the surface, then the overall or effective carrier lifetime is the same as the carrier lifetime in the bulk material. On the other hand, if the surface conditions are such that no carriers exist at the surface, the effective carrier lifetime will be reduced from its bulk value. The previous analysis of the PIN diode carrier density and impedance assumed the former case (constant throughout). Insight into the impact of the surface can be gained by performing a three-dimensional geometric analysis of the relationship between the effective carrier lifetime that includes effects of the surface and the bulk carrier lifetime. Instead of the carriers being uniformly distributed throughout the I-region, the boundary conditions at the surface of the PIN diode can be selected so that no carriers existed at the outside surface (so-called infinite surface recombination velocity) [9, 36]. The one-dimensional carrier transport equation in (3.13) has to be expanded to look at the other two dimensions using a cylindrical coordinate system as shown in Figure 3.16. The carrier transport equation then becomes

$$\frac{\partial n(x,r)}{\partial t} = -\frac{n(x,r)}{\tau_{\text{bulk}}} + D_a \frac{\partial^2 n(x,r)}{\partial x^2} + D_a \frac{1}{r}\frac{\partial}{\partial r}\left(r \frac{\partial n(x,r)}{\partial r}\right) \tag{3.31}$$

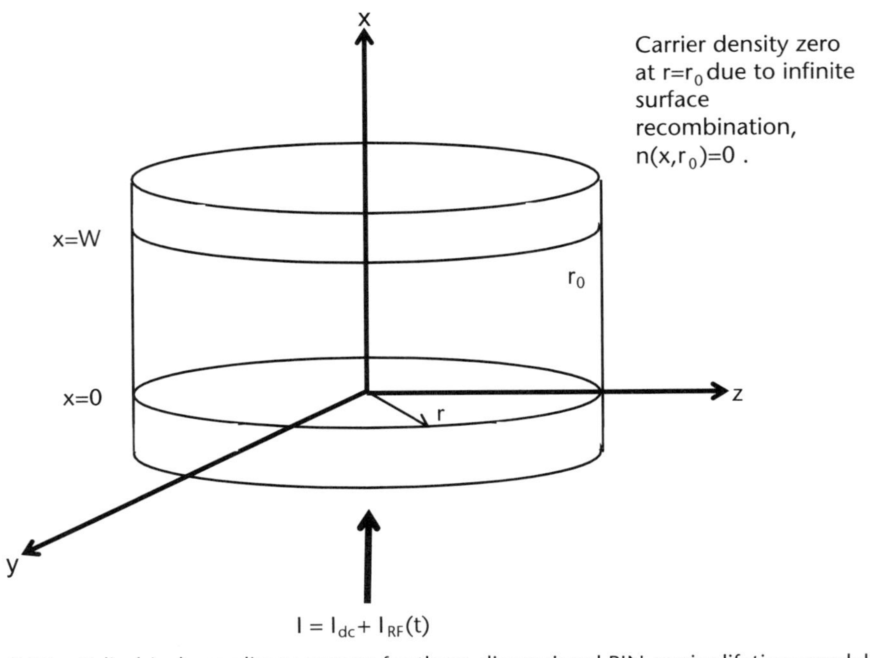

Figure 3.16 Cylindrical coordinate system for three-dimensional PIN carrier lifetime modeling.

A solution based on the Fourier-Bessel series was used in [18, 36], with the previous boundary conditions as described in (3.18) and with the additional boundary condition of infinite surface recombination (carrier density goes to zero at $r = r_0$). The resulting form for the I-region carrier density can be written as

$$n_I(x,r) = \sum_{m=1}^{\infty} A_m J_0(a_m r)\left[D_p \cosh\left(\tfrac{W+2x}{2\gamma_m}\right) + D_n \cosh\left(\tfrac{W-2x}{2\gamma_m}\right)\right] \tag{3.32}$$

where the subscript of n_I indicates either dc (n_0) or ac (n_{RF}) conditions. The other parameters in the equation are $\alpha_m = z_m/r_0$ where the z_m are the mth zeroes of the zeroth order Bessel function $[J_0(a_m r)]$, the A_m are defined by the boundary conditions in (3.18), $\gamma_m^2 = L_a^2/(1 + L_a\alpha_m)^2)$, and $L_a = \sqrt{D_A \tau_{bulk}}$ is the ambipolar diffusion length. Integrating $n_I(x,r)$ over the entire volume of the I-region provides a measure of the total charge, and using the relationship for the stored charge ($Q = I_{dc}\tau$) for both the ideal surface and zero carrier surface conditions allows a ratio of the effective to bulk carrier lifetime to be defined [36]:

$$\frac{\tau_{eff}}{\tau_{bulk}} = \frac{\displaystyle\sum_{m=1}^{\infty} \frac{\sinh(W/\gamma_m)}{z_m^2\sqrt{1+(L_a\alpha_m)^2}\left(D_n + D_p\cosh(W/\gamma_m)\right)}}{\displaystyle\sum_{m=1}^{\infty} \frac{\sinh(W/\gamma_m)\sqrt{1+(L_a\alpha_m)^2}}{z_m^2\left(D_n + D_p\cosh(W/\gamma_m)\right)}} \tag{3.33}$$

Figure 3.17 shows the effective carrier lifetime τ_{eff}/τ_{bulk} versus normalized diode diameter (r_0/L_a) and width (W/L_a) for silicon PIN diodes calculated using (3.33). The first 11 terms of the Fourier-Bessel series ($m = 11$) were used in the computations. The results indicate that smaller geometry devices (small normalized I-region width W/L_a and radius, r_0/L_a) show significantly reduced effective carrier lifetimes compared with the bulk silicon value due to the reduction in the total I-region stored charge. Lower carrier lifetimes tend to drive the impedance of the PIN diode to higher values, as well as shift the impedance-frequency characteristic to higher frequencies.

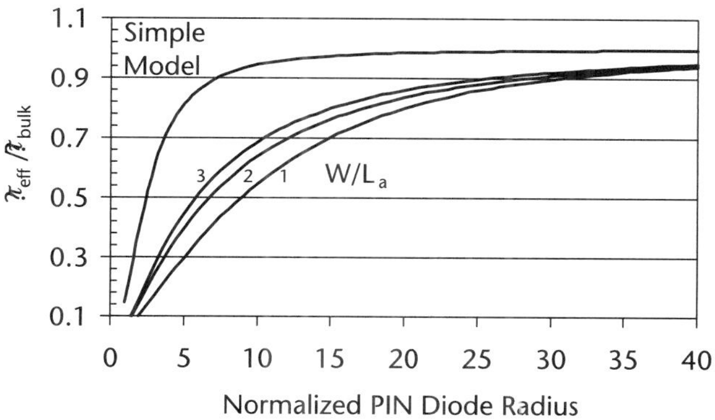

Figure 3.17 Effective carrier to bulk lifetime ratio in PIN diodes as a function of normalized PIN diode radius (r_0/L_a) using normalized I-region width (W/L_a) as a parameter.

3.5 PIN Diode Impedance as a Function of Frequency

In Section 3.2.2, there were two definitions provided for the PIN diode resistance. The first one was based on the slope of the dc *I-V* characteristic and is valid only at very low frequencies (3.6). The other definition was the high-frequency resistance based on I-region conductivity modulation created by the dc forward current (3.5):

$$R_S = \frac{2V_T}{I_{\text{dc}}} \tag{3.34a}$$

$$R_S = \frac{W^2}{\left(\mu_n + \mu_p\right) I_{\text{dc}} \tau} \tag{3.34b}$$

Both equations described purely resistive behavior of the PIN diode, except of course for the reactive components that make up the interconnections or the package. The equations though only tell part of the story. Traditional understanding of the PIN diode resistance is that RF resistance is much smaller than the low-frequency value from the dc *I-V* curve. No mention, however, is made of what happens between the low- and high-frequency limit. Two completely different PIN diodes biased at the same dc forward current can exhibit two completely different resistance-frequency relationships. Take, for example, two PIN diodes dc biased at 10 mA dc forward current with typical I-region characteristics, Diode1 with an I-region thickness of 5 μm with 100 nS carrier lifetime and Diode2 with I-region thickness 250 μm and 2 μs carrier lifetime (the first could be a switching device, the second could be used in attenuator applications). Both have the same low-frequency resistance, 5.2Ω. The RF resistance for Diode1 from (3.34b) is 0.2Ω, much lower than the low-frequency value and in line with traditional understanding. The RF resistance for Diode2, however, is 24Ω, nearly five times higher than the low-frequency resistance. Clearly, there is some interesting PIN diode behavior going on between the lower and upper frequency limits. One of the key factors to explore is the difference in the two PIN diodes' recombination length, L_a, defined in the mathematical analysis section. Note for Diode1 L_a = 22 μm, whereas for Diode2, L_a = 100 μm. For Diode1, W is less than L_a; conversely, *W* is greater than L_a for Diode2. This phenomenon and dependence on the W/L_a ratio can be studied using the in-depth mathematical analysis of the previous section to provide insight into what is occurring with the PIN diode.

3.5.1 PIN Diode Impedance Versus Frequency: Mathematical Analysis

To study this impedance-frequency phenomenon, the detailed expressions in Section 3.4.1 will be used with silicon PIN diodes and the behavior will be based on the characteristics of the two example diodes, Diode1 and Diode2. Extensions will then be made for other device characteristics. Using the characteristics of Diode1 and Diode2 and the impedance expressions in (3.24) and (3.27), Figure 3.18 shows the resistance behavior as a function of frequency for the two diodes. In this figure, the contact resistance and bond inductance were not included, and the resistive part of the total diode impedance is defined as the real part of the total impedance,

RE{Z_{PI} + Z_{IN} + Z_I}. The same low-frequency resistance of 5.2Ω is noted, and the high-frequency resistance shows the clear split between Diode1 decreasing with frequency and Diode2 increasing with frequency. In Figure 3.19, the series reactive part of the impedance is defined as the imaginary part of the total impedance, IM{Z_{PI} + Z_{IN} + Z_I}. From Figure 3.19, the reactance curve shows that there is a reactance minimum at a frequency of approximately $1/2\pi\tau$ (1.6 MHz) for the thin diode Diode1 that is not caused by any package effects but rather is a consequence of the process of conductivity modulation in the I-region and its interaction with the junction regions by both the RF and dc currents, with the RF current leading the voltage in the vicinity of this frequency. Even more interesting is the reactive behavior of the thick diode, Diode2. At a frequency below $1/2\pi\tau$ (80 kHz), the PIN diode exhibits inductive behavior, changing back to capacitive behavior above $1/2\pi\tau$ and then decaying toward a zero value at high frequencies. Like Diode1, the inductive behavior is a consequence of the I-region conductivity/junction effect and not of any packaging effects. At low frequencies, the RF voltage leads the RF current to create the inductive effect, switching at medium frequencies to the RF current leading the voltage and then completely decaying. The quality factor Q (not the stored charge, although the two share the same symbol) of the PIN diodes is defined in (3.35) as the energy stored in the PIN diode relative to that dissipated by the diode and is plotted in Figure 3.20, illustrating that Q is less than unity for both diodes over the entire frequency range.

$$Q = \frac{IM(Z_{\text{PIN}})}{RE(Z_{\text{PIN}})} = \frac{X_{PI} + X_I + X_{\text{IN}}}{R_{PI} + R_I - R_{\text{IN}}} \tag{3.35}$$

This low Q implies that the PIN diodes would be very poor active inductors over any frequency range. However, this phenomenon has been mentioned so that engineers using PIN diodes and measuring their characteristics note that this is natural behavior of the devices and not a potential flaw in either measurements or fabrication.

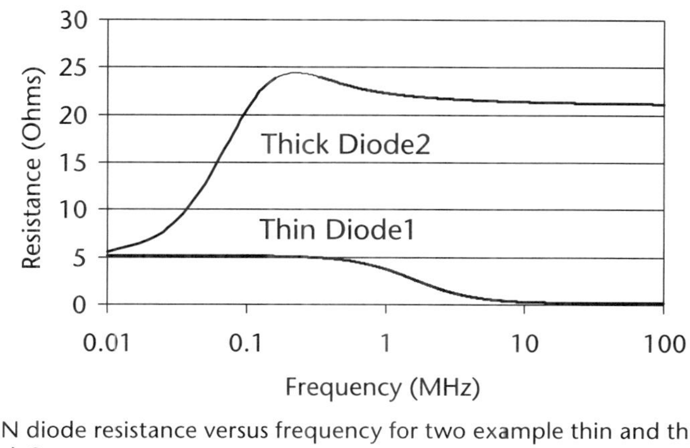

Figure 3.18 PIN diode resistance versus frequency for two example thin and thick PIN diodes, Diode1 and Diode2.

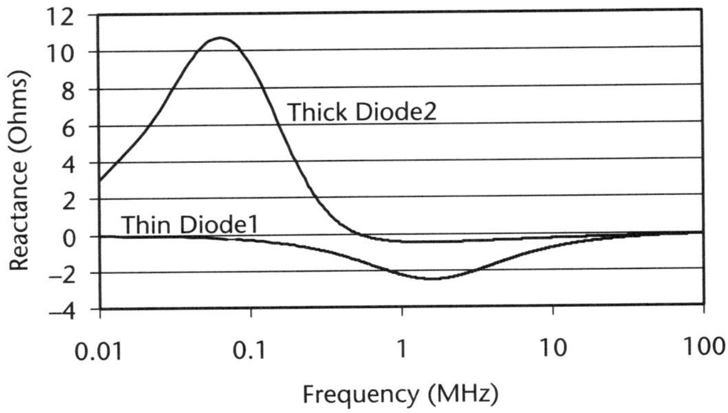

Figure 3.19 PIN diode reactance versus frequency for two example thin and thick PIN diodes, Diode1 and Diode2.

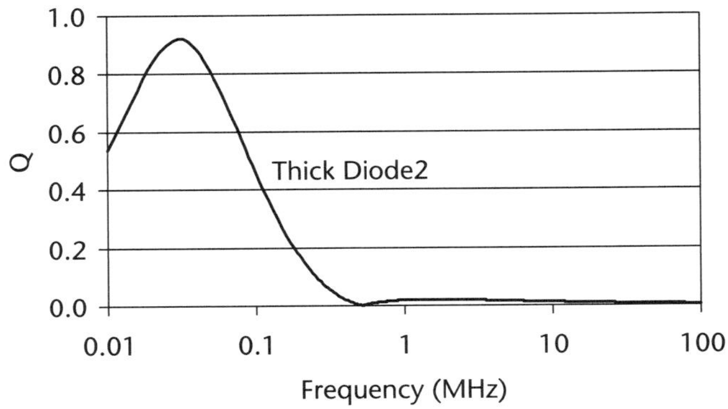

Figure 3.20 Quality factor Q of thick example PIN diode Diode2. Note that the quality factor of the PIN diode is less than unity at all frequencies, making it a poor active inductor.

A more general set of plots showing the impedance behavior versus frequency can be performed if two unitless parameters are defined: the I-region width normalized to the recombination length, W/L_a, and the radian frequency–carrier lifetime product $\omega\tau$. The impedance relationships in (3.24) and (3.27) were used to plot PIN diode impedance components and shows resistive and reactive components of a Si PIN diode (Figures 3.21 and 3.22) and a GaAs PIN diode (Figures 3.23 and 3.24) in the range $0.01 < W/L_a < 4$ and $0.01 < \omega\tau < 10$, both with a dc forward current I_{dc} = 10 mA (no contact resistance R_c or bond wire inductance L_{bond} included in these calculations, but they could be easily incorporated). The figures show in all the cases that for small $\omega\tau$, the PIN diode resistance goes to the junction governed resistance $2V_T/I_{dc}$ limit of 5.2Ω at this current with zero reactance. As $\omega\tau$ increases, the influence of the I-region becomes apparent with diodes of small W/L_a values exhibiting a decrease in resistance to a fraction of an Ohm. The reactance characteristic for small W/L_a diodes is purely capacitive with a minimum in the vicinity of $\omega\tau = 1$. The capacitive reactance decreases toward zero for large $\omega\tau$. Note that for PIN diodes that exhibit $W/L_a = 2$ there is little or no reactive component; the resistance shows little variation with frequency for this same W/L_a value, implying a constant resistance with frequency. For W/L_a greater than approximately two,

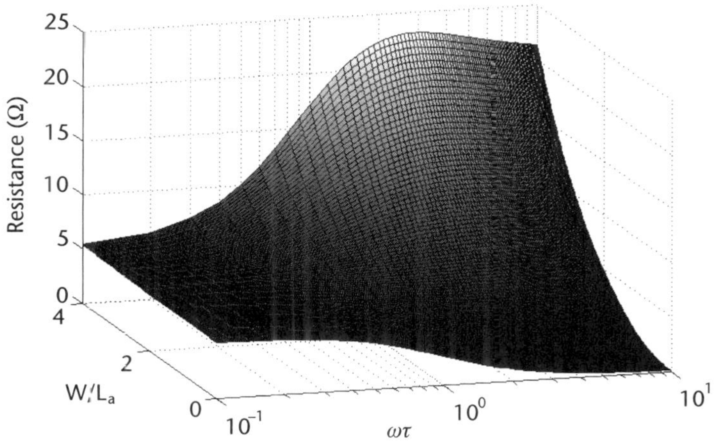

Figure 3.21 PIN diode resistance versus normalized values W/L_a and $\omega\tau$ for a Si PIN diode biased at 5 mA.

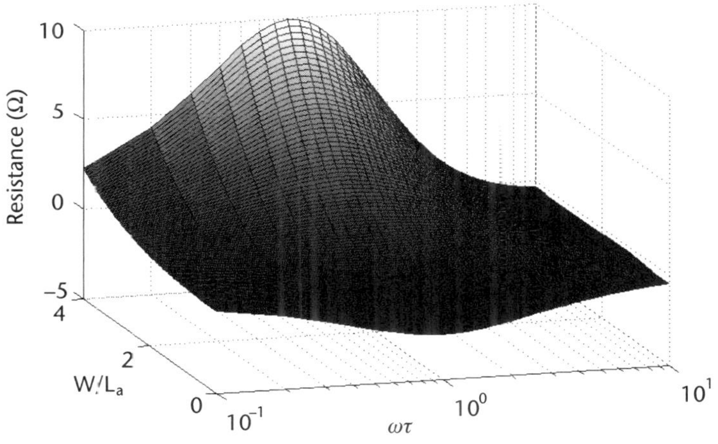

Figure 3.22 PIN diode reactance versus normalized values W/L_a and $\omega\tau$ for a Si PIN diode biased at 5 mA.

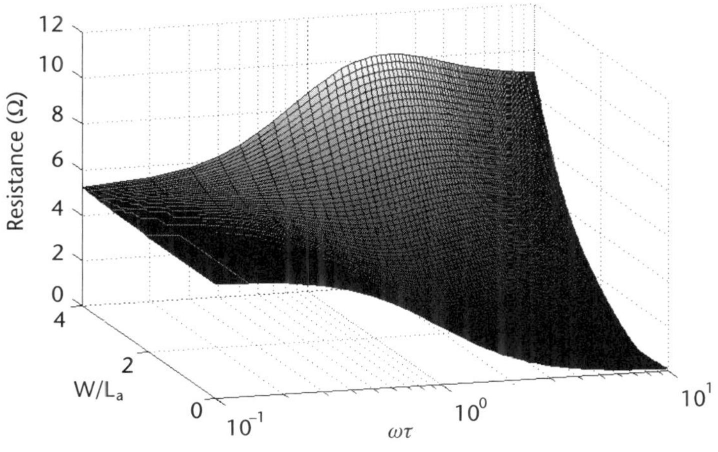

Figure 3.23 PIN diode resistance versus normalized values W/L_a and $\omega\tau$ for a GaAs PIN diode biased at 5 mA.

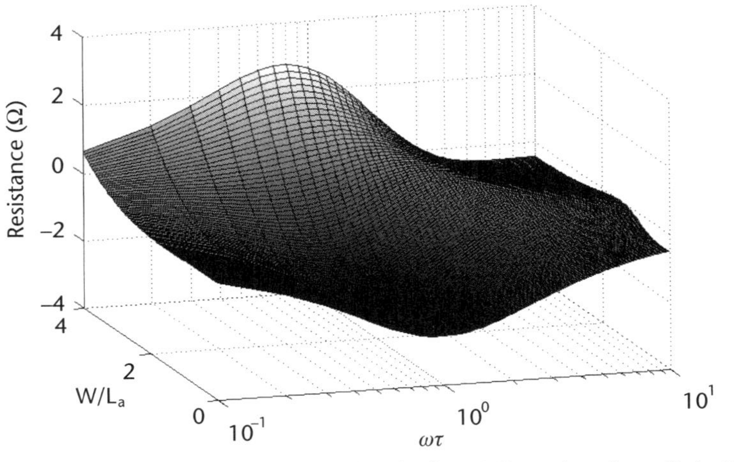

Figure 3.24 PIN diode reactance versus normalized values W/L_a and $\omega\tau$ for a GaAs PIN diode biased at 5 mA.

the PIN diode resistance increases with increasing $\omega\tau$ with the reactance characteristic exhibiting inductive reactance for $\omega\tau$ values near unity, with a changeover to a small capacitive reactance at higher $\omega\tau$ before dying out completely. The GaAs PIN diode shows similar impedance behavior with its silicon counterpart, although the overall values of resistance and reactance are smaller by virtue of GaAs's higher electron mobility [27, 28].

3.5.2 Carrier Lifetime Measurement

Traditionally, the PIN diode carrier lifetime has been based on measurements of the charge storage time defined as (Figure 3.25) [10]

$$\tau = T_D \ln\left(1 + \frac{I_{dc}}{I_R}\right) \tag{3.36}$$

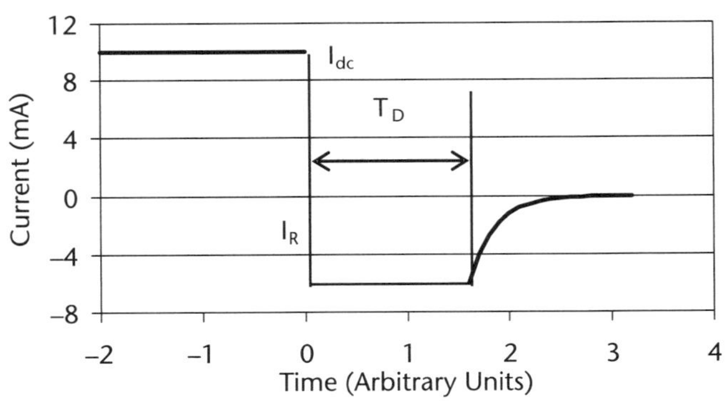

Figure 3.25 Measuring charge storage in a PIN diode to obtain the I-region carrier lifetime.

where T_D is the charge storage time delay, I_{dc} is the dc forward current, and I_R is a reverse current removing the I-region stored charge. If the currents are chosen properly, carrier lifetime τ is the same as charge storage delay, T_D. This occurs if the device has 10 mA dc forward current and is then switched to 6 mA reverse current.

There is another way, however, to obtain a measurement of the PIN diode carrier lifetime, and that is noting the reactance curves in Figures 3.22 and 3.24. Note that for PIN diodes with W/L_a less than unity, the capacitance minimum occurs near $\omega\tau = 1$. Because that capacitance minimum frequency F_τ can be easily measured, the carrier lifetime becomes [29]

$$\tau = \frac{1}{2\pi F_\tau} \tag{3.37}$$

For example, if the capacitance minimum occurs at 1.5 MHz, the I-region carrier lifetime obtained with the RF method is approximately 100 nS.

This section has shown that for relatively small RF signals, the PIN diode behaves as a linear impedance; that is, the impedance does not change with applied RF signal. A closer analysis of (3.21), however, shows that the approximation $1/(1 + x) \approx 1 - x$ for small x is not a good assumption as x approaches unity. A more accurate approximation for $1/(1 + x)$ as x approaches unity would be $1 - x + x^2 - x^3$. This more accurate approximation for $1/(1 + x)$ contains quadratic and higher terms, providing the basis for a nonlinear impedance characteristic; the ramifications of this nonlinearity will be the focus of the next chapter.

3.5.3 Effects of Temperature on PIN Diode Impedance

There are several parameters that make up the PIN diode's electrical characteristics that are dependent on temperature. The obvious first one is the thermal voltage, kT/q, where T is in Kelvin (0.0258 V at 290K or 17°C). Two other parameters, the mobility and the carrier lifetime, are also temperature dependent. The temperature dependence of mobility has been studied by many researchers over the decades, with the hole and electron mobilities decreasing with increasing temperature following an approximate $T^{-2.0}$ temperature dependence over the temperature range that covers most PIN diode applications [9]:

$$\mu_a(T) \approx \mu_a(290\text{K})\left(\frac{T}{290}\right)^{-2.0} \tag{3.38}$$

More exact formulations for the mobility temperature dependence can be found in the research literature (for example, [37]). The mobility-temperature dependence is somewhat more complicated for GaAs; the mobility temperature dependence is approximately $T^{-1.0}$ and $T^{-2.1}$ for electrons in n-type material and holes in p-type material, respectively. The mobility-temperature equation above, however, is still valid for the individual carrier mobilities with a change of the respective 290K mobility value or exponent.

The temperature dependence of the carrier lifetime is not as well understood, with most investigators relying on measurements and analysis on gold-doped devices. Based on these measurements and work performed by other investigators on silicon devices [30–35], carrier lifetime has been found to increase with increasing temperature. The temperature dependence of carrier lifetime may be modeled by the following exponential expression:

$$\tau(T) \approx \tau(290\text{K})\left(\frac{T}{290}\right)^{m} \tag{3.39}$$

The value of the m-exponent can vary over a wide range, with a variety of factors influencing the carrier lifetime temperature characteristic. Some of the important ones are the diode geometry (specifically the I-region width and diameter) and the type of material used to passivate the diode surface. Studies have shown [34, 35] that this coefficient can vary between approximately 0.75 and 2.2 for passivations used in typical PIN diode manufacturing processes (oxide, glass, and nitride [34]).

The result of these two temperature-varying parameters is that the PIN diode RF resistance can have either a positive or negative temperature coefficient, governed primarily by the carrier lifetime temperature coefficient since the mobility temperature dependence is fixed for a given semiconductor type. Using the traditional form of the PIN diode RF resistances, this temperature dependent resistance can be written as

$$R_S(T) \approx R_S(290\text{K})\left(\frac{T}{290}\right)^{2.0-m} \tag{3.40}$$

The ratio $R_S(T)/R_S(290\text{K})$ is plotted in Figure 3.26 and shows that the PIN diode RF resistance shows a positive temperature coefficient (increasing resistance with temperature) for m-values of 2.0 or less and a slight negative temperature coefficient

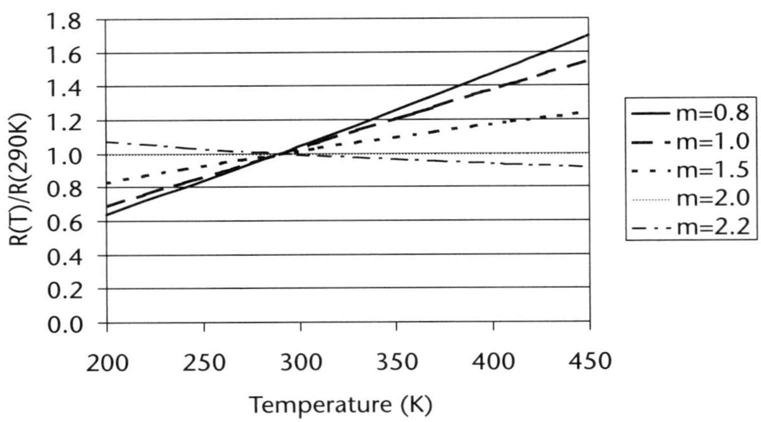

Figure 3.26 Plot of PIN diode resistance ratio (3.40) as a function temperature for various values of the carrier lifetime temperature coefficient m. The resistance ratio is normalized to the 290K resistance.

for m-values greater than 2.0. The mobility and carrier lifetime temperature expressions can also be easily integrated into the more detailed integral equation forms for the PIN diode resistance described in (3.24) through (3.27). This varying PIN diode RF resistance especially impacts their use in attenuator circuits since high values of resistance are usually required for these applications [34]. These applications are the focus of a later chapter.

3.6 PIN Diode Reverse Bias Modeling

In Section 3.2.3, the case was made for a detailed resistance-capacitance model for the PIN diode under zero or reverse bias that divided the I-region into two distinct regions, one where there were no free carriers (swept region) and one where there were still charge carriers (unswept region). A model in terms of the full I-region width W and the boundary between the two distinct regions, W_D, was developed and is repeated in Figure 3.27 and with (3.8) and (3.9) ($W_D < W$):

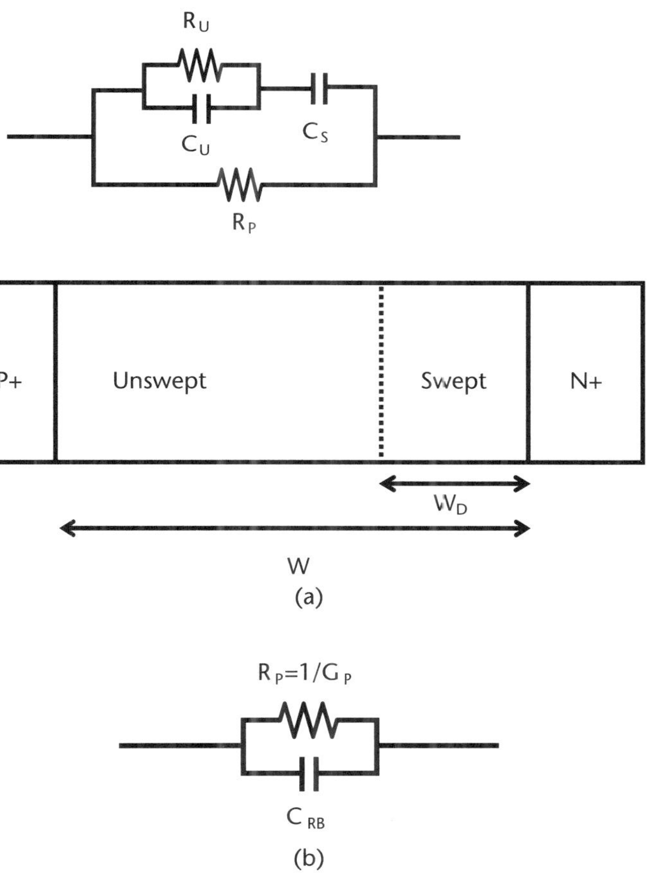

Figure 3.27 (a) Reverse bias lumped element equivalent circuit for PIN diode showing the physical origin of the various components. (b) Equivalent circuit for the reverse bias PIN diode impedance.

$$R_U = \rho_I \frac{W - \sqrt{\frac{2\varepsilon_r\varepsilon_0}{qn_I}\left(V_{bi} - V\right)}}{A}; \quad C_U = \varepsilon_r\varepsilon_0 \frac{A}{W - \sqrt{\frac{2\varepsilon_r\varepsilon_0}{qn_I}\left(V_{bi} - V\right)}};$$

$$C_S = \varepsilon_r\varepsilon_0 \frac{A}{\sqrt{\frac{2\varepsilon_r\varepsilon_0}{qn_I}\left(V_{bi} - V\right)}} \tag{3.41}$$

Note from (3.3) that the dielectric relaxation frequency F_D can be written in terms of the two "unswept" region terms:

$$F_D = \frac{1}{2\pi\rho\varepsilon_r\varepsilon_0} = \frac{1}{2\pi R_U C_U} \tag{3.42}$$

For frequencies below F_D, the full equivalent circuit for the PIN diode should be used. For frequencies above F_D, the capacitance becomes the geometric capacitance of the device given by $\varepsilon_r\varepsilon_0 A/W$.

The impact of these components can be observed (Figure 3.28) by means of an example using a W = 100 μm silicon PIN diode with background carrier density of 10^{12} cm^{-3} carriers (approximately 100 Ω-m) in the "unswept" region, a built-in junction potential V_{bi} = 0.7V, a diameter of 250 μm and the semiconductor values in Table 3.1. The corresponding dielectric relaxation frequency is approximately 15 MHz. The figures plot the three reverse bias components as a function of voltage out to −10 V. The curves show that the resistance R_u decreases with increasing reverse bias and is effectively zero at approximately −6.8V, with its shunting capacitance C_u going to very large values at the same voltage. Note that this voltage, −6.8V, is the full I-region depletion voltage (or all the carriers swept out of the I-region), and it is at this point that the expression for R_u and C_u are no longer valid, since $W = W_D$. Figure 3.28(b) shows the resulting RF conductance and capacitance at 1 MHz and 100 MHz. The conductance goes toward zero at −6.7V; however, in an actual PIN diode, there will always be some shunt resistance R_P (usually due to small leakage currents along the mesa sidewall surface due to the semiconductor-passivation interface) that give a non-zero floor to the overall diode conductance. The capacitance measured at 1 MHz (where commercial devices are typically measured for manufacturer's specification sheets) is much higher at lower voltages, a consequence of the interaction between the swept and unswept regions. The 100 MHz capacitance, however, is essentially constant for all voltages since 100 MHz is well above the F_D of 15 MHz. The 1 MHz capacitance is the same as the 100 MHz capacitance at the punchthrough voltage of −6.7V.

The analysis presented for the reverse bias PIN diode assumed an abrupt junction that allowed the depletion boundary W_D to be easily defined as proportional to the square root of the applied voltage $V^{1/2}$:

$$W_D = \sqrt{\frac{2\varepsilon_r\varepsilon_0}{qn_I}\left(V_{bi} - V\right)} \tag{3.43}$$

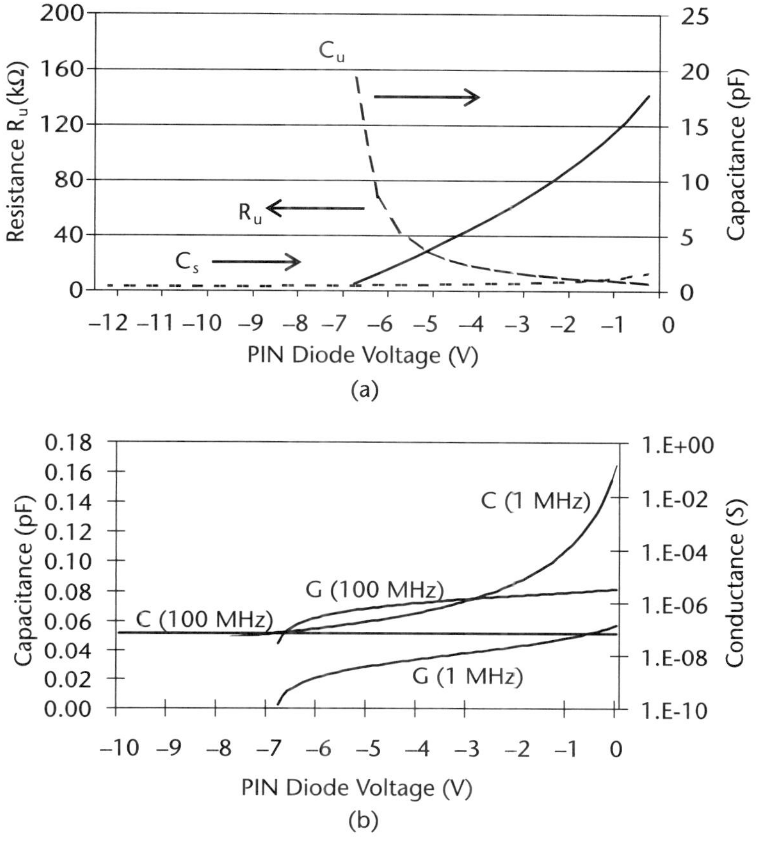

Figure 3.28 (a) Reverse bias resistance R_P and capacitances components C_u and C_s of an example PIN versus PIN diode voltage. (b) Total shunt conductance G_P and capacitance C_{RB} at 1.0 MHz and 100 MHz.

The single-sided abrupt junction approximation, however, is an idealization as the PI and IN junctions, which are usually not abrupt (although close if an MBE or MOCVD process are used) but rather exhibit diffusion "tails" that are the result of the diffusion fabrication process. The result is that the W_D boundary may not exhibit the $V^{1/2}$ relationship but may exhibit more complicated behavior. For example, an analysis of a linear or graded PN junction device shows that the depletion boundary exhibits a voltage dependence of $V^{1/3}$ instead of the $V^{1/2}$ relationship previously discussed [9]. Dopant diffusion tails can be described in terms of complementary error functions, $erfc(x)$ or e^{-x^2}dependencies, greatly complicating the derivation of a closed form mathematical solution for W_D. Since there are actually two junctions, both with potential diffusion tails, even above the now so-called punchthrough voltage, there may still be a small change in reverse bias capacitance with voltage (even well above F_D). For the general case, the depletion width can always be described as a Taylor series:

$$W_D(V) = \sum_{k=0}^{\infty} w_k V^k \tag{3.44a}$$

where the w_k coefficients can be found from measurements of C_{RB} at reverse bias since the reverse capacitance can be written as

$$C_{RB} = \frac{\varepsilon_r \varepsilon_0 A}{W_D(V)} = \sum_{k=0}^{\infty} c_k V^k \tag{3.44b}$$

The w_k can be determined from the shape of the C_{RB}-V curve using standard Taylor series approaches (derivatives calculated from the C_{RB}-V curve around a given reverse bias voltage V_0):

$$\begin{aligned} \frac{dC_{RB}}{dV} &\approx \frac{C_{RB}(V_0 + \Delta V) - C_{RB}(V_0)}{\Delta V}; \\ \frac{d^2 C_{RB}}{dV^2} &\approx \frac{C_{RB}(V_0 + \Delta V) - 2C_{RB}(V_0) + C_{RB}(V_0 - \Delta V)}{(\Delta V)^2} \end{aligned} \tag{3.45}$$

where a second-order central form was used for the second derivative and ΔV is a small voltage difference about V_0. While the seemingly small changes in C_{RB} at high voltages may appear to have minimal impact on the linear characteristics of the device, the next chapter will show that the C-V characteristic has a large influence on PIN diode nonlinear behavior in reverse bias. Irrespective of the W_D-V relationship by abrupt, linear, or complicated diffusion tails, the overall discussion of the reverse bias characteristic is still valid under these circumstances; it is just that the actual voltage dependence on the equivalent circuit parameters will be quite different.

References

[1] Uhlir, A., Jr., "The Potential of Semiconductor Diodes in High Frequency Communications," *IRE Proc.*, June 1958, pp. 1099–1115.

[2] Coale, F. S., "A Switch Detector Circuit," *IRE Trans. on Microwave Theory and Techniques*, Vol. MTT-3, December 1955, pp. 59–61.

[3] Garver, R. V., "Theory of TEM Diode Switching," *IRE Trans. on Microwave Theory and Techniques*, May 1961, pp. 224–238.

[4] Leenov, D., J. H. Forster, and N. G. Cranna, "PIN Diodes for Protective Limiter Applications," *ISSCC Dig. Papers*, 1961, pp. 84–85.

[5] Leenov, D., "The Silicon PJN Diode as a Microwave Radar Protector at Megawatt Levels," *IEEE Trans. on Electronic Devices*, 1964, pp. 53–61.

[6] Garver, R. V., *Microwave Diode Control Devices*, Dedham, MA: Artech House, 1976.

[7] Mortenson, K. E., "Microwave Semiconductor Control Devices," *Microwave J.*, May 1964, pp. 49–57.

[8] White, J. F., *Microwave Semiconductor Engineering*. New York: Van Nostrand Reinhold Co., 1982 [previously published by Artech House in 1977 under the title *Microwave Semiconductor Control*].

[9] Sze, S. M., *Physics of Semiconductor Devices*, 2nd ed., New York: John Wiley, 1981.

[10] Doherty W. E., and R. Joos, *PIN Diode Circuit Designers Handbook*, Watertown, MA: Microsemi Corp., 1998.

[11] Walston, J., "SPICE Circuit Yields Recipe for PIN Diode," *Microwaves & RF*, 1992, pp. 78–89, 1992.

[12] Jang, B.-J., I.-B. Yom, and S.-P. Lee, "An Enhanced PIN Diode Model for Voltage-Controlled PIN Diode Attenuator," *Proc. 33rd European Micro. Conf.*, 2003, pp. 231–234.

[13] *MicroSim PSpice A/D Reference Manual*, Hillsboro, OR: Microsim Corp., 1998.

[14] White, J., "Origins of High-Power Diode Switching," *IRE Trans. on Microwave Theory and Techniques*, Vol. 32(9), 1984, pp. 1105–1117.

[15] Ryder, R., in a talk at the IEEE NEREM show in Boston, MA, circa 1965 (from White, 1984).

[16] Ghandhi, S. K., *Semiconductor Power Devices*, New York: John Wiley, 1977.

[17] Varshney, R., D. Roulston, and S. Chamberlain, "Some Properties Concerning the a.c. Impedance of PIN and P-N-N+ diodes," *Solid-State Electronics*, Vol. 17, No. 7, 1974, pp. 699–706.

[18] Aitchison, J. M., and F. Berz, "The Effect of Surface Recombination on Pin Diodes," *Solid-State Electronics*, Vol. 24, 1981, pp. 795–804.

[19] Ghandhi, S. K., *Semiconductor Power Devices*, New York: John Wiley and Sons, 1977.

[20] Caverly, R., and G. Hiller, "The Small Signal ac Impedance of Gallium Arsenide and Silicon PIN Diodes," *Solid-State Electronics*, Vol. 33(10), October 1990, pp. 1255–1263.

[21] Shockley, W., *Electrons and Holes in Semiconductors*. Princeton, NJ: Van Nostrand, 1950.

[22] Lindmayer, J., and C. Y. Wrigley, *Fundamentals of Semiconductor Devices*. Princeton, NJ: Van Nostrand, 1965.

[23] Neudeck, G. W., *The PN Junction Diode*. Reading, MA: Addison-Wesley, Modular Series on Solid State Devices, 1983.

[24] Shur, M., *Physics of Semiconductor Devices*. Englewood Cliffs, NJ: Prentice-Hall, 1990.

[25] Gatard, E., R. Sommet, P. Bouysse, and R. Quéré, "An Improved Physics-Based Formulation of the Microwave p-i-n Diode Impedance," *IEEE Micro. Wire. Comp. Lett.*, Vol. 17(3), 2007, pp. 211–213.

[26] Strollo, A.G.M., and P. Spirito, "A New PiN Diode Modelling Approach for Power Electronic PSPICE Simulations," in *Proc. 25th Annu. IEEE Power Electron. Spec. Conf.*, Vol. 1, June 1994, pp. 52–58.

[27] Gopinath, A., "Comparison of GaAs MESFET and GaAs p-i-n Diodes as Switch Elements," *IEEE Electron Device Letters*, Vol. 6, No. 10, October 1984, pp. 505–506.

[28] Abid, Z., and A. Gopinath, "Impedance and Switching of GaAs p-i-n Diodes," *IEEE Trans. on Microwave Theory and Techniques*, Vol. 38, No. 10, October 1990, pp. 1526–1528.

[29] Caverly, R., "RF Technique for Determining the Ambipolar Carrier Lifetime in Pin RF Switching Diodes," *Electronics Letters*, Vol. 34(23), 1998, pp. 2277–2278.

[30] Gamal, S., H. Morel, and J. P. Chante, "Carrier Lifetime Measurement by Ramp Recovery of p-i-n Diodes," *IEEE Trans. Electron Devices*, Vol. 37, No. 8, August 1990, pp. 1921–1924.

[31] Dudeck, I., and R. Kassing, "Determination of the Temperature Dependence of the Capture Cross-Sections of the Gold Acceptor Level and of the Temperature of Current Filaments in Silicon pin Diodes," *Phys. Stat. Sol.(a)*, Vol. 49, 1978, pp. 153–161.

[32] Bassett, R., "Observations on a Method of Determining the Carrier Lifetime in p-i-n Diodes," *Solid-State Electronics*, Vol. 12, 1969, pp. 385–391.

[33] Wilson, P., "Recombination in Silicon p-i-n Diodes," *Solid-State Electronics*, Vol. 10, 1967, pp. 145–154.

[34] Caverly, R., and G. Hiller, "The Temperature Dependence of Silicon PIN Diode Attenuators," *Proc. 1993 IEEE International Microwave Symp.*, Vol. 2, June 1993, pp. 553–556.

[35] Xiaochun, G., and W. Sihan, "Relationship Between Resistance and Temperature of Microwave Pin Diode," *Semiconductor Technology*, Vol. 33, No. 12, 2008, pp. 1066–1069.

[36] Caverly, R., and N. Jain, "Modeling GaAs PIN Diodes for Microwave and Millimeter Wave Applications," *Digest 24th European Microwave Conference*, Vol. 2, 1994, pp. 1622–1627.

[37] Jacoboni, C., C. Canali, G. Ottaviani, and A. A. Quaranta, "A Review of Some Charge Transport Properties of Silicon," *Solid-State Electronics*, Vol. 20, 1977, pp. 77–89.

[38] Application Note, "GaAs PIN Diode Chips," MACOM Technology Solutions, Inc., 2011 http:// http://www.macom.com/ [accessed 8-6-2015, http://cdn.macom.com/datasheets/GaAs_pindiodechips.pdf].

[39] Bellantoni, J. V., D. C. Bartle, D. Payne, G. McDermott, S. Bandla, et al., "Monolithic GaAs p-i-n Diode Switch Circuits for High-Power Millimeter-Wave Applications," *IEEE Trans. on Microwave Theory and Techniques*, Vol. 37, No.12, December 1989, pp. 2162, 2165.

CHAPTER 4

Modeling PIN Diodes—Nonlinear and Time Domain Behavior

4.1 Introduction

Any electronic component, if driven with a large enough signal, will generate spurious signals due to nonlinear behavior inherent to the device. For passive elements such as resistors, capacitors, and inductors, any nonlinearities are sufficiently small so that in most applications the nonlinearities can be ignored, and these elements can be treated as a constant resistance or reactance. Active devices such as diodes and transistors, on the other hand, can have parameters that vary with voltage, current, or time and give rise to harmonic and intermodulation signals that cause distortion in the system. PIN diodes are susceptible to this phenomenon as well. The first applications of PIN diodes were at microwave frequencies where the operating frequency was much higher than the carrier lifetime, and so nonlinear behavior was not a major concern in the design. As applications of PIN diodes started to occur more often at lower frequencies, this nonlinear behavior became more important and therefore needed to be explored to see what diode parameters influenced the distortion and how it could be reduced. This exploration requires the use of a full mathematical description of the diode, and this section presents the mathematical basis for a PIN diode nonlinear model.

4.2 PIN Diode Forward Bias Distortion

4.2.1 Detailed Mathematical Modeling

In Chapter 3, the dc and RF charge densities were derived from the semiconductor transport equations and provided insight into the frequency-impedance properties of the PIN diode. The main contributor to the observed frequency-impedance phenomenon was the interaction of I-region conductivity modulation with its current-voltage characteristics and that of the two junctions. This same conductivity modulation by the RF signal makes the impedance a function of time and therefore generates unwanted harmonically related signals. There are a number of ways to explore the nonlinear behavior in solid-state devices, but the most common techniques that have been applied to PIN and other diodes use Taylor [1, 2] or Volterra series [3] expansions or numerical methods [4] to perform the necessary mathematics. The approach taken in this chapter will tend toward the Taylor series approach.

To investigate this nonlinear behavior of the PIN diode, the same modeling approach is used as with the detailed impedance modeling, but with two important extensions to the theory outline in Chapter 3:

- Harmonic and intermodulation frequencies have to be modeled, and so a multi-tone excitation is needed.
- The Taylor series expansion needs to use higher order terms to model the nonlinear behavior.

Equation (3.17) can be easily modified to included N multiple tones with the same dc forward current I_{dc}:

$$I(t) = I_{\text{dc}} + \sum_{k=1}^{N} RE\left[I_k e^{j\omega_k t}\right] = I_{\text{dc}} + \sum_{k=1}^{N} I_k \cos\left(\omega_k t\right) \tag{4.1}$$

where RE implies taking the real part of the expression. As before, I_{dc} supports the dc charge density n_0, and now the I_k RF tones support the n_k RF charge densities (the RF charge density n_{RF} in (3.19) for the single tone now becomes n_k for the multiple RF tone case). The I-region voltage described in (3.23) can be updated with the additional RF charge densities and the associated RF tones as

$$V_I(t) = -\frac{I(t)}{qA\left(\mu_n + \mu_p\right)} \int_{-W/2}^{W/2} \frac{dx}{n_0(x)\left[1 + \sum_{k=1}^{N} RE\left[\frac{n_k(x)}{n_0(x)} e^{j\omega_k t}\right]\right]} \tag{4.2}$$

If the term $\sum_{k=1}^{N} RE\left[\left(n_k(x)/n_0(x)\right)e^{j\omega_k t}\right]$ is less than unity as before, then (4.2) can be again expanded into a Taylor series, but instead of only taking the first term in the series, the I-region voltage drop can be generalized to as many orders as needed to model the various harmonic and intermodulation products using the expansion $1/(1 + x)$ is $1 - x + x^2 - x^3 \ldots$. The I-region voltage relationship can then be written as

$$V_I(t) = -\frac{1}{qA\left(\mu_n + \mu_p\right)} \int_{-W/2}^{W/2} \frac{dx}{n_0(x)} \left(I_{dc} + \sum_{k=1}^{N} I_k \cos\left(\omega_k t\right)\right)\left(\sum_{l=0}^{\infty} (-1)^l \left[\sum_{k=1}^{N} RE\left[\frac{n_k(x)}{n_0(x)} e^{j\omega_k t}\right]\right]^l\right) \tag{4.3}$$

The voltage drop across the two junctions can be determined in a similar fashion to (3.25):

$$\begin{aligned} V_J(X,t) &= \frac{kT}{q} \ln\left(\frac{1}{n_i} n_0(x)\left[1 + \sum_{k=1}^{N} RE\left[\frac{n_k(x)}{n_0(x)} e^{j\omega_k t}\right]\right]\right) \\ &\approx \frac{kT}{q} \sum_{l=1}^{\infty} \frac{(-1)^l}{l} \left[\sum_{k=1}^{N} RE\left[\frac{n_k(x)}{n_0(x)} e^{j\omega_k t}\right]\right]^l \end{aligned} \tag{4.4}$$

where X is $-W/2$ for the PI junction and $W/2$ for the IN junction. The term $(kT/q)\ln[n_0(x)/n_i]$ has been neglected since this term has no time variation and is the term associated with the applied dc bias. As in the linear case, the total PIN diode voltage is just the sum of the voltage expressions in (4.3) and (4.4), with the two distinct junction voltages:

$$V(t) = V_I(t) + V_J\left(-\frac{W}{2},t\right) + V_J\left(\frac{W}{2},t\right) \tag{4.5}$$

The ratio of the RF to dc charge density is common in both (4.3) and (4.4), and so a new variable, α_k, can be introduced to ease a bit of the mathematical burden:

$$\begin{aligned}\alpha_k(x) &= \frac{n_k(x)}{n_0(x)} = |\alpha_k(x)|e^{j\Theta_k} \Rightarrow \sum_{k=1}^{N} RE\left[\frac{n_k(x)}{n_0(x)}e^{j\omega_k t}\right] \\ &= \sum_{k=1}^{N}|\alpha_k(x)|\cos\left(\omega_k t + \Theta_k\right)\end{aligned} \tag{4.6}$$

Gretsch showed that the lth power of (4.6), again common in (4.3) and (4.4), can be written as [1]

$$\begin{aligned}&\left[\sum_{k=1}^{N}|\alpha_k(x)|\cos\left(\omega_k t + \Theta_k\right)\right]^l \\ &= \frac{2l!}{m!n!\cdots}\left[\left(\frac{\alpha_1(x)}{2}\right)^m\left(\frac{\alpha_2(x)}{2}\right)^n\cdots\right]\cos\left\{\left[m\left(\omega_1 + \Theta_1\right) \pm n\left(\omega_2 + \Theta_2\right)\cdots\right]t\right\}\end{aligned} \tag{4.7}$$

where $l = m + n + \ldots$ describes the permutations of the frequency components and allows the calculation of fundamental as well as harmonic and intermodulation products at any frequency for N tones. For example, for a single tone ($N = 1$, all others zero), the first three terms $l = 1, 2, 3$ in (4.7) yield the following frequencies, either the fundamental or the harmonics of ω_1:

$$\begin{aligned}&\left(\frac{\alpha_1}{2}\right)\cos\left\{\left[\left(\omega_1 + \Theta_1\right)\right]t\right\} && l = 1, m = 1 \\ &2\left(\frac{\alpha_1}{2}\right)^2\cos\left\{\left[2\left(\omega_1 + \Theta_1\right)\right]t\right\} && l = 2, m = 2 \\ &2\left(\frac{\alpha_1}{2}\right)^3\cos\left\{\left[3\left(\omega_1 + \Theta_1\right)\right]t\right\} && l = 3, m = 3\end{aligned} \tag{4.8}$$

The PIN diode I-region voltage drop given in (4.3) can be generalized with the introduction of the term α_k and (4.7) as

$$V_I(t) = -\frac{1}{qA\left(\mu_n + \mu_p\right)} \sum_{l=0}^{\infty} (-1)^l \int_{-W/2}^{W/2} \frac{I_{\rm dc} + \sum_{k=1}^{N} I_k \cos\left(\omega_k t\right)}{n_0(x)} \left[\sum_{k=1}^{N} \left|\alpha_k(x)\right| \cos\left(\omega_k t + \Theta_k\right)\right]^l dx \tag{4.9}$$

$$= -\sum_{l=0}^{\infty} \beta_l \left[I_{\rm dc} + \sum_{k=1}^{N} I_k \cos\left(\omega_k t\right)\right] \cos\left\{\left[m\left(\omega_1 + \Theta_1\right) \pm n\left(\omega_2 + \Theta_2\right)\cdots\right] t\right\}$$

where $l = m + n + \ldots$, and

$$\beta_l = (-1)^l \frac{2l!}{m!n!\cdots} \frac{1}{qA\left(\mu_n + \mu_p\right)} \int_{-W/2}^{W/2} \frac{1}{n_0(x)} \left\{\left[\left(\frac{\alpha_1(x)}{2}\right)^m \left(\frac{\alpha_2(x)}{2}\right)^n \cdots\right]\right\} dx \tag{4.10}$$

Using the same approach with (4.4) and (4.7) for the junction voltage gives

$$V_J(X,t) = \frac{kT}{q} \sum_{l=1}^{\infty} \frac{(-1)^l}{l} \left[\sum_{k=1}^{N} \left|\alpha_k(x)\right| \cos\left(\omega_k t + \Theta_k\right)\right]^l \tag{4.11}$$

where X is $-W/2$ for the PI junction and $W/2$ for the IN junction.

4.2.2 PIN Diode Distortion at High Frequencies

In many applications, operation of the PIN diode occurs at frequencies where the $\omega\tau$ product is significantly greater than unity. This can be as low as 1 MHz for PIN diodes with carrier lifetimes of a few microseconds or as high as 1000 MHz with carrier lifetimes of a nanosecond or less. Simplifying (4.9) and (4.11) makes it easier to see the parameters that govern the nonlinearities. To perform this simplification, the expression for α_k in (4.6) can be approximated for small W/L_a as [5]

$$\alpha_k(x) \approx \frac{I_k}{I_{\rm dc}} \frac{W}{L} \frac{1}{\sqrt{\omega\tau}} e^{\frac{2x-W}{2L_a}\sqrt{\frac{\omega\tau}{2}}} \tag{4.12}$$

The α_k in (4.12) will be less than unity even if the RF current is larger than the dc forward current since the exponential is very small everywhere (except for x = W/2, where it is unity) and $\omega\tau$ is greater than one. An example will help illustrate this point. For a PIN diode with τ = 2 μs and $W/L_a = 1$ (W approximately 75 μm), the approximation is valid for RF signals up to about 11 times the forward bias current level at 10 MHz; thus, for 100 mA of forward bias, the model applies to RF drives of up to 1.1A. For the large $\omega\tau$ case, the β_l from (4.10) can be written as

$$\beta_l = (-1)^l \frac{2l!}{m!n!\cdots} \frac{1}{qA\left(\mu_n + \mu_p\right)} \left(\frac{1}{I_{\rm dc}} \frac{W}{2L} \frac{1}{\sqrt{\omega\tau}}\right)^l \left(I_1^m I_2^n \ldots\right) \int_{-W/2}^{W/2} \frac{1}{n_0(x)} e^{\frac{2x-W}{2L_a}\sqrt{\frac{\omega\tau}{2}}(l)} dx \tag{4.13}$$

Note that the previous equations show temperature-dependent parameters, namely mobility and carrier lifetime, and so the effects of temperature on nonlinear behavior can be modeled using the expressions presented earlier in Chapter 3 ((3.38) and (3.39)).

The single-tone test as described by the preceding equations only gives information on the fundamental and harmonic response. Although the harmonically generated signals are important, they do not tell the whole story because the nonlinear behavior of the PIN diode (or any nonlinear device) can generate intermodulation products if two or more signals excite the PIN diode. Of the intermodulation terms, it is the odd-order terms that introduce in-band distortion products and are the more important of the two. The simplest way to describe these intermodulation products (as described earlier in Chapter 2) is through the use of the two-tone test. Using as few as two excitation frequencies increases the number of frequency components substantially and complicates the analysis. For the first three orders of l (fundamental, second, and third order nonlinearity), the permutations of m and n based on a given value of l are as follows:

$l = 1$: $m = 1$ or $n = 1$;
$l = 2$: $m = 2$, $n = 0$; $m = 0$, $n = 2$; $m = n = 1$;
$l = 3$ -> $m = 3$, $n = 0$; $m = 0$, $n = 3$; $m = 2$, $n = 1$; $m = 1$, $n = 2$

for a total of the two fundamental signals and 10 other signals when taking into account the impact of the sum and differences (±) in (4.7). The frequency components that need to be tracked in the analysis are

$$\begin{aligned}&\omega_1, \omega_2, \omega_1 + \omega_2, \omega_1 - \omega_2, 2\omega_1, 2\omega_2, 3\omega_1, 3\omega_2, 2\omega_1\\&+ \omega_2, 2\omega_1 - \omega_2, 2\omega_2 + \omega_1, 2\omega_2 - \omega_1\end{aligned} \quad (4.14)$$

The generated phases are of the same form as the generated frequencies in (4.7) but with the ω_k replaced by the corresponding Θ_k.

Up until this point, the analysis has been exact (except for possibly the truncation of the infinite series expansion). For nonlinear circuit analysis of the PIN diode, an approximation has to be made: the PIN diode voltage defined by (4.9) and (4.11) is generated in response to the current excitation $I(t)$, (4.1). In the linear case, the current-voltage relation (or the diode voltage drop) is related through the PIN diode impedance, Z_{PIN} as defined in (3.5) or (3.28). The generator supplying the excitation current in the linear case was assumed to be a pure single frequency source. In the nonlinear case, the distortion signals are generated by the PIN diode's nonlinear response, but the origin of the excitation is still from pure single frequency sources. From a circuit perspective, two equivalent circuits now have to be created, one with the pure independent sources (Figure 4.1(a)) and the other with no independent sources but with the "PIN diode current generator" from (4.9) and (4.11) generating the harmonic and intermodulation signals (Figure 4.1(b)). This dual equivalent circuit approach is similar to that followed in the harmonic balance technique where the circuit is divided into its linear and nonlinear parts, with

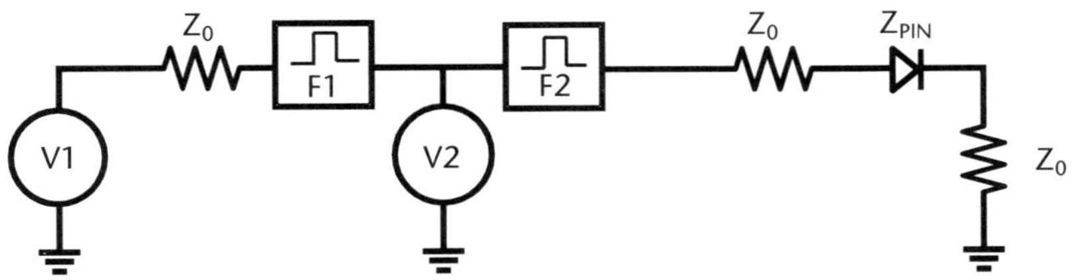

(a)

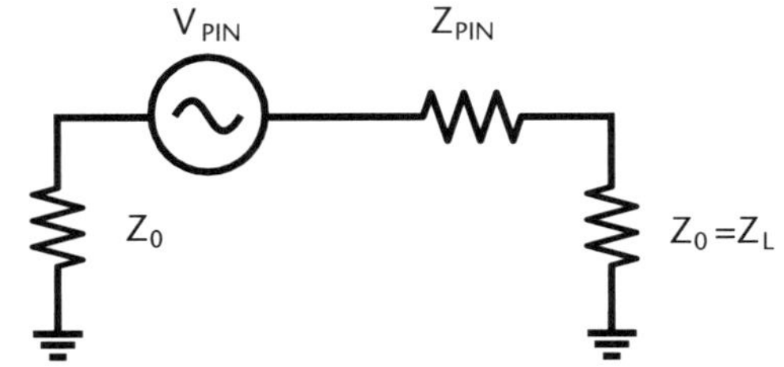

(b)

Figure 4.1 (a) Linear excitation circuit with two single frequency sources of source impedance Z_0 and ideal bandpass filters tuned to the source frequency. (b) Nonlinear excitation circuit showing the nonlinear voltage source V_{PIN} that generates the harmonic and intermodulation signals.

the linear part defining the excitation levels for the nonlinear part. Note that in Figure 4.1, the source impedance of the PIN diode current generator, V_{PIN}, is the linear impedance of the PIN diode, Z_{PIN}.

From Figure 4.1(b), the voltage from the PIN diode distortion generator, V_{PIN}, dropped across the load impedance, Z_L, can be computed using voltage division:

$$V_L(t) = V_{\text{PIN}}(t)\frac{Z_L}{Z_0 + Z_L + Z_{\text{PIN}}} \tag{4.15}$$

Note that in this nonlinear equivalent circuit diagram, no passive components associated with the PIN diode (shunt capacitance, bond wire inductance) have been added; these components can be wrapped into the equivalent circuit during simulation. The location of these components in the equivalent circuit is important for accurate frequency response modeling and so the bond wire inductor L_{bond} can be placed in series with Z_{PIN}. The PIN diode impedance Z_{PIN} used in this equivalent circuit is calculated based on the linear response, but any reactive elements must have their reactance computed based on the nonlinear distortion frequency component under study. For example, the reactance of L_{bond} at distortion component $2\omega_2 - \omega_1$ would be $j(2\omega_2 - \omega_1)L_{\text{bond}}$.

The nonlinear components generated by the PIN diode can be studied by looking at the same diode examples from Chapter 3, Diode1 and Diode2, where thin Diode1 exhibits an I-region thickness of 5 μm with 100 nS carrier lifetime and thick Diode2 with I-region thickness 250 μm and 2 μs carrier lifetime. Both PIN diodes are biased at 10 mA dc forward current and excited by an RF current of 1 mA. For this example, the focus will be on the fundamental, second, and third order single-tone nonlinear responses ω_1, $2\omega_1$, and $3\omega_1$. Figure 4.2(a) shows the magnitude of the second and third order harmonic voltages generated across PIN diodes Diode1 and Diode2 as a function of frequency; Figure 4.2(b) shows the resistance of these two devices over the same frequency range. Similarities can be seen between the linear and nonlinear characteristics at low frequencies; junction effects dominate in the low-frequency limit and the linear and second and third order nonlinear responses converge to identical values. The second and third order distortion voltages generated by the PIN diodes begin to roll off rapidly at lower frequencies in Diode2 than in Diode1, since this device exhibits a higher carrier lifetime, and I-region conductivity modulation effects begin to decrease at the lower frequency. The Diode2 resistance transition also occurs at a lower frequency than Diode1 due to the same effect. In both nonlinear cases, however, at frequencies significantly higher than $1/\tau$, both diodes exhibit monotonically decreasing generated distortion voltages, with junction effects being negligible and the I-region dominating the distortion response. At 100 MHz with an RF current of 1 mA, the linear voltage drop

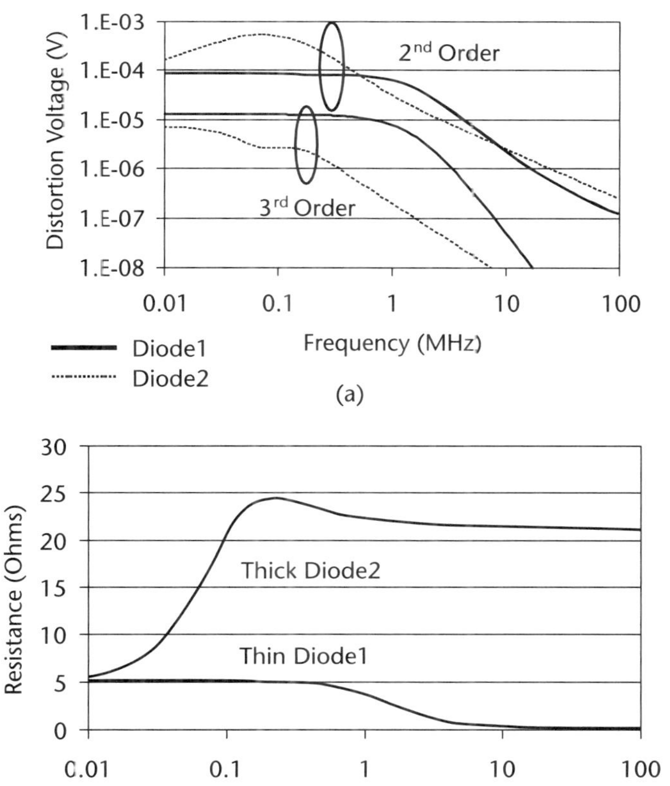

Figure 4.2 (a) Magnitude of the second and third order harmonic voltages generated across PIN diodes Diode1 and Diode2 as a function of frequency. (b) Resistance of Diode1 and Diode2 over the same frequency range.

Table 4.1 Summary of Example PIN Diode Distortion Responses at 100 MHz

	Diode1	*Diode2*	*Diode1*	*Diode2*
Linear Voltage	100 μV	21 mV	—	—
2nd Order Voltage	0.12 μV	0.25 μV	60 dBc	98 dBc
3rd Order Voltage	65 pV	210 pV	123 dBc	160 dBc

across Diode1 (Diode2) is approximately 100 μV (21 mV). The second and third order distortion voltage drops across Diode1 (Diode2) are approximately 0.1 μV (0.25 μV) and 65 pV (210 pV), respectively, many orders of magnitude less than the fundamental voltage dropped across the diodes. Table 4.1 summarizes these data in terms of voltages and dB below carrier (dBc), where the carrier is the associated fundamental voltage drop.

The key factor to note in this discussion is that these generated distortion voltages will eventually find their way to the circuit load (Z_L in Figure 4.1(b) for example) and will manifest themselves as load distortion voltages that show up as spurious signals in the final load output. For the engineer concerned with the distortion introduced into a circuit by solid-state devices such as PIN diodes, the distortion intercept points (IP), primarily the second and third order ones (IP2, IP3), are of interest. Since the voltages generated by the PIN diode have been defined for the various orders of distortion in (4.9) and (4.11), using the nonlinear equivalent circuit shown in Figure 4.1(b) provides a basis to calculate the IP at the load ($Z_L = Z_0$). To illustrate this calculation, example Diode1 and Diode2 will be used again under the same RF (1 mA) and dc currents (10 mA) used to determine the values in Table 4.1. In both cases, the RF current of 1 mA through the load $Z_L = Z_0 = 50\Omega$ provides a load power of 25 μW (−16 dBm) that is constant with frequency (as long as the RF current remains constant). The intercept points are calculated using the expressions derived in Chapter 2, IP2 = $2P_F - P_{2D}$, IP3 = $3/2P_F - 1/2P_{3D}$, where P_F is the fundamental load power (−16 dBm in this case) and P_{2D} (P_{3D}) is the second (third) order distortion power at the load and are shown in Figure 4.3.

4.2.2.1 Approximate Distortion Intercept Point Model

In this section, the discussion will be limited to operation at frequencies well above $\omega\tau$, where a number of simplifying assumptions can be made that lead to closed form solutions for important distortion parameters. Most PIN diodes used in control circuits, especially switch circuits, exhibit W/L_a ratios less than unity to keep the resistance low, which implies there is little variation in dc charge density throughout the I-region. If there is little dc charge variation across the I-region, then the approximation $n_0(x) \approx n_0$ can be made so the β_1 in (4.13) can be further simplified. In addition, an investigation of (4.13) shows that β_0 is simply the high-frequency RF resistance, R_S, of the PIN diode:

$$\beta_0 = \frac{W}{qA(\mu_n + \mu_p)n_0} = R_S \Rightarrow n_0 = \frac{W}{qA(\mu_n + \mu_p)R_S} \tag{4.16}$$

which greatly simplifies the β_l expression:

$$\beta_l \approx (-1)^l \frac{2(l-1)!}{m!n!\cdots} R_S \frac{L_a}{W} \sqrt{\frac{2}{\omega\tau}} \left(\frac{1}{I_{dc}} \frac{W}{2L_a} \frac{1}{\sqrt{\omega\tau}} \right)^l \left(I_1^m I_2^n \ldots \right) \tag{4.17}$$

for l > 0. Specific choice of the expansion parameter l provides a means of determining the distortion components specifically related to the second and third order intermodulation (or any other component, depending on how far out one wishes to take the value of l). The simplest approach is to assume a two-tone test for the analysis, with the further assumption that RF currents are equal $I_1 = I_2 = I_{RF}$, a common measurement setup for two-tone tests. For the second order intermodulation distortion component (frequency $F_a + F_b$ as shown in Figure 2.13), the value of l to choose is $l = 1$ with $m = 1$, $n = 0$, and for third order intermodulation distortion component (frequency $2F_a \pm F_b$ or $2F_b \pm F_a$ as shown in Figure 2.13), choose $l = 2$ with $m = n = 1$.

(a)

(b)

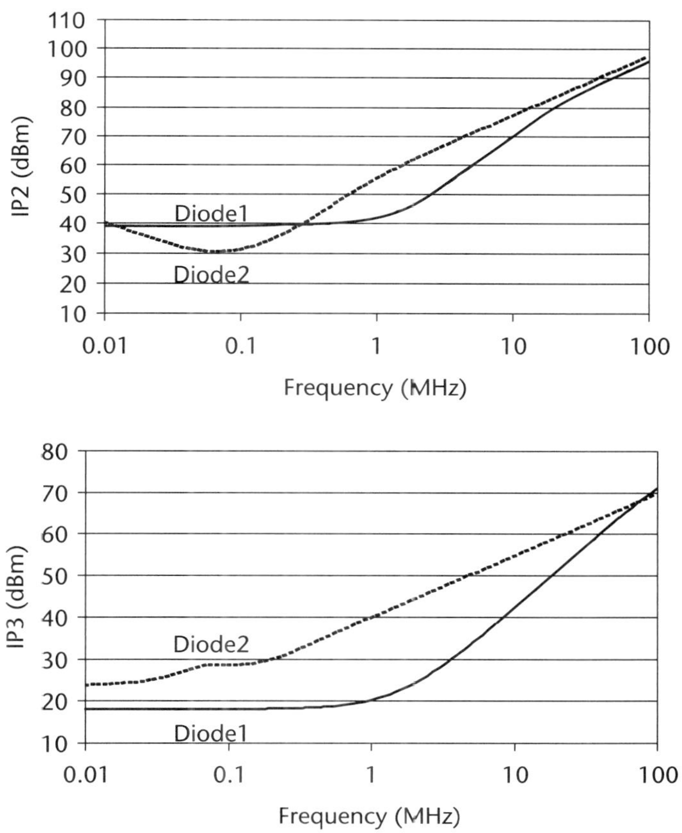

Figure 4.3 (a) IP2 and (b) IP3 for example PIN diodes Diode1 and Diode2 as a function of frequency.

The associated β_l for the intermodulation terms are

$$\beta_1 \approx -\frac{2(0)!}{1!0!\cdots} R_S \frac{L_a}{W}\sqrt{\frac{2}{\omega\tau}}\left(\frac{1}{I_{dc}}\frac{W}{2L_a}\frac{1}{\sqrt{\omega\tau}}\right)^1 I_{RF} = -R_S \frac{I_{RF}}{I_{dc}}\frac{\sqrt{2}}{\omega\tau} \tag{4.18}$$

$$\beta_2 \approx \frac{2(1)!}{1!1!\cdots} R_S \frac{L_a}{W}\sqrt{\frac{2}{\omega\tau}}\left(\frac{1}{I_{dc}}\frac{W}{2L_a}\frac{1}{\sqrt{\omega\tau}}\right)^2 I_{RF}^2 = R_S \frac{I_{RF}^2}{I_{dc}^2}\frac{W}{L_a}\frac{1}{\sqrt{2}(\omega\tau)^{3/2}}$$

The associated second (third order) distortion voltages generated by the PIN diode and then dropped across the load $Z_L = Z_0$ at frequency $\omega_1 + \omega_2$ $(2\omega_1 + \omega_2)$ can now be written

$$V_{2\text{-IMD}} = -R_S \frac{I_{RF}^2}{I_{dc}}\frac{\sqrt{2}}{\omega\tau}\frac{Z_0}{R_S + 2Z_0} \tag{4.19}$$

$$V_{3\text{-IMD}} = R_S \frac{I_{RF}^3}{I_{dc}^2}\frac{W}{L_a}\frac{1}{\sqrt{2}(\omega\tau)^{3/2}}\frac{Z_0}{R_S + 2Z_0}$$

Intermodulation or harmonic distortion components for any higher order can be computed in a similar fashion. Since the peak fundamental voltage drop at the load is $Z_0 I_{RF}$, the fundamental, second and third order load powers can be computed as

$$P_{F\text{-L}} = \frac{(Z_0 I_{RF})^2}{2Z_0} = \frac{Z_0}{2} I_{RF}^2$$

$$P_{2\text{-IMD-L}} = \frac{1}{2Z_0}\left(R_S \frac{I_{RF}^2}{I_{dc}}\frac{\sqrt{2}}{\omega\tau}\frac{Z_0}{R_S + 2Z_0}\right)^2 \tag{4.20}$$

$$P_{3\text{-IMD-L}} = \frac{1}{2Z_0}\left(R_S \frac{I_{RF}^3}{I_{dc}^2}\frac{W}{L_a}\frac{1}{\sqrt{2}(\omega\tau)^{3/2}}\frac{Z_0}{R_S + 2Z_0}\right)^2$$

Load-referenced IP2 and IP3 can now be written with the aid of (4.20) by substituting the I_{RF} term in $P_{F\text{-L}}$ into the two other IMD expressions to yield

$$\text{IP2} = \frac{Z_0^3}{4}\left(\frac{\omega Q}{R_S}\right)^2\left(\frac{R_S + 2Z_0}{Z_0}\right)^2 \approx Z_0^3\left(\frac{2\pi F Q}{R_S}\right)^2 \tag{4.21}$$

$$\text{IP3} = \frac{Z_0^2}{2}\left(\frac{\omega Q}{R_S}\right)^{1.5}\sqrt{\frac{kT}{q}}\left(\frac{R_S + 2Z_0}{Z_0}\right) \approx Z_0^2\left(\frac{2\pi F Q}{R_S}\right)^{1.5}\sqrt{\frac{kT}{q}}$$

where the far right term (4.21) is the case for R_S much smaller than $2Z_0$ and where the I-region stored charge Q has replaced the $I_{dc}\tau$ term. Since R_S is proportional to W^2/Q, the QF/R_S ratio is also proportional to $F(Q/W)^2$. Figure 4.4 shows a graph of these expressions for IP2 and IP3 in dBm as a function of the parameter QF/R_S for R_S much less than $2Z_0$ ($Z_0 = 50\Omega$) at room temperature ($kT/q = 0.0258$V). The

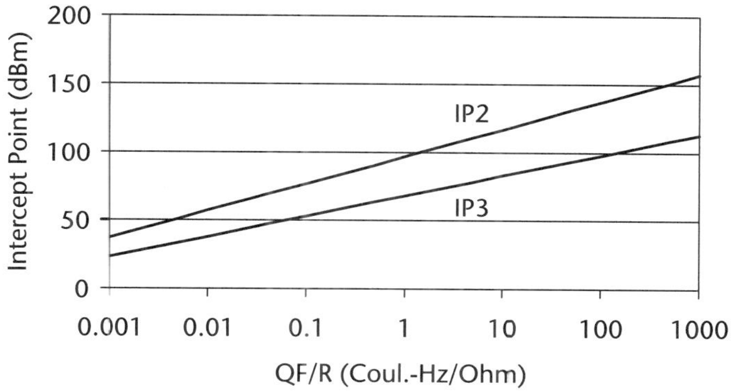

Figure 4.4 IP2 and IP3 in dBm plotted from (4.21) as a function of the QF/R_S ratio.

figure shows that the distortion intercept points increase (or absolute level of distortion decreases) as the ratio QF/R_S increases, indicating that absolute distortion levels as measured at the load will

- Decrease with increasing frequency for fixed Q/R_S;
- Decrease with a reduction in PIN diode resistance R_S and/or increasing the dc stored charge Q at a given frequency.

This relationship between IP2 and IP3 versus QF/R has been verified by experimental measurements and shows very good agreement over a wide range of QF/R values [5].

Looking back at the example PIN diodes Diode1 and Diode2, the 10 mA bias level provided high-frequency resistance of approximately 0.25Ω and 21Ω, respectively, for the two devices. The stored charge for PIN diodes Diode1 and Diode2 can be computed from their carrier lifetimes and the 10 mA dc bias level as 10^{-9} Coulombs (1 nC) and 20×10^{-9} Coulombs (20 nC), respectively. For operation at 100 MHz, the QF/R_S ratios for Diode1 and Diode2 are 0.4 and 0.095 C-Hz/Ω, respectively. The resulting intercept points for these two devices are shown in Table 4.2.

The thinner PIN diode (Diode1) has higher IP2 and IP3 when measured at the load, even though it has lower stored charge than the thicker PIN diode (Diode2), by virtue of its lower resistance due to the thinner I-region (Q/W ratio higher). For any PIN diode, doubling I_{dc} will both double Q and halve R_S, increasing IP2(IP3) by 12 dB (9 dB). Increasing the dc forward current cannot be done without bound, however, since diode burnout will occur above a certain threshold. This QF/R_S ratio and its influence on more complex control circuits will be investigated further in later chapters.

Table 4.2 Intercept Points for Example PIN diodes Diode1 and Diode2

	QF/R_S (C-Hz/Ω)	IP2 (dBm)	IP3 (dBm)
Diode1	0.40	89	62
Diode2	0.095	76	52

4.3 PIN Diode Reverse Bias Distortion

As described in Chapter 3, the capacitance of the ideal PIN diode was generally assumed constant for all reverse bias voltages beyond the punchthrough voltage. However, in actual PIN diodes, the boundaries between the I-region and the heavily doped end regions are not exactly abrupt, but rather exhibit a continuously increasing doping concentration that approaches the end region doping concentration. These so-called diffusion tails cause the reverse biased PIN diode's capacitance to continue to decrease for voltages beyond punchthrough, independent of frequency, since the depletion layer will advance into the end regions with increasing reverse bias voltage. Even with fabrication processes such as MBE or MOCVD, where these tails are practically nonexistent, there will still be a small increase in depletion width with increasing reverse bias voltage. In general, thicker PIN diodes tend to exhibit flatter *C-V* profiles beyond punchthrough than thinner diodes because the depth that the depletion boundary extends into the end region is a smaller portion of the overall device thickness in the thicker devices. This results in a lower rate of capacitance change with reverse voltage for thick PIN diodes.

In Chapter 3, the reverse bias capacitance-voltage (C_{RB}-V) characteristic of the PIN diode was mathematically represented by a Taylor series in (3.44). This difference in series choice reflects that the PIN diode is reverse biased at some dc reverse voltage V_{RB}:

$$C_{RB} = \sum_{k=0}^{\infty} c_k \left(V - V_{RB}\right)^k = c_0\left(V_{RB}\right) + c_1\left(V - V_{RB}\right)^1 + c_2\left(V - V_{RB}\right)^2 \ldots \quad (4.22)$$

where the c_k are based on the derivatives of the C_{RB}-V characteristic:

$$c_k = \frac{1}{k!}\frac{d^k C_{RB}}{dV^k}\bigg|_{V=V_{RB}} \quad (4.23)$$

The rate of capacitance change, as well as the higher order derivatives, with reverse voltage is highly dependent on the end-region doping concentration profiles, making it impractical to attempt to mathematically quantify since these profiles are highly dependent on manufacturing processes that may not be consistent between various fabrication processes, process runs, or facilities. However, the important characteristics required for estimating the reverse bias distortion may be experimentally determined from measured *C-V* data using the following expressions for the first two derivatives of the C_{RB}-V data:

$$\frac{dC_{RB}}{dV} \approx \frac{C_{RB}\left(V_{RB} + \Delta V\right) - C_{RB}\left(V_{RB}\right)}{\Delta V};$$
$$\frac{d^2C_{RB}}{dV^2} \approx \frac{C_{RB}\left(V_{RB} + \Delta V\right) - 2C_{RB}\left(V_{RB}\right) + C_{RB}\left(V_{RB} - \Delta V\right)}{(\Delta V)^2} \quad (4.24)$$

where a second-order central form was used for the second derivative and ΔV is a small voltage difference about V_0. Higher order derivatives can also be calculated using a similar numerical approach. While changes in C_{RB} at high voltages may appear small, the variation in reverse bias capacitance with applied RF voltage can generate harmonic and intermodulation signals that introduce distortion into the circuit, in a similar way as the reverse C-V characteristic does in the classic varactor diode. The corresponding equivalent circuit for the reverse biased PIN diode is shown in Figure 4.5(a), where L_{bond} and C_P are the standard PIN diode parasitic/package parameters, $C_{RB}(V)$ is variable with voltage, and R_{RB} models any loss in the device [6]. In varactors, this R_{RB} is an important parameter since it governs the quality factor of the varactor and acts as an additional resistive-reactive load in the circuit. This resistance, however, is not of concern in most PIN diode control circuits (but it is of great concern in varactors), and so the focus on the modeling will be strictly on the C_{RB}-V characteristic. The analysis of the distortion generated by the reverse biased PIN diode from this point on assumes that the diode is in a series connected circuit (Figure 4.1(b) with Z_{PIN} the capacitive reactance at the reverse bias voltage) with the nonlinear equivalent circuit shown in Figure 4.5(b). The same approach used for modeling the forward bias nonlinearities is used here as well: the linear voltage drop across the reverse bias PIN diode's capacitance is the driving function for the generation of the distortion currents with a separate nonlinear equivalent circuit using a dependent distortion current generator (I_{RB}).

(a)

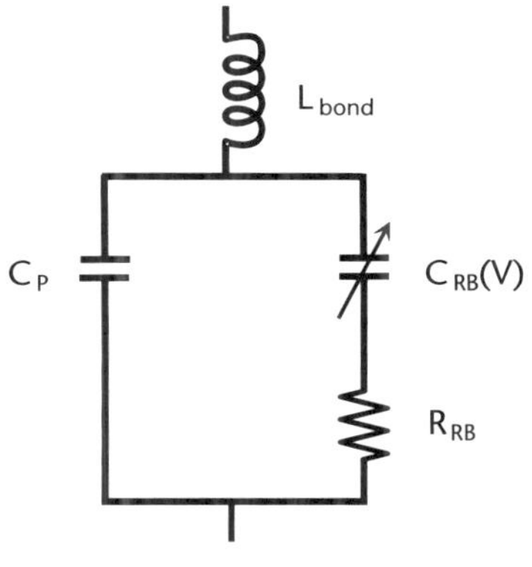

(b)

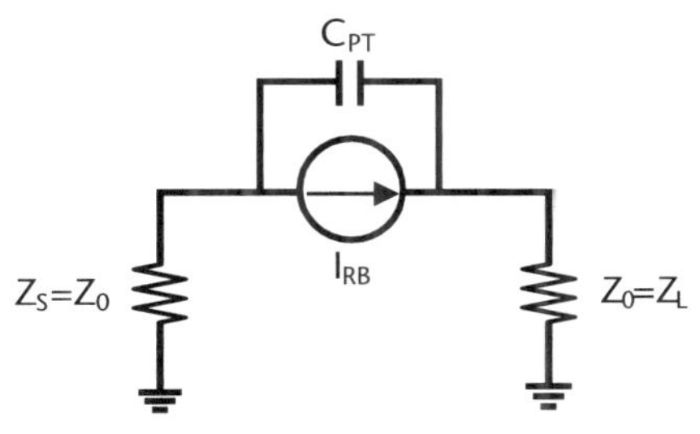

Figure 4.5 (a) Reverse bias equivalent circuit. (b) Nonlinear current generator equivalent circuit (L_{bond} and C_P removed).

These distortion currents can be derived from the fundamental definition of current, namely the change in charge with time. For the PIN diode capacitance, the product of the capacitance and the voltage is the charge $Q_{cap} = CV$. The RF voltage, $V_{RF}(t)$, swings around the reverse bias voltage, V_{RB}, so the time-varying charge associated with the PIN reverse bias capacitance can be written as

$$I_{RB}(t) = \frac{dQ_{cap}(t)}{dt}; \quad Q_{cap}(t) = C_{RB}\left[V_{RF}(t)\right]V_{RF}(t) = \sum_{k=0}^{\infty} c_k V_{RF}(t)^{k+1} \tag{4.25}$$

For N multiple signal sources, the RF voltage dropped across the PIN diode can be written as

$$V_{RF}(t) = \sum_{n=1}^{N} V_n \cos(\omega_n t)$$

in which case the reverse bias current can be modified as

$$I_{RB}(t) = \frac{dQ_{cap}}{dt} = \frac{d}{dt}\left[\sum_{k=0}^{\infty} c_k V_{RF}(t)^{k+1}\right] = \left\{\sum_{k=0}^{\infty} c_k \frac{d}{dt}\left[\sum_{n=1}^{N} V_n \cos(\omega_n t)\right]^{k+1}\right\} \tag{4.26}$$

For a two-tone test ($N = 2$) with $V_1 = V_2 = V_{RF}$, this current becomes

$$I_{RB}(t) = \sum_{k=0}^{\infty} c_k V_{RF} \frac{d}{dt}\left[\cos(\omega_1 t) + \cos(\omega_2 t)\right]^{k+1} \tag{4.27}$$

Using (4.27), the first three terms (k = 0, 1, 2) in the summation provide the fundamental and second ($\omega_2 \pm \omega_1$) and third order ($2\omega_2 \pm \omega_1$, $2\omega_1 \pm \omega_2$) intermodulation current products for two closely spaced frequencies:

$$\begin{aligned}
I_{RB\text{-}fund} &= \omega\left(c_0 V_{RF} + c_2 V_{RF}^3\right) \approx \omega c_0 V_{RF} \approx \omega C_{PT} V_{RF} \\
I_{RF\text{-}2nd} &= \left(\omega_2 \pm \omega_1\right) c_1 V_{RF}^2 \\
I_{RF\text{-}3rd} &= \left(2\omega_2 \pm \omega_1\right) \frac{3}{4} c_2 V_{RF}^3
\end{aligned} \tag{4.28}$$

where for $k = 0$, the c_0 coefficient is approximated by the punchthrough capacitance (C_{PT}) and only the amplitudes of the currents are shown. The value of V_{RF} in (4.27) is the amplitude of the fundamental voltage across the reverse biased PIN diode since that is the signal controlling the generation of the distortion components in the device. Using traditional circuit analysis techniques for the second and third order generated distortion currents and the equivalent circuit in Figure 4.5(b), IP2 and IP3 can be written for this circuit as

$$\text{IP2} = \frac{1 + \left[2(\omega_2 \pm \omega_1) Z_0 C_{PT}\right]^2}{2c_1^2 (\omega_2 \pm \omega_1)^2 Z_0^3} \approx \frac{1}{2c_1^2 (\omega_2 \pm \omega_1)^2 Z_0^3} \tag{4.29}$$

$$\text{IP3} = \frac{2}{3} \frac{\sqrt{1 + \left[2(\omega_2 \pm \omega_1) Z_0 C_{PT}\right]^2}}{(2\omega_2 \pm \omega_1) c_2 Z_0^2} \approx \frac{2}{3} \frac{1}{(2\omega_2 \pm \omega_1) c_2 Z_0^2}$$

where the second expression holds if the capacitive reactance of C_{PT} is much larger than the load impedance (usually the case for high isolation in a series connected switch circuit with the PIN diode in its reverse bias state). These expressions indicate that the intercept point decreases (distortion increases) as frequency increases, a direct consequence of dependence on the time-varying capacitance (d/dt). Since thicker I-region PIN diodes tend to exhibit flatter *C-V* characteristics, thicker PIN diodes are expected to perform with better distortion properties than thinner diodes. Published information on values of c_1 for representative thin and thick commercial PIN diodes have shown this value to be in the range of approximately 10^{-15} F/V and 10^{-16} F/ at 10V reverse bias (approximate punchthrough voltage for the thick diode) for the diodes at 1000 MHz [7]. The measured reverse bias distortion reported in [7] was 73 dBm (61 dBm) for the thick (thin) PIN diode at 1000 MHz and 10V reverse bias, in good agreement with the modeling approach shown here. The increase in distortion with increasing frequency was also noted in this paper [7], again validating this approach to modeling the reverse bias distortion in PIN diodes.

4.4 Minimum Reverse Bias in High-Power Applications

A common question a PIN diode circuit designer asks is the amount of reverse bias needed to hold off a certain peak RF signal level. The traditional answer to this question requires a look at the PIN diode's current-voltage characteristics, including reverse breakdown, and the RF signal itself. For safe operation of the PIN diode, the instantaneous voltage across the PIN diode (both RF and dc) should not exceed its avalanche breakdown voltage (V_{BR} in Figure 4.6), where high reverse current densities may cause diode failure, and should also never swing into forward conduction where large forward currents could cause similar failure.

To fully keep the PIN diode in reverse bias during the entire RF cycle, the "safest" choice of reverse bias voltage is halfway between zero bias (0V) and the reverse breakdown voltage, V_{BR}, thus limiting the maximum RF voltage swing to $V_{BR} = 2V_{RF}$. For high-power applications, large applied reverse bias voltages may not be available or are expensive to implement, and so the $V_{3R}/2$ dc reverse bias voltage is a conservative but potentially costly design choice.

The real PIN diode, however, does not show this instantaneous turn-on time, as the ideal rectifier diode as described above in real microwave and RF applications. There is a finite amount of time before the diode starts conducting in the forward direction after the RF signal swings above zero bias. The time it takes to fully turn the PIN diode on in the forward direction is the time required for charge carriers

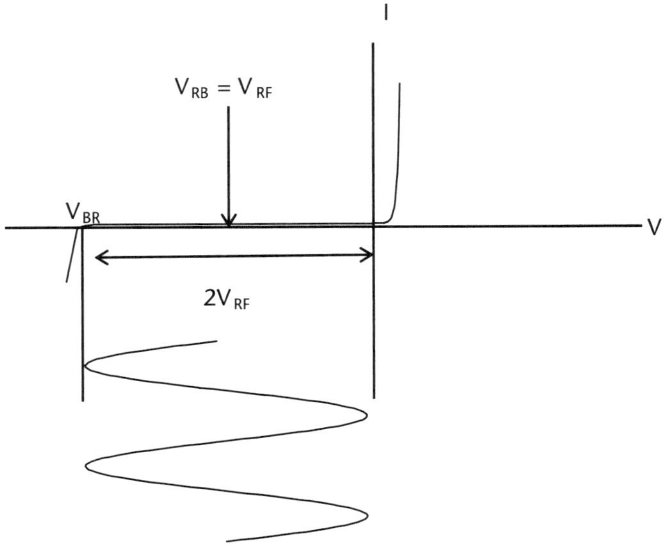

Figure 4.6 The traditional reverse bias point for maximizing the RF swing is half the breakdown voltage of the PIN diode, $V_{BR}/2$.

to move from the heavily doped end regions and completely fill the I-region during the forward cycle of the applied RF signal. Depending on the frequency, the peak RF voltage level ($|V_{RF}|$), duty cycle (D), and the I-region thickness (W), this time may be too short (especially for short duty cycles where the RF is only on for a short period of time) for the carriers to completely traverse the I-region (the so-called transit time), preventing the PIN diode from fully entering its conducting state [8–10], in which case the diode will then act as a lossy capacitor.

An investigation looking at the characteristics of the PIN diode reverse bias behavior in the presence of an applied RF signal has shown that this $V_{BR}/2$ bias point can be relaxed depending on a complex interaction between the I-region width W, the I-region transit time, magnitude of the RF signal, and duty cycle [11]. In the PIN diode under reverse bias, conduction and displacement currents both flow through the device [8]:

$$J(t) = 2n_0 q v_{\text{drift}} + \varepsilon \frac{dE(t)}{dt} \tag{4.30}$$

where $J(t)$ is the total current density, n_0 is the I-region carrier density (where the density of holes equals the density of electrons), v_{drift} is the drift velocity of the charge carriers (assuming equal hole and electron drift velocities), ε is the dielectric permittivity, and $E(t)$ is the electric field. Further, the drift velocity is related to the electric field through the ambipolar mobility, $v_{\text{drift}} = \mu_a |E|$ and has a maximum of approximately 10^5 m/s in silicon (the so-called saturation velocity, v_{sat}). From a charge control perspective, the total current as indicated by (4.30) contains a dc term ($2n_0qv$) and an ac term ($\varepsilon dE(t)/dt$) that are supported by a dc and RF charge in the I-region:

$$I = I_{dc} + I_{RF} = \frac{Q_D}{T} + \frac{dQ_D}{dt} \quad (4.31)$$

where Q_D contains both dc and RF components, Q_{dc} and Q_{RF} ($Q_D = Q_{dc} + Q_{RF}$), and T is the I-region transit time defined as $W/2v_{drift}$ [8]. If an $e^{j\omega t}$ variation is assumed for the time-varying portion, the dc and RF currents can be written as

$$I_{dc} = \frac{Q_{dc}}{T}; \quad I_{RF} = Q_{RF}\frac{1 + j\omega T}{T} \quad (4.32)$$

From the discussion in Chapter 3, the voltage-current relationship in the PIN diode is related to the stored charge by $W^2/2\mu_a Q$, so the corresponding dc and RF voltages can be written as [11, 12]

$$\begin{aligned} V_{dc} &= \frac{W^2}{2\mu_a Q_{dc}} I_{dc} = \frac{W^2}{2\mu_a T} \\ V_{RF} &= \frac{W^2}{2\mu_a Q_{RF}} I_{RF} = \frac{W^2}{2\mu_a}\frac{1 + j\omega T}{T} \end{aligned} \quad (4.33)$$

The quantity of interest in this discussion is the required dc reverse voltage to the RF peak voltage:

$$|V_{dc}| = \frac{1}{\sqrt{1 + (\omega T)^2}}|V_{RF}| = \frac{1}{\sqrt{1 + \left(\pi F W/v_{drift}\right)^2}}|V_{RF}| \quad (4.34)$$

The drift velocity increases linearly with electric field until it reaches its saturation velocity and may be approximated using the following expression [13]:

$$v_{drift} = \frac{2\mu_a E}{1 + \sqrt{1 + \left(2\mu_a E/v_{sat}\right)^2}} \quad (4.35)$$

The designer of high-power microwave systems using PIN diodes is usually interested not in the CW power but in the pulsed power with duty cycle $D = T_D/T_P$, which is the ratio of the pulse duration (T_D) to the pulse period (T_P). For an operating frequency greater than $1/T_D$, the rms value of the electric field E can be approximated as [12],

$$E = \frac{|V_{RF}|}{W}\sqrt{D} \quad (4.36)$$

Combining (4.35) and (4.36) into (4.34) provides for a measure of the required dc reverse voltage needed for a specific RF voltage:

$$
\begin{aligned}
\left|V_{dc}\right| &== \frac{\left|V_{RF}\right|}{\sqrt{1+\left(\pi F W / v_{drift}\right)^{2}}} \\
&= \frac{\left|V_{RF}\right|}{\sqrt{1+\left(\frac{\pi F W^{2}}{2\mu_{a}\left|V_{RF}\right|\sqrt{D}}\left[1+\sqrt{1+\left(\frac{2\mu_{a}\left|V_{RF}\right|\sqrt{D}}{W v_{sat}}\right)^{2}}\right]\right)^{2}}}
\end{aligned}
\tag{4.37}
$$

From (4.35), if the electric field in the I-region is low enough so that velocity saturation does not occur ($|V_{RF}|/W < v_{sat}$), (4.37) can be simplified:

$$
\left|V_{dc}\right| == \frac{\left|V_{RF}\right|}{\sqrt{1+\left(\frac{\pi F W^{2}}{\mu_{a}\left|V_{RF}\right|\sqrt{D}}\right)^{2}}}
\tag{4.38}
$$

The relationship between the dc to RF voltage can be seen more clearly by plotting (4.38) for the required dc reverse voltage $|V_{dc}|$ as a function of the RF signal amplitude $|V_{RF}|$ using $FW^2/\sqrt{D}$ as a parameter (Figure 4.7). The utility of (4.38) and Figure 4.7 can best be seen by use of an example. Consider a 1 kW pulsed ($|V_{RF}|$ = 316V in a 50Ω system) using a 100 μm PIN diode with a 10% duty cycle at 1000 MHz. For this example, $FW^2/\sqrt{D}$ = 31.6, and so extrapolating between the $FW^2/\sqrt{D}$ = 10 and 100 curves in Figure 4.7 shows that only approximately 95V of reverse bias is needed to hold off 316V of RF in this 10% duty cycle application. Under these same conditions but using a 10 μm PIN diode ($FW^2/\sqrt{D}$ = 0.31) requires that the dc reverse bias voltage be the same as the peak RF voltage swing (316V).

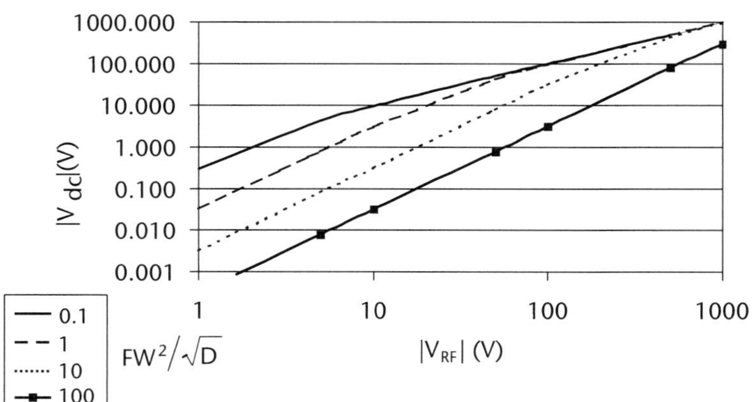

Figure 4.7 DC reverse voltage $|V_{dc}|$ as a function of applied RF voltage ratio $|V_{RF}|$ with the term $FW^2/\sqrt{D}$ as a parameter.

These relationships indicate that thicker PIN diodes require lower dc reverse voltages and would appear preferable to thinner PIN diodes. Thicker PIN diodes may have higher forward bias resistance and so will have increased insertion loss and may switch at a slower speed.

4.5 Time Domain Models

One of the biggest advances in microwave and RF circuit analysis and design has been the use of CAD tools. A number of microwave and RF circuit tools such as Microwave Office, Advanced Design System (ADS), SPICE (and its many variants), and others have greatly simplified the old methods of circuit analysis and trying to invert $N \times N$ matrices by hand. These tools have many built-in models for the commonly used components in microwave and RF circuit design: resistors, capacitors, inductors, rectifier diodes, transistors of various flavors, and independent and dependent sources, to name just a few. Conspicuously absent until the last few years was a standard PIN diode model that was based on the underlying physics of the device rather than empirical models such as the Walston or Jang models as outlined earlier in Chapter 3. This section will present a review of a time-domain model of the PIN diode suitable for use in SPICE-like simulators. The isothermal version of this model has been implemented in a number of CAD tools; a more recently derived electrothermal model will be included in the presentation.

4.5.1 SPICE Model—Isothermal

4.5.1.1 Forward Bias

In Chapter 3, a detailed analysis of PIN diode behavior was covered and began with the traditional semiconductor transport equation:

$$\frac{\partial^2 n(x,t)}{dx^2} = \frac{n(x,t)}{D_a \tau} + \frac{1}{D_a}\frac{\partial n(x,t)}{dt} \tag{4.39}$$

where $n(x,t)$ is the one-dimensional carrier density in the I-region, τ is the I-region ambipolar carrier lifetime, D_a is the ambipolar diffusion constant, and t and x were time and position variables, respectively. While these expressions are important for mathematical modeling, they are too complex to be easily integrated into a time-domain model. An interesting approach for the solution of (4.39) was introduced by Strollo [14] for modeling the dc and transient characteristics of power PIN diodes. Subsequent analysis has shown that with additional effort, this approach is also successful in modeling the microwave and RF behavior of the PIN diode [15, 16]. The translation of these analytic models to a form more suitable for simulation using time domain simulator elements requires that the boundary conditions for solving (4.39) be altered from those typically used in the above closed form analytical solutions [15, 16]. Assuming an $e^{jwt} = e^{st}$ time variation for all quantities (s is the Laplace Transform variable), the boundary conditions for solving (4.39) are described in (4.40) and shown in Figure 4.8:

$$\left.\frac{dn(x)}{dx}\right|_{x=0} = -\frac{b}{1+b}\frac{1}{qAD_a}I(t) \quad \left.\frac{dn(x)}{dx}\right|_{x=x_m} = 0 \tag{4.40}$$

where b is the electron to hole mobility ratio (μ_n/μ_p), x_m is the location of the minimum I-region charge density computed from the transcendental equation $\sinh(x_m/L_a) = b\sinh((W - x_m)/L_a)$, and the other symbols have their usual meanings. For equal hole and electron mobilities ($b = 1$), the minimum I-region charge density occurs in the middle of the I-region, $W/2$. Using the solution technique advanced by Strollo [14], the Laplace Transform for the charge density in the I-region may be written as

$$n(x,s) = \lambda(s)\frac{b}{1+b}\frac{1}{qAD_a}I_1\frac{\cosh\left(\dfrac{x - x_m}{\lambda(s)}\right)}{\sinh\left(\dfrac{x_m}{\lambda(s)}\right)} \tag{4.41}$$

where $\lambda(s)^2 = D_a\tau/(1 + s\tau)$ and $s = j\omega$. As in the previous derivations, the I-region dc charge density $n_{dc}(x,0)$ is supported by a dc bias current $I_1 = I_{dc}$ ($\lambda(s) = L_a$ at dc), whereas the I-region RF charge density $n_{RF}(x,s)$ is supported by the RF current $I_1 = I_{RF}$. These two currents also flow through the two end regions; the junction voltage drops are modeled using traditional PN junction SPICE-based simulator elements [17].

By rearranging the variables in (4.31) and noting the e^{st} temporal variation, the I-region stored charge-current (Q-I) relationship in the s-domain, $H(s)$, may be written in the form of a transfer function as

$$H(s) = \frac{I(s)}{Q(s)} = \frac{D_a}{x_m^2}\frac{x_m}{\lambda(s)}\tanh\left(\frac{x_m}{\lambda(s)}\right) \tag{4.42}$$

where $I(s)$ and $Q(s)$ are the current and I-region stored charge in the s-domain, respectively. Unfortunately, the mathematical form for the Q-I relationship in (4.42)

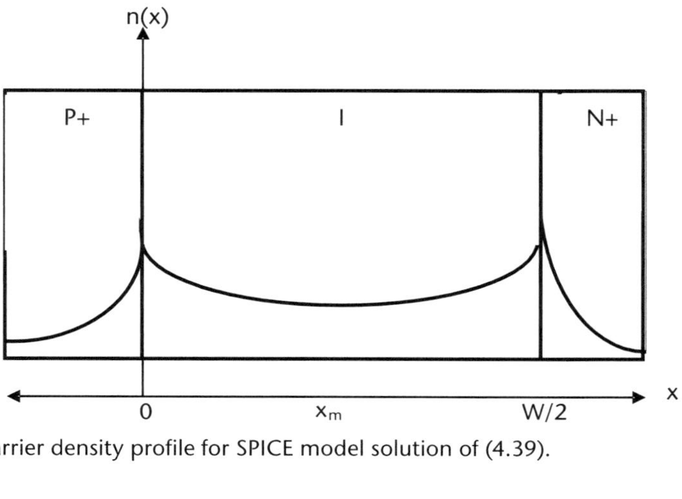

Figure 4.8 Carrier density profile for SPICE model solution of (4.39).

is not in a form that can be directly implemented in time domain circuit simulators such as SPICE. However, Strollo used a unique mathematical expansion based on what is called the Pade approximation for the general function $x_m/\lambda(s)\tanh[x_m/\lambda(s)]$ [14], which allows the Q-I relationship $H(s)$ to be written as

$$H(s) = \frac{I(s)}{Q(s)} = \cfrac{1}{Z + \cfrac{1}{\cfrac{3}{T} + \cfrac{1}{5Z + \cfrac{1}{\cfrac{7}{T} + \dots}}}} \tag{4.43}$$

where the variables $Z = \tau/(1 + s\tau)$ and $T = x_m^2/D_a$. This form for the Q-I transfer function $H(s)$ can be modeled as a lumped element equivalent circuit by noting that the current $I(s)$ is the response to a stored charge $Q(s)$. The form for Z and T suggests that these variables may be used to model $H(s)$ using a parallel resistance-reactive element (for Z) and a resistance element (for T) as part of a ladder-type network relating the $Q(s)$ excitation to the $I(s)$ response. Figure 4.9 shows an RF ladder network structure as an example of how this $H(s)$ relationship out to fourth order for the charge storage region in the PIN diode can be modeled in a circuit simulator.

The circuit simulator chosen for this discussion is based on the standard SPICE implementation. The SPICE netlist file format and variables differ from those used in standard mathematics and so SPICE variables will all be capitalized to distinguish them from other mathematical forms.

The source $E(s)$ is governed by current continuity through the PN junctions, which also flows through the I-region; $E(s)$ is implemented in SPICE as a controlled source directly dependent on this PN junction current (Figure 4.10(b)). The total current through the PN junction is modeled with dependent source GPIN in Figure 4.10(a); the voltage across this element (the voltage between points C and B) governs the voltage source Ej in Figure 4.10(b). The product $I(s)H(s)$ is then the stored charge in the I-region, $Q(s)$. The I-region modulated conductivity is coupled to the main time domain model via the controlled current source GMOD. A shunt inductor of high value in parallel with GMOD (LRMOD) ensures dc continuity through the I-region but has minimal effect in RF simulations. Since the I-region stored charge $Q(s)$ is calculated from the equivalent circuit in Figure 4.9(a), the scaling factor for

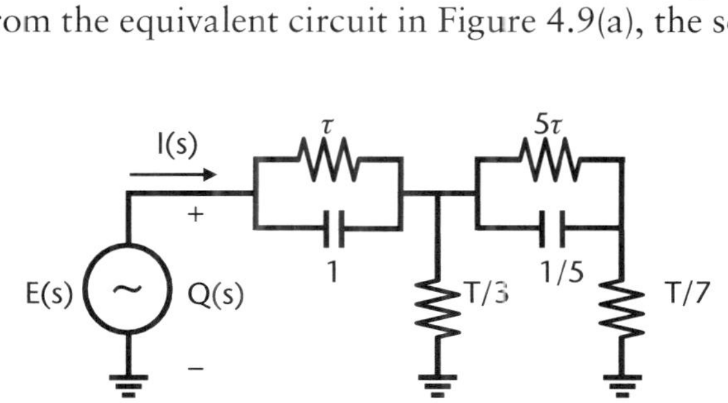

Figure 4.9 4th order approximation to the transfer function $H(s) = I(s)/Q(s) = Ij/Q(s)$ and its mathematical formulation. This equivalent circuit is coupled to the controlled current source GMOD in the main time domain model for the I-region (Figure 4.10(a)).

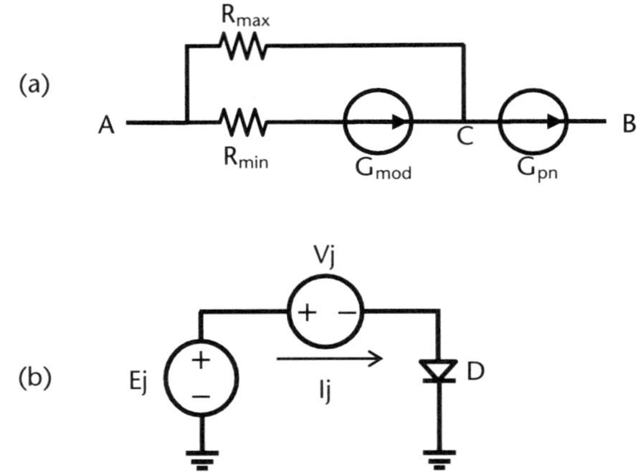

Figure 4.10 (a) Main RF equivalent circuit of the PIN diode model showing the controlled sources for the I-region conductivity modulation (GMOD) and the PN junction (GPN). (b) PN junction modeled as a conventional diode D with the corresponding diode current *Ij* acting as a source for the *H*(*s*) I-region stored charge equivalent circuit (Figure 4.9).

this dependent source in SPICE-based simulators is $[W^2/2\mu_a\tau]^{-1}$. Two resistors are added to the model to represent the minimum resistance of the I-region at high dc bias (RMIN) and the zero bias resistance (RMAX). The full PIN diode model with parasitic elements is shown in Figure 4.11. Additional accuracy in describing the transfer function in (4.42) can be obtained by using higher order terms in (4.43) and increasing the corresponding number of SPICE circuit elements. Extensive analysis using this model has shown that little additional accuracy is achieved by using expansion of (4.43) above 8th order. Some versions of SPICE do not allow the advanced calculations required to specify the GMOD controlled current source as shown in Table 4.3; a series of native interactive SPICE controlled sources such as those found in the online repository file 'SPICEfile-1.xls' can also be used. This spreadsheet also computes the various elements of the I-region transfer function H(s) for specific diode types.

Table 4.3 shows the SPICE model netlist for this example structure with Figure 4.12 showing the results of a SPICE simulation of the frequency-dependent impedance of a thin PIN diode (SPICE netlist 'netlist4-1.net' in the file repository). The simulation parameters are a dc bias of 10 mA, an I-region thickness of 10 μm with 1.0μS carrier lifetime, and package parasitics included (see table for details). The resistance and reactance can be computed as the real and imaginary part of the diode

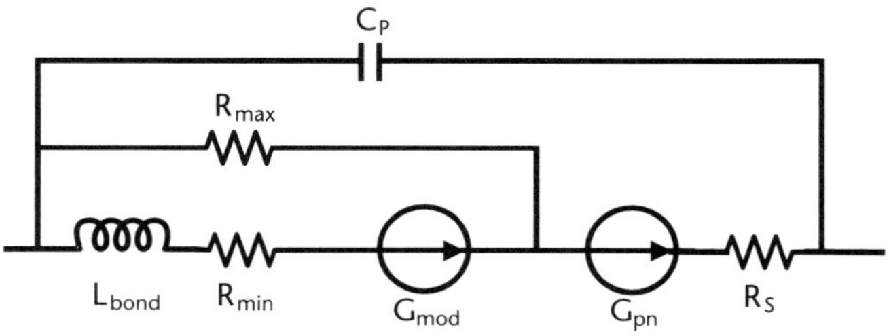

Figure 4.11 PIN diode forward bias model with parasitic elements shown.

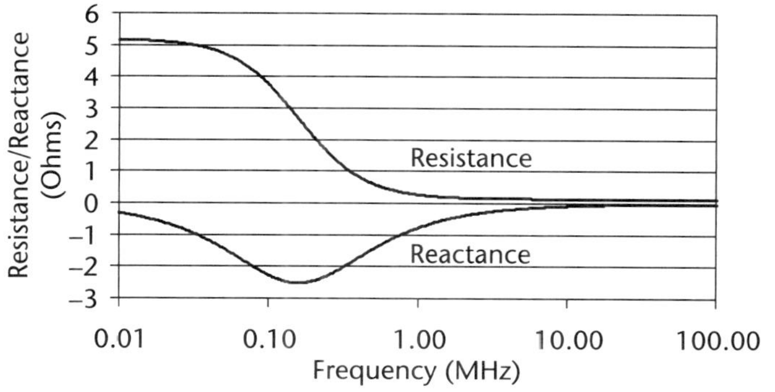

Figure 4.12 Results of SPICE frequency sweep modeling (.ac) of an example PIN diode $W = 10$ μm, $\tau = 1$ μs PIN diode at 10 mA dc forward current.

Table 4.3 SPICE netlist for PIN diode subcircuit model ('netlist4-1.net')

```
*EXAMPLE PIN SPICE FILE USING THE PIN DIODE SUBCIRCUIT
* CURRENT SOURCE FOR AC AND DC EXCITATION AS PART OF AN AC SWEEP
I1 0 1 DC 0.01 AC 0.001
X1 1 0 PIN
.AC DEC 100 1E4 1E9
.OPTIONS GSHUNT=1E-7
* SOURCE NETLIST FOR AC AND DC EXCITATION AS PART OF A TRANSIENT
*SIMULATION
*V1 1 0 SIN(1.2 0.1 1.5E5)
*RS 1 2 100
*X1 2 0 PIN
*.TRAN 100NS

* SUBCIRCUIT FOR PIN DIODE
.SUBCKT PIN 9 21 PARAMS:
+ RMIN = 0.01 RMAX= 50K
* MUA – AMBIPOLAR MOBILITY
+ W = 10U TAU0=100N MUA=0.05
+ LBOND = 0.5NH CP=0.1PF RS=0.01
+ N=2
.PARAM VI={W*W/2/MUA}
.PARAM T={W*W/4/.0258/MUA}

* THE "RCONVERGE" RESISTORS ARE PURELY TO AID IN SPICE CONVERGENCE
RCONVERGE1 12 20 1E7
RCONVERGE2 11 12 1E7
```

(continues)

Table 4.3 SPICE netlist for PIN diode subcircuit model ('netlist4-1.net') *(Continued)*

```
RMIN 10 11 {RMIN}
GMOD 11 12 VALUE={(V(11,12)*V(2,3)/VI)}
LRMOD 11 12 1E9
LBOND 9 10 {LBOND}

GPN 12 20 VALUE={I(VS2)}
RMAX 10 12 {RMAX}
RS 20 21 {RS}
CP 9 21 {CP}

EJ 30 0 VALUE={V(12,20)}
VS1 30 31 0
DIN 31 0 DIN
* THE NEXT LINE MODELS THE PN JUNCTION. USE SPICE MODEL PARAMETERS
*HERE [19].
.MODEL DIN D N={N}
E1 1 0 VALUE={I(VS1)}
VS2 1 2 0

* 4TH ORDER APPROXIMATION FOR BASE REGION
RP1 2 3 {TAU0}
CP1 2 3 {1}
RS1 3 0 {T/3}
RP2 3 4 {5*TAU0}
CP2 3 4 {1/5}
RS3 4 0 {T/7}
.ENDS
```

voltage (V(9,21)) divided by the diode current based on the associated SPICE node voltages and branch currents. The results clearly show that the reactance minimum as described in Chapter 3 (Figure 3.19 for example) occurs with rapidly decreasing resistance. The reactance minimum at approximately 150 kHz is also in agreement using the RF technique of carrier lifetime determination in (3.37), $\tau = 1/2\pi F_\tau$.

Figure 4.13 shows the results of a SPICE transient analysis of this same PIN diode using the circuit elements in Table 4.3. The simulations show the PIN diode voltage and current using a 100Ω signal source with a dc value of 1.2V (approximately 80 mA of dc bias current) and RF amplitude of 0.1V. Two simulation frequencies are shown: 15 MHz and 1500 MHz. These two frequencies where chosen to show the following:

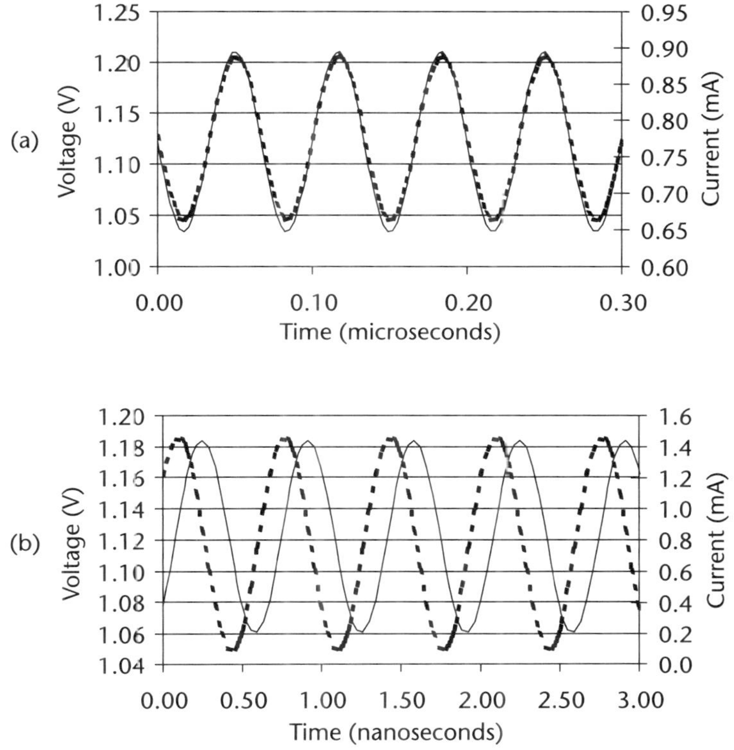

Figure 4.13 Simulated PIN diode voltage (bold lines) and current (dotted lines) responses using a SPICE simulator for (a) 15 MHz and (b) 1500 MHz.

- At 15 MHz, Figure 4.13(a) shows primarily resistive behavior since the PIN diode voltage and current are in-phase.
- At 1500 MHz, the capacitive reactance of element C_P starts to become apparent in Figure 4.13(b), and the capacitive phase shift is noted in the PIN diode voltage-current response.

The SPICE netlist file 'netlist4-1.net' can be used to test the simulations by the reader.

Modeling of various switch and attenuator circuit topologies using this SPICE approach will be a topic in later chapters.

4.5.2 SPICE Model—Electrothermal

In Chapter 3, the two factors determining the PIN diode resistance-temperature characteristic were shown to be the carrier mobility and the carrier lifetime temperature dependences. Knowing the heating characteristics of the PIN diode due to an applied dc bias, large RF signal, or a combination of the two aids the design engineer in determining important characteristics about the PIN diode's survival in high-power applications. The SPICE model introduced in the previous section describes PIN diode behavior in the time domain and can be modified to include thermal effects. The temperature rise is based on the instantaneous voltage and current measured at the terminals of the PIN diode, with the product of these two terms a measure of the input power (P_{therm}). In PIN diodes, the standard parameter used to model the temperature rise is the thermal resistance (Θ_{therm}) with units of °C/W. As the units imply, the thermal resistance provides a metric to compute the amount of temperature rise per watt of applied power. The temperature rise, however, is not instantaneous but takes some finite amount of time to respond to the applied power. The standard parameter describing this finite temperature rise time (or fall time, if the power is removed) is the thermal time constant. Since Θ_{therm} describes the thermal resistance, an analog to the standard RC time constant can be applied if one defines the thermal capacity γ_{therm}, yielding the product $\gamma_{\text{therm}} * \Theta_{\text{therm}}$ as the thermal time constant. The thermal time constant is a function of many factors, including the PIN diode geometry and how well the die is thermally attached to the package [18]. However, using this concept of a thermal time constant allows a thermal circuit to be created as an electrical circuit analog where the excitation of this thermal circuit is an independent source modeling the applied power and a standard single-pole thermal circuit modeling the thermal response of the PIN diode. Figure 4.14 shows this thermal model using a so-called Foster form to compute the temperature rise above the ambient temperature [19, 20] (set at 27°C since this is the default temperature used in SPICE simulations and modeled as independent voltage source V_{ambient}) [17]. Homogenous heating of the PIN diode throughout the device is assumed.

The two PIN diode regions, PN junctions and conductivity modulated I-region, have different thermal behaviors and have to be modeled separately. The PN junction thermal variation is calculated using the standard SPICE diode equation (4.44) and parameters at temperature T (T_{nom} is 27°C by default in SPICE) [17]. A similar dynamic thermal modeling technique as described in [20] is employed using a voltage-controlled current source GTHERM to alter the overall current flow (4.45) through the PIN diode at fixed diode voltage V_d based on the instantaneous temperature rise from the thermal circuit in Figure 4.14:

$$I(T) = IS\left(\frac{T}{T_{nom}}\right)^{XTI/N} e^{\left[\left(\frac{T}{T_{nom}}-1\right)\cdot\left(\frac{qE_{\text{Gap}}}{NkT}\right)\right]}\left(e^{\frac{qV_d}{NkT}}-1\right) \tag{4.44}$$

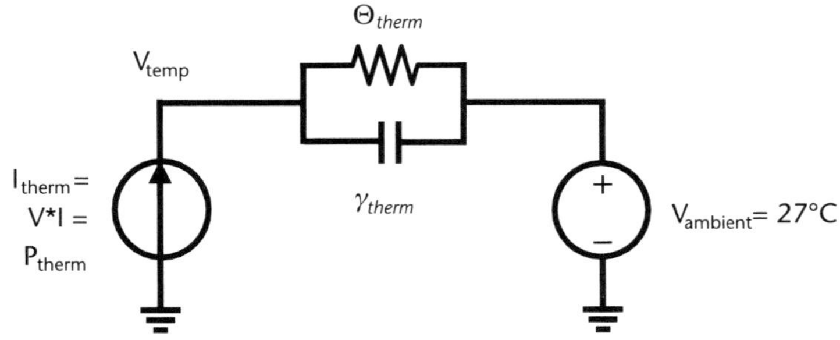

Figure 4.14 Single-pole Foster [29] thermal impedance model with voltage V_{temp} corresponding to the instantaneous device temperature.

where IS is the PIN diode 27°C reverse saturation current; the remaining symbols are the usual SPICE parameters for the PN junction [17]. The diode D1 is the same PN junction device that is a component in the current time-domain model; controlled source GJUNCTION is added to modulate the D1 current as a function of the instantaneous PIN diode temperature. The current difference equation (4.45) can be directly implemented in SPICE using the difference between (4.44) at T_{nom} and the instantaneous temperature value, T (Figure 4.15):

$$\Delta I(T) = G_{\text{junction}}(T) = I(T) - I\left(T_{nom}\right) \tag{4.45}$$

I-region thermal modeling follows the approach in [21] by using the temperature dependence of the I-region ambipolar mobility $\mu_a(T)$ and carrier lifetime $\tau(T)$ [21]:

$$\mu_a(T) = \mu_a\left(T_{nom}\right)\left(\frac{T}{273 + T_{nom}}\right)^{-2.3}; \quad \tau(T) = \tau\left(T_{nom}\right)\left(\frac{T}{273 + T_{nom}}\right)^{mtau} \tag{4.46}$$

where mtau is the carrier lifetime temperature coefficient, T is in Celsius, T_{nom} = 27°C and the other symbols have their usual meanings. Similar for the PN junction thermal modeling, the instantaneous PIN diode temperature T is computed using the Foster thermal circuit in Figure 4.14. The mobility and carrier lifetime temperature dependencies are then coupled to the current time domain model through the element GXFERTEMP that then modifies the conductivity modulation in the I-region, GMOD, by use of the temperature-scaling factor based on the mobility and carrier lifetime temperature dependences in (4.46); (i.e., mtau-2.3). This temperature scaling factor is computed in the SPICE equivalent circuit netlist shown in Table 4.4; the command PWR raises the temperature difference to the mtau-2.3 power. The I-region conductivity modulation factor GMOD is modified from that in the isothermal case shown in Table 4.3 with the addition of a voltage element $V(1000,0)$ that scales the I-region characteristics according to the mobility-carrier lifetime temperature dependence computer using the GXFERTEMP circuit.

Figure 4.16 shows the time domain (transient analysis, SPICE command .TRAN) results for the same diode whose responses are shown in Figures 4.12 and 4.13 with a thermal resistance of 50 °C/W and a thermal time constant of 100 nS at

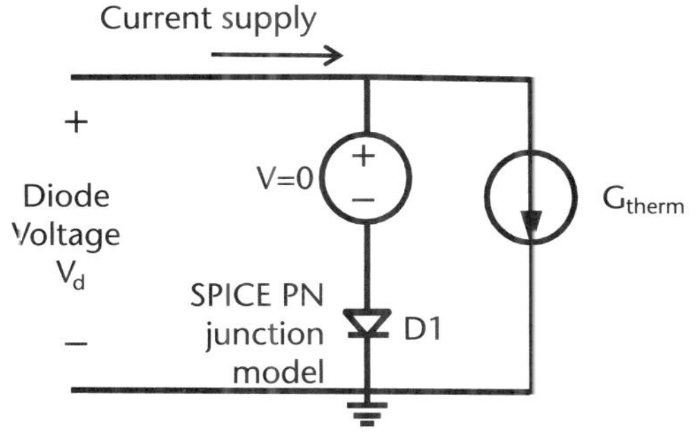

Figure 4.15 Modified PN junction component of the PIN diode model, showing the controlled current source that models the temperature rise in the junction dynamically.

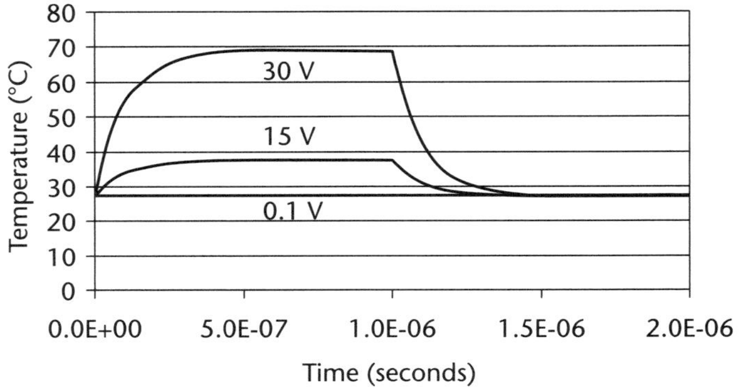

Figure 4.16 PIN diode temperature increase and decrease with application of a 1.0 μs RF pulse of different amplitudes.

Table 4.4 SPICE Netlist Elements for I-Region Thermal Modeling

```
* THESE CIRCUIT ELEMENTS IMPLEMENT THE FOSTER MODEL IN FIG. 4.14
* AS A FUNCTION OF THE VOLTAGE ACROSS THE PIN DIODE AND CURRENT
* THROUGH IT -> POWER
* THERMAL PARAMETER LIST – SEE TABLE 4.3 FOR ISOTHERMAL PARAMETER
* DEFINITIONS
+ THETATHERM=50 THERMAL_TIME_CONSTANT=100NS TAMBIENT=27

*CIRCUIT FOR INSTANTANEOUS POWER COMPUTATION AS V(9,21)*I(VMEAS)
* VMEAS MEASURES THE CURRENT THROUGH THE PIN DIODE SUBCIRCUIT TO
*COMPUTE POWER
VMEAS 99 9 DC 0
* GTHETA IS THE ‘THERMAL HEAT SOURCE’ FOR THE THERMAL MODEL
* AND IS V*I ACROSS/THROUGH DIODE AS V(99,21)*I(VMEAS)
GTHETA 0 VTEMP VALUE={ABS(V(99,21)*I(VMEAS))}
RTHETA VTEMP VTEMP1 {THETATHERM}
CTHETA VTEMP VTEMP1 {THERMAL_TIME_CONSTANT/THETATHERM}
+ IC={TAMBIENT}
* VTHETA FIXES THE AMBIENT TEMPERATURE
VTHETA VTEMP1 0 DC {TAMBIENT}

* THESE TWO LINES PROVIDE THE SCALING FOR THE I-REGION TEMPERATURE
*DEPENDENCE
GXFERTEMP 0 1000 VALUE={1E-9*PWR(EXP((V(VTEMP,0)-TAMBIENT)/108),(MTEMP2))}
RXFERTEMP 1000 0 1E9

* MODIFIED GMOD ELEMENT
GMOD 11 12 VALUE={(V(11,12)*V(1000,0)*V(2,3)/VI)}
```

500 MHz. To show a significant temperature rise at the same 1.2V dc bias voltage (approximately 80 mA of dc bias current), RF voltage swings of 0.1, 15, and 30V are simulated. The RF signal is applied at time $t = 0$ and then removed after 1.0 μs to show both the thermal rise and fall due to the applied RF signal. The 100 nS thermal time constant was chosen for purely illustrative purposes; typical thermal time constants are usually larger, especially for large packaged devices for high power applications. The thermal modeling results are shown in Figure 4.16 for all three RF drive signals. The 0.1V signal indicates no significant heating of the device, whereas the 30V signal shows the device heating up to approximately 68°C. The 15V RF drive signal causes the PIN diode to heat up to about 38°C. In all cases, the thermal rise and fall times are consistent with the 100 nS thermal time constant value listed in the simulation setup. The SPICE netlist file 'thermal1.net' is available to test the simulations by the reader.

4.5.3 Comments on SPICE Simulations

This section provides a few notes on interpretations of the SPICE results as well as the setup for the simulation.

4.5.3.1 SPICE Options

SPICE provides a number of options to aid in convergence. In almost all PIN diode circuits, the various components in the system rarely exceed 1 MΩ in value, either resistive or reactive. This relatively low impedance can be exploited in two different ways: one, by allowing 10 MΩ resistors to bridge various circuit nodes to help in convergence and two, by using the .OPTIONS command in SPICE to set the nodal conductance 'GSHUNT to 10^{-8} S. These two techniques have been used in the simulations shown in this section.

4.5.3.2 Electrothermal Modeling

The frequency sweep .ac analysis is performed assuming small signal conditions (i.e., small ac or RF variations) around the dc bias point. In fact, SPICE reference manuals state that the SPICE .ac analysis operates by "linearizing the circuit around the bias point" [17]. Nonlinear effects should be studied using the transient analysis [17]. In electrothermal simulations, the only thermal effects that will be observed are the small temperature variations about the dc bias point due to small signals and not the PIN diode temperature rise due to any large signal effects. True thermal effects due to large signals can only be observed with .dc and transient analyses using this model. This is not usually a problem because the temperature rise is most important under large signal conditions anyway.

References

[1] Gretsch, W., "The Spectrum of Intermodulation Generated in a Semiconductor Diode Junction," *IEEE Proc.*, Vol. 54, No. 11, November 1966, pp. 1528–1535.

[2] Heiter, G., "Characterization of Nonlinearities in Microwave Devices and Systems," *IEEE Trans. on Microwave Theory and Techniques*, Vol. MTT-21, No. 12, December 1973, pp. 797–805.

[3] Albrecht, C., and F. Jansen, "Numerical Analysis of Nonlinear Small-Signal Distortion in p-n Structures," *IEEE Trans. on Electron Devices*, Vol. ED-24, No. 2, February 1977, pp. 91–98.

[4] Reiss, W., "Nonlinear Distortion Analysis of p-i-n Diode Attenuators Using Volterra Series Representation," *IEEE Trans. on Circuits Syst.*, Vol. CAS-31, No. 6, June 1984, pp. 535–542 (see also W. Reiss, "Volterra Series Representation of a Forward Biased p-i-n Diode," *IEEE Trans. on Electron Devices*, Vol. ED-28, December 1981, p. 1495).

[5] Caverly, R., and G. Hiller, "Distortion in p-i-n Diode Control Circuits," *IEEE Trans. on Microwave Theory and Techniques*, Vol. MTT-35(5), May 1987, pp. 492–501. (See also G. Hiller and R. Caverly, "Predict PIN-Diode Switch Distortion: Microwaves and RF," Vol. 25. No. 1, January 1986, p. 111.)

[6] Application Note 200824(A): Varactor Diodes, Skyworks Solutions, Inc., 2008, pp. 1–6.

[7] Caverly, R., and G. Hiller, "Distortion in Microwave and RF Switches by Reverse Biased PIN Diodes," *Proc. 1989 IEEE Intl. Microwave Symp.*, 1989, pp. 1073–1076.

[8] Lucovsky, G., R. Schwarz, and R. Emmons, "Transit Time Considerations in p-i-n Diodes," *J. Appl. Phys.*, Vol. 35, No. 3, Pt. 1, March 1964, p. 622.

[9] Ward, A., "Calculations of High-Current Characteristics of Silicon Diodes at Microwave Frequencies," *Harry Diamond Lab. Tech. Rpt. HDL-TR-2057*, October 1984. (See also A. Ward, "Calculations of Second Breakdown in Silicon Diodes," *Harry Diamond Laboratories Tech. Rpt. HDL-TR-1978*, August 1982.)

[10] Ward, A., J. Deppe, and R. Carver, "Spike Leakage, Limiting and Rectification in Silicon PIN Diodes," *High Power Microwave Tech. for Defense Appl. Conf.*, December 1986.

[11] Caulton, M., A. Rosen, P. Stabile, and A. Gombar, "p-i-n Diodes for Low-Frequency, High-Power Switching Applications," *IEEE Trans. on Microwave Theory and Techniques*, Vol. MTT-30(6), June 1982, pp. 876–881.

[12] Caverly, R., and G. Hiller, "Establishing the Minimum Reverse Bias for a p-i-n Diode in a High-Power Switch," *IEEE Trans. on Microwave Theory and Techniques*, Vol. 38, No. 12, 1990, pp. 1938–1943.

[13] Wang, S., *Fundamentals of Semiconductor Theory and Device Physics*, Englewood Cliffs, NJ: Prentice Hall, 1989.

[14] Strollo, A., "A New SPICE Model of Power P-I-N Diode Based on Asymptotic Waveform Evaluation," *IEEE Trans. Power Electronics*, Vol. 12, No. 1, January 1997, pp. 12–20.

[15] Caverly, R., and M. Quinn, "A SPICE Model for Simulating the Impedance-Frequency Characteristics of High Frequency PIN Switching Diodes," *Proc. 1999 IEEE ISCAS*, 1999, pp. VI-282–VI-285.

[16] Caverly, R., "Microwave and RF PIN Diode Model for Time Domain Simulation," *IEEE Trans. on Microwave Theory and Techniques*, Vol. 60, No. 7, July 2012, pp. 2158–2164.

[17] *MicroSim PSpice A/D Reference Manual*, Hillsboro, OR: Microsim Corp., 1998.

[18] Thermal equivalent circuit models, App Note AN2008-03, Industrial Power, Infineon, edition 2008-06-16, 2008. (See also "High Frequency Transistor Primer, Part III: Thermal Properties," HP App Note, 1998.)

[19] Caverly, R., "Time-Domain Electrothermal Circuit-Level Modeling of Microwave and RF PIN Diodes," *2012 IEEE Intl. Microwave Symp.*, June 2012, pp. 1–3.

[20] März, M., and P. Nance, "Thermal Modeling of Power-Electronic Systems," *Fraunhofer Ins. Integrated Circuits Thermal System Modeling*, 2000, pp. 1–22.

[21] Caverly, R., and G. Hiller, "The Temperature Dependence of Silicon PIN Diode Attenuators," *Proc. 1993 IEEE Intl. Micro. Symp.*, Vol. 2, June 1993, pp. 553–556.

CHAPTER 5

Modeling MOSFET Control Devices

5.1 Introduction

Up until the mid-1990s, the terms microwave and CMOS were not usually used in the same sentence. At the time, the geometries of digital CMOS devices were such that the capacitances were too high and gate widths too wide to be of use at the high frequencies microwave and RF engineers typically used. The main microwave semiconductor technology was GaAs for not only RF control but also for full system design. One of the first papers that looked at RF CMOS in applications beyond power amplification changed the way microwave and RF design engineers looked at the role of CMOS at microwave and RF frequencies [1]. The use of RF CMOS since that time has increased dramatically with full transceivers integrating both analog and digital functions now available on a single SoC (see [1–4] for overviews). RF control devices based on MOS technology were a part of this silicon revolution, with papers on MOS devices as RF switches starting to appear in the late 1990s [5, 6]. One of the big advantages, besides the ease of integration with digital CMOS circuitry, is that CMOS switches require only positive control voltages as opposed to GaAs MESFET switches, which require negative control voltages or that the circuit be at dc float (as discussed in the next chapter) [7]. In CMOS technology, the bulk of the current flow is primarily majority carrier based as opposed to minority carrier based, as in the case of the PIN diode. Since the electron mobility (in n-channel MOS devices) is higher than the hole mobility (in p-channel MOS devices), the n-channel MOS transistor (nMOS) is almost exclusively used for RF and microwave control, and so the n-channel MOS transistor will be the focus of the modeling described in this chapter.

5.2 Review of CMOS Technology

This section provides a review of CMOS technology, including the physical structure and technology scaling.

5.2.1 The CMOS Physical Structure

The fabrication details of CMOS integrated circuit (IC) technology and their relationship with modeling of fabricated MOSFETs can be found in the literature because of the widespread interest in CMOS for digital IC design (see for example [8–13]). For the n-channel MOSFET, highly doped drain and source-diffused regions of

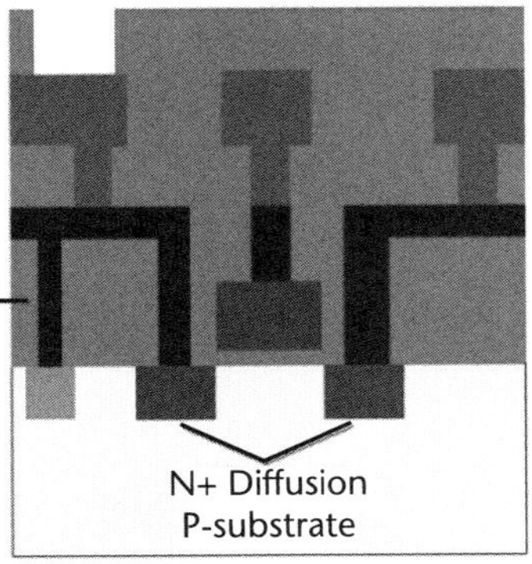

Figure 5.1 Cross section of n-channel FET in a conventional bulk CMOS process with a single poly layer (two or more may be available in CMOS processes) and two levels of metal (seven or more metal layers may be available in CMOS processes) (after [18]).

n-type material are introduced into a p-type substrate, physically separated by a gate material grown over a thin insulating layer sometime referred to as the thin oxide or thinox (Figure 5.1). This gate material was originally aluminum (Al) in early MOSFETs. Over the years of geometry reduction, the original Al gate material was replaced by highly doped polycrystalline silicon (polysilicon, or poly) in order to tolerate the higher fabrication temperatures and to improve long-term reliability[1] [14, 15]. However, metallic gates over high dielectric insulating material are gaining increased attention again at the 45 nm feature size since they are of lower resistivity and therefore can improve switching speeds [14, 15]. The thin oxide can be silicon dioxide or other material with good insulating and lattice matching properties, but the term oxide is still used when discussing this insulating material. In addition to the components comprising the MOSFET itself, a number of metal layers above the wafer surface of increasing thickness provide low resistivity interconnections. The dc current carrying capability of these metal layers depends on the metal layer cross section, with dc current densities on the order of 2 mA/μm^2 typically assumed for reliability [16]; studies show that this maximum current density can be approximately 25 to 30 times higher for RF signals [17]. Duty cycles will also affect this maximum current density, with low duty cycles allowing higher current densities. Vias are used to connect adjoining metal layers, with contacts used to connect semiconductor layers or metal layers to other semiconductor layers. The process parameters describing the important characteristics (primarily resistance and capacitance) of the various layers are provided by the fabrication houses, specific to their propriety processes. Representative process parameters for a 180 nm CMOS process are illustrated in Table 5.1. Knowledge of these process parameters by the device modeler is crucial for successful and accurate modeling. More-detailed process parameters can often be obtained from the IC manufacturer, but in many cases nondisclosure agreements are involved.

1 To this day, the term metal is still used to describe the MOSFET, although a more accurate term may be SOS (semiconductor oxide semiconductor) or SIS (semiconductor insulator semiconductor); these terms may also be seen in the literature.

Table 5.1 Sample set of process and other MOS fabrication parameters for a 180 nm CMOS IC process

Resistance Characteristics

	N+ layer	*P+ layer*	*Poly*	*Metal1*	*Metal2*	*Metal3*	*Units*
Sheet Resistance (R_S)	5.0	4.0	5.0	0.1	0.1	0.05	Ohms/sq
Contact Resistance (R_c)	7.0	7.0	6.0	2.0	4.0		Ohms

Capacitance Characteristics

	N+ layer	*P+ layer*	*Poly*	*N-well*	*Units*
Area (substrate)	800	1200	200		aF/um^2
Area (N^+)			7600		aF/um^2
Area (P^+)			7400		aF/um^2
CGDO	5.0E-10				F/m
CGSO	5.0E-10				F/m
CJ	8.0E-4				F/m^2
CJSW	2.0E-10				F/m

The primary fabrication technology for digital CMOS relies on a bulk technology; the majority of the silicon used is purely for mechanical and thermal purposes. This bulk CMOS technology unfortunately adds unwanted parasitic characteristics, primarily capacitive, that dramatically influence the RF behavior in control applications. Silicon on insulator (SOI) technology utilizes a thin epi layer of silicon grown on an insulating substrate (glass or sapphire are usually used), which reduces many of the unwanted bulk capacitive effects that negatively impact bulk CMOS but at the expense of reduced thermal and power handling capacity. Minimum SOI feature size also tends to lag digital CMOS by about a generation [technical reviewer, private communication].

5.2.2 Technology Scaling

The phrase "180 nm CMOS technology" or "65 nm CMOS technology" refers to the minimum feature size available in the technology and is almost always synonymous with the MOSFET gate length, L. The primary indicator of MOSFET frequency response is also the gate length. The transit time needed for an electron to move from the source to the drain over distance L is dependent on the electron mobility μ_n and the applied electric field E and is given in (5.1a). This transit time cannot be decreased indefinitely by increasing the electric field; in silicon, the drift velocity v_{drift} saturates (v_{sat}) at approximately 10^5 m/s, requiring reductions in gate length L to reduce transit time. A quick calculation of frequency response would look at the inverse of the transit time (at least dimensionally):

$$\tau = \frac{L}{\mu_n E} = \frac{L}{v_{\text{drift}}} = \frac{L^2}{\mu_n V_{\text{DS}}} \tag{5.1a}$$

$$f = \frac{1}{2\pi\tau} = \frac{v_{sat}}{2\pi L} = 15.9 \text{ GHz}/\mu\text{m} \tag{5.1b}$$

While a crude measure of frequency response, (5.1b) indicates significant improvements are garnered by reducing the gate length. Reducing L, however, also means varying many of the fabrication process parameters such as layer thickness, layer separations, and doping concentrations, another reason why the device modeler needs the very specific technology process parameters for accurate modeling.

5.3 Current-Voltage (I-V) Characteristics of the nMOSFET RF Control Device

For RF control applications using the MOSFET, the RF signal to be controlled is assumed to flow from the drain to the source with the gate acting as the device control node. The gate terminal is connected to a dc bias source, while the drain connection can either be a dc or RF or combined source (Figure 5.2(b)). The basic nMOSFET device structure is shown in Figure 5.2(c) with the standard current and voltage polarities indicated. When V_{GS} exceeds a certain voltage termed the threshold voltage, V_T, the region under the gate material "inverts" by having more mobile electrons than the holes introduced by the substrate doping, creating a thin conducting channel directly under the gate that allows current I_{DS} to flow when a drain-source voltage, V_{DS}, is applied. Otherwise, the two back-to-back PN junctions created by the two n-type diffused regions and the p-type substrate act to prevent any such current flow regardless of polarity of V_{DS}. With V_{GS} below V_T, the nMOSFET has no conducting channel and theoretically acts as a dc open circuit. However, there is RF current flow between the drain and source through the device capacitance, a topic for the next section. Because a conducting channel needs to be created, the device is often referred to as an enhancement mode device. The threshold voltage V_T is determined during the MOS fabrication process and can be estimated using the expression [19]

$$V_T = \frac{t_{ox}}{\varepsilon_{ox}}\sqrt{2\varepsilon_{Si} q N_{sub}\left(2\phi_B + \left|V_{SB}\right|\right)} + 2\phi_B \tag{5.2}$$

where $\phi_B = kT/q \ln(N_{sub}/n_i)$, t_{ox} is the oxide thickness under the gate (the thin oxide), N_{sub} is the substrate or well doping (m^{-3}), k is Boltzmann's constant, T is the temperature, q is the electronic charge, n_i is the intrinsic carrier concentration, and ε_{Si} and ε_{ox} are the dielectric permittivities of silicon and the insulating oxide, respectively. The true nMOSFET is actually a four-terminal device, with the gate-drain-source (G, D, S) connections making up the traditional three-terminal device concept, with the fourth node, often referred to as the bulk or substrate node (B), being the connection between the conducting channel region and the substrate. In conventional amplifier circuits, for example, the source and bulk are tied together, usually at the ground node, and so V_{SB} is zero. However, there are many RF control

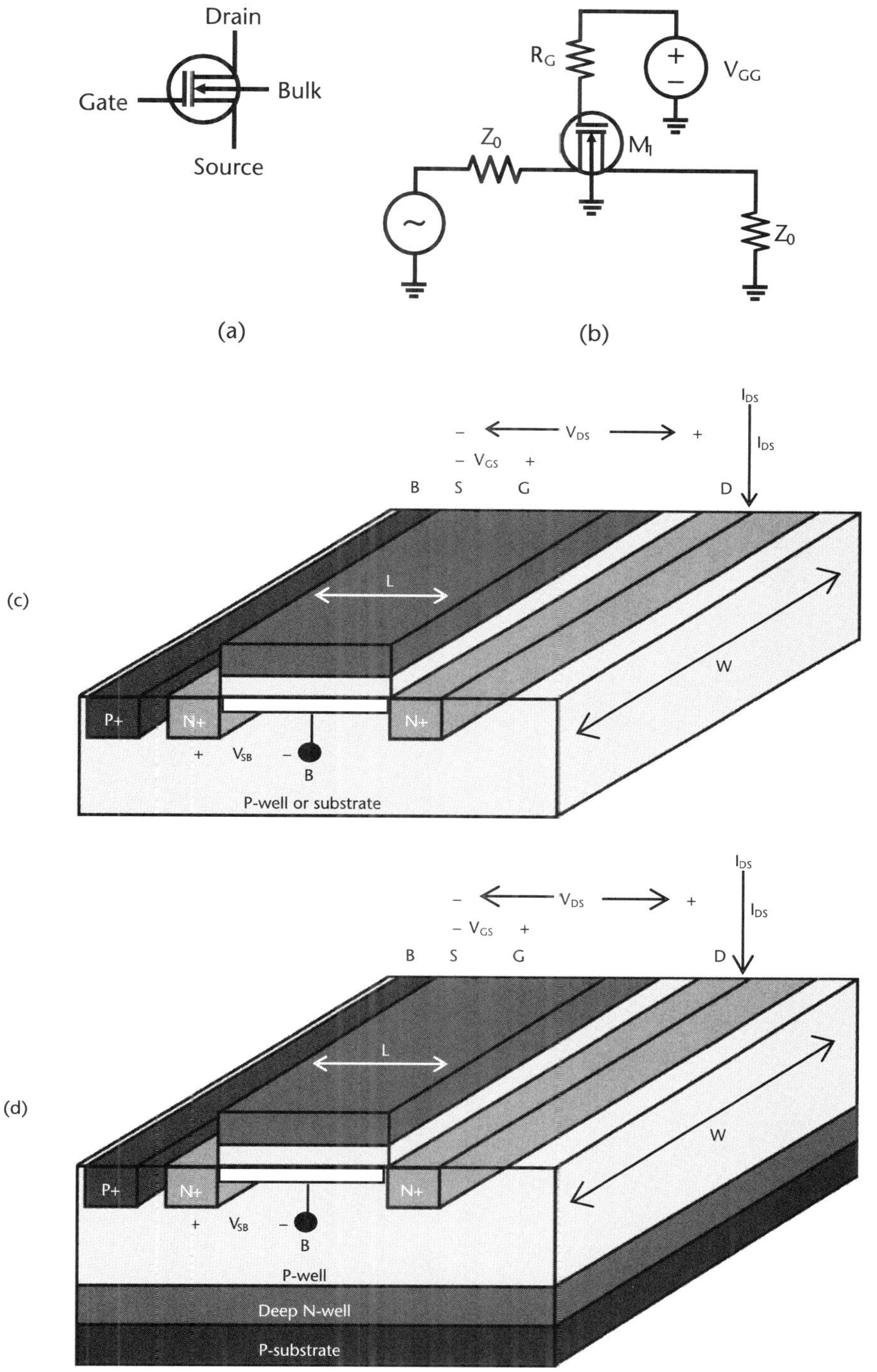

Figure 5.2 (a) MOSFET circuit symbol, (b) standard series-connected MOSFET in a microwave and RF control application, (c) cross section of bulk n-channel MOSFET fabrication technology, and (d) triple-well n-channel MOSFET fabrication technology showing channel region, device width W, the four nodes (G, D, S, B), and voltage polarities (after [18]).

applications where the source and substrate are not at the same potential, and so bulk or substrate effects can greatly influence RF behavior. These substrate impacts are discussed in Section 5.4.4.

5.3.1 I-V Characteristics

From the device modeler's perspective, the threshold voltage, V_T, is set by the process and fabrication house and is typically 10 to 20% of the target operating voltage. The threshold voltage usually tracks with the gate length L reduction. For example, in 180 nm CMOS, the threshold voltage is approximately 0.4V, whereas the threshold voltage is approximately 0.3V for 65 nm CMOS [56].

A simple explanation for the I-V characteristics for the nMOSFET can be developed from a charge control perspective. Once the gate-source voltage, V_{GS}, rises above V_T, an inversion layer under the gate forms, with the charge in this inversion layer dependent on the capacitance making up the oxide layer and the average drain-source voltage in the channel, $Q_{gate} = (\varepsilon_{ox}/t_{ox}) \cdot WL(V_{GS} - V_T - 0.5V_{DS})$. This charge is supported under drift conditions in this inverted region (not diffusion as in the case of PN junctions) by the product of the drain current I_{DS} and the gate transit time (5.1a) $Q_{gate} = I_{DS}\tau = I_{DS}L^2/\mu_n V_{DS}$. Rearranging terms allows the current-voltage expression to be written as

$$I_{DS} = KP\frac{W}{L}V_{DS}\left(V_{GS} - V_T - 0.5V_{DS}\right) \quad KP = \mu_n C_{ox} = \frac{\mu_n \varepsilon_{ox}}{t_{ox}} \tag{5.3}$$

where μ_n is the mobility of the dominant charge carrier in the induced channel and KP is the intrinsic transconductance, another parameter set by the process and not under the designer's direct control (other than choosing a particular fabrication process). The factor W/L in (5.3), which is under the device designer's control, is often referred to as the aspect ratio of the device and can be quite large in devices used for RF control, as opposed to unity aspect ratios often used for digital logic gates where the critical characteristic is a voltage level near one of the power rails, representing the digital bit.

Intrinsic transconductance values for KP are inversely proportional to the gate length L since for reductions in L, fabrication processes dictate that the oxide thickness t_{ox} also must reduce. Typical values of KP for 180 nm and 130 nm CMOS are 150 and 280 μA/V^2, respectively [56]. For small values of V_{DS}, 5.4 shows that I_{DS} increases linearly (so-called *linear* region of operation), the main focus region of operation for the microwave and RF control device modeler.

5.3.2 RF On-State Resistance

At this point in the general derivation of the I-V characteristics for an nMOSFET device [8–12], the discussion starts to look at device behavior as V_{DS} begins to approach $V_{GS} - V_T$. However, operation in the so-called saturation region, so critical for the rich variety of MOS-based active circuit applications such as amplifiers and oscillators, is not the normal operating region for the RF control case, which is confined primarily to the linear operating region where V_{DS} will always be less

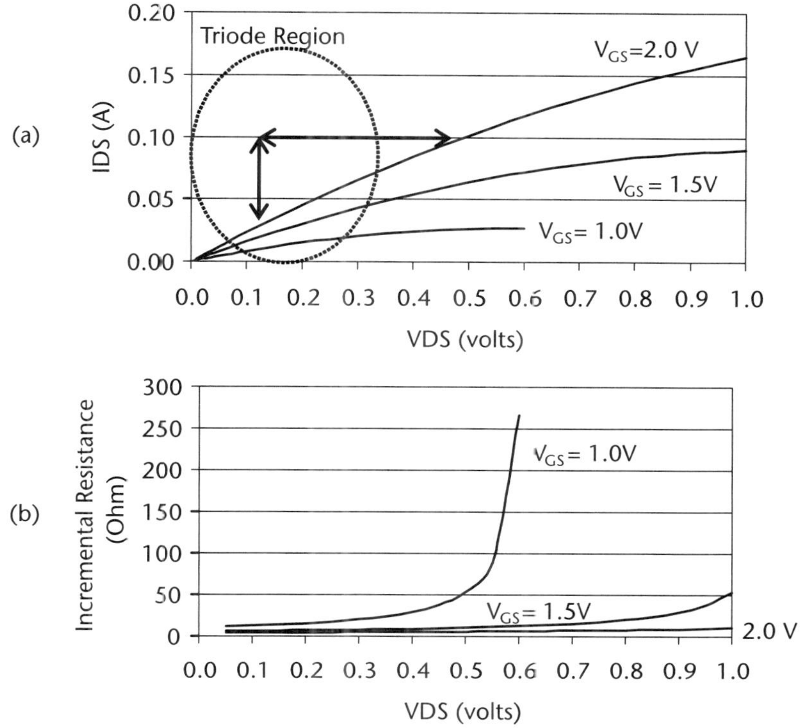

Figure 5.3 (a) Example I_{DS}-V_{DS} curves for various gate voltages. The triode region or linear region of operation is indicated. (b) Dynamic resistance for the three gate voltages.

than saturation voltage, approximately $V_{GS} - V_T$. When operated in this linear or so-called triode[2] region, the nMOSFET acts as a voltage-controlled resistor with the gate being the controlling voltage. Figure 5.3 shows the I-V characteristics of a 180 nm nMOSFET as a function of V_{GS} using (5.2) and the previously mentioned device characteristics and W/L = 1000; the triode region is marked on the figure. The dynamic resistance of the device can be found by looking at the change in V_{DS} over I_{DS} ($\Delta V_{DS}/\Delta I_{DS}$ or the slope of the I-V curve) at fixed gate voltage as indicated in Figure 5.3(a). For this condition, the change in voltage is approximately 0.3V with a corresponding change in current of 0.06A, yielding a slope of 0.3/0.06 V/A or 5Ω. Figure 5.3(b) shows the dynamic resistance over the three gate operating voltages shown in Figure 5.3(a); note that the dynamic resistance is relatively constant for small V_{DS} but increases as V_{DS} increases toward the onset of saturation (which ideally occurs at $V_{DS} = V_{DS\text{-sat}} = V_{GS} - V_T$) [18]. While at first this high-resistance regime looks attractive for use in attenuator circuits, for example, operation of MOS devices as control devices near the "knee" point in the dynamic resistance should be avoided. Figure 5.3(b) shows that very small changes in V_{DS}, such as during the RF

2 The term *triode region* is a holdover from vacuum tube terminology. In a vacuum tube, the output current (from the high-voltage plate) was controlled by a voltage applied to the two other terminals: the cathode (where the electrons originated) and the grid. The resulting curves were linear in nature, and since three terminals were used, the term triode was created.

voltage swing, can cause large changes in the resistance, making this resistance a function of time and therefore a highly nonlinear region of operation, which introduces spurious distortion signals into the circuit. This nonlinear operation will be discussed in a Section 5.5.3.

Using the same concept as the above example, a mathematical relationship for the channel resistance, R_C, can be described as the on-state RF resistance, $R_C = R_{ON}$, of the nMOSFET can be derived by looking at the change of I_{DS} with respect to V_{DS} in (5.3), which shows the dependence of the various fabrication parameters on the MOSFET resistance:

$$R_C = R_{ON} = \left(\frac{\partial I_{DS}}{\partial V_{DS}}\right)^{-1} = \frac{1}{\frac{W}{L}KP\left(V_{GS} - V_T\right)} \tag{5.4}$$

In RF switch applications, the on-state resistance R_{ON} should be small, usually less than a few Ohms, typically around 5Ω, for low insertion loss in series-connected 50Ω circuits (5Ω yields an insertion loss of approximately 0.4 dB in a 50Ω system). Using the typical values of V_T and KP mentioned above for 180 nm CMOS requires that the aspect ratio, W/L, be greater than approximately 900 for a dc gate bias of V_{GS} = 2.0V (five times V_T). This is physically a very large device (in digital CMOS, W/L is usually unity) exhibiting a gate periphery of nearly 0.3 mm, which often requires that the device be divided into a number of parallel elements with multiple gate fingers. This multiple gate finger layout of the device will have a positive impact on the overall capacitance of the device, a topic for a later section in this chapter. The channel resistance is also a function of V_{GS}, which is in the RF path, potentially making the channel resistance a time-varying parameter, which has ramifications in terms of nonlinear operation, a topic for Section 5.5.3.

The gate material exhibits resistive characteristics that, while not directly in the RF path, will have an eventual influence on MOSFET switching speed. The sheet resistance R_{SH} of the gate material can be found from manufacturer process parameters such as those shown in Table 5.1. For the case of a polysilicon gate material, the value of gate sheet resistance shown in the table is 5.0 Ω/square; the gate resistance can then be computed as the product of the sheet resistance and the aspect ratio, $R_G = R_{SH}(W/L)$. For the example above, a poly gate resistance can be computed as R_G = 5.0(900) or about 4.5 KΩ. R_G in typical MOSFET control devices, however, is considerably less than this value due to improvement in gate material technology (metallic for sub-90 nm devices, for example) or through the use of multiple gate fingers. Multiple gate finger technology will be discussed in Section 5.4 since this layout technology has an impact on the overall MOSFET capacitance.

5.3.3 Bulk Resistance

Figure 5.2 shows the bulk node connection B at the channel-substrate interface. In the figure, this node is shown as floating but in actual MOSFET fabrication, a bulk or substrate contact is required. This substrate connection usually consists of a diffused region of highly doped p-type material introduced near the MOSFET source contact. The p-material provides the surface connection, but there exists a finite path from the bulk node under the MOSFET channel through the medium resistivity substrate to the actual bulk contact, resulting in the bulk node resistance, R_B [20]. Depending on the geometry of the substrate connection chosen, this resistance can vary from a few Ohms to several hundred [20].

5.3.4 RF Off-State Resistance

If the dc gate voltage, V_{GS}, is less than threshold voltage, the two PN junctions that make up the source and drain connections are back-to-back PN junction diodes and therefore impede current flow regardless of the polarity of the RF signal (or dc level) between drain and source, resulting in an effective open circuit between the two nodes. The value of the channel resistance R_C is now infinite and the device is considered OFF. However, as will be seen in Section 5.4, there are still RF paths around this channel due to device capacitances that can be lumped together into a single off-state capacitance value, C_{OFF}. A frequently cited figure of merit (FOM) combines the on- and off-state characteristics into broadband switch cutoff frequency, $F_C = 1/2\pi R_{ON}C_{OFF}$ (or the associated broadband switch time constant, $T_C = R_{ON}C_{OFF}$). With typical on-state resistance of 5Ω and off-state capacitances of a few tenths of a pF, F_C values for MOSFETs can reach into hundreds of GHz range (with associated time constants in the femtosecond range). It should be noted that this F_C is not a measure of the operational upper frequency limit but rather is an FOM that allows comparison between different devices. In this case, the higher FOM device will be usable at a higher frequency than the lower FOM device.

5.4 Detailed Capacitance Characteristics

5.4.1 Intrinsic Device Capacitance Origin

The various capacitances that are an inherent part of the device structure are the main factor limiting MOSFET RF performance (Figure 5.4). These MOSFET capacitances scale with device gate width W, and since W is relatively large in MOSFET control devices to keep the on-state resistance small, the capacitances can be quite large. The main capacitances of interest in modeling the MOSFET for RF control are the gate oxide (C_{gc}) and gate overlap (C_{ovr}) capacitances that make up the MOSFET gate source and capacitances, C_{GS} and C_{GD}, as shown in (5.5):

$$\begin{aligned} C_{ovr} &= C_{\mathrm{GX0}} \cdot W \\ C_{gc} &= \frac{\varepsilon_{ox}}{t_{ox}} \cdot WL \\ C_{\mathrm{GS}} &= C_{\mathrm{GD}} = C_{ovr} + C_{gc}/2 \end{aligned} \tag{5.5}$$

where the X is either drain or source [8]. For the example device presented in Section 5.3.2 and using the process parameters in Table 5.1 with a simple SiO_2 insulator

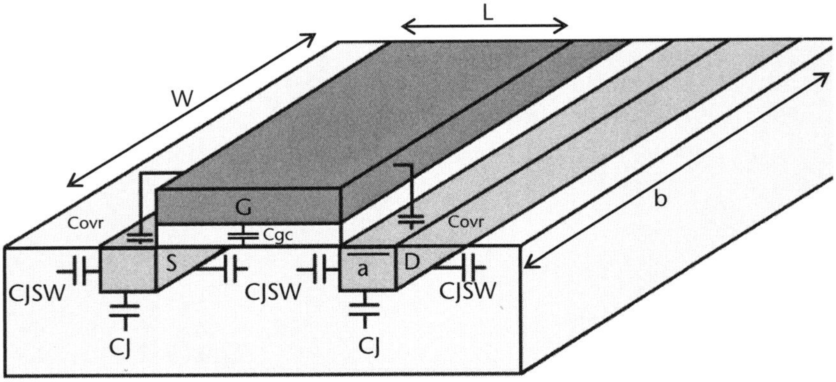

Figure 5.4 Capacitances in the MOS structure (after [18]).

($\varepsilon_{ox} = 4\varepsilon_0$), the gate-channel capacitance C_{gc} is approximately 0.23 pF. Assuming no overlap capacitance (C_{ovr}), C_{GS} and C_{GD} are approximately 0.12 pF each.

The drain and source n-type diffused regions in the p-type substrate are PN junctions and define the drain-bulk and source-bulk capacitances (C_{DB} and C_{SB}, respectively). Since the RF path is through the drain to the source, these capacitances primarily govern the upper frequency limit of the MOSFET as an RF control device. The PN junction nature of these capacitances makes them dependent on both the RF voltage and any dc voltages that appear on the respective node. This voltage dependence is seen in the capacitance expression in (5.6). The PN junction surrounds the diffused region on four sides, yielding a capacitance dependent not only on the area of the diffused region (ab in Figure 5.4) but also the perimeter ($2a + 2b$). The area (perimeter) term combines with the process area capacitance per unit area (length) parameter CJ (sidewall $CJSW$) to form the drain bulk C_{DB} and source bulk C_{SB} diffusion capacitance contributions (Figure 5.4):

$$C_{DB(SB)} = (ab) \cdot \frac{CJ}{\sqrt{1 - \frac{|V_{DB(SB)}|}{V_{bi}}}} + 2(a+b) \cdot \frac{CJSW}{\sqrt{1 - \frac{|V_{DB(SB)}|}{V_{bi}}}} \tag{5.6}$$

where V_{bi} is the bulk PN junction built-in potential, $kT/q \ln(n_{\text{diffusion}} N_{\text{sub}}/n_i^2)$ (approximately 0.6 to 0.7V in silicon). For small signal RF modeling, the associated voltage $V_{DB(SB)}$ can be assumed zero so that the capacitance is simply an area or a perimeter times the appropriate capacitance parameter, $C_{DB(SB)} = (ab)CJ \cdot +2(a+b)CJSW$. For the example RF MOS switch, the width W is the same as dimension b, approximately 160 μm. Integrated circuit design rules often specify that the diffused regions (dimension a) be approximately 2.5 times the gate length, in this case, approximately 0.45 μm for a 180 nm gate length [21]. For the example MOSFET, the drain/source area is then approximately 75×10^{-12} m^2 and the drain/source perimeter is 321 μm. Since these capacitances and their related voltage dependences are in the RF path, distortion signals will be generated, a topic of Section 5.5.4.

5.4.2 Multiple Gate Fingers

These capacitances can be reduced by the device manufacturer by using a merged source-drain and multifingered gate layout. The multifinger gate allows two transistors to "share" a source and drain, thereby reducing both the area and the periphery (Figure 5.5 shows a two-finger MOSFET with this source-drain sharing). Design rules for CMOS require a large area be reserved for source and drain diffusions (diffusion dimension Y in Figure 5.5 is approximately 2.5 times the gate length [21]) so that metal contacts can be attached; these large area (and perimeter) contacting regions significantly increase the capacitance at the source and drain nodes. The area requirements for merging the source and drain connections are significantly reduced and therefore reduce the nodal capacitance. Designs that require low R_{ON} need increased values of W and exhibit larger capacitances, and hence have poorer RF performance unless the layout is optimized to reduce the parasitic capacitance.

The reduction in area and perimeter can be seen with the aid of the following example. Assume that the source and drain regions in Figure 5.5(a) have a horizontal length of X and a vertical dimension Y as indicated in the figure. The total drain and source area and perimeter are then $2XY$ and $4(X + Y)$, respectively. By reducing the horizontal dimension for the two gate-finger MOSFET in Figure 5.5(b), the total source area becomes $2(XY/2) = XY$ and source perimeter becomes $2(2X/2 + 2Y) = 2(X + 2Y)$. In other words, there is little improvement in the source characteristics since $Y \ll X$; the drain characteristics, however, are a different story. The total drain area and perimeter is now $XY/2$ and $2(X/2 + Y) = X + 2Y$. The total drain and source area and perimeter for the two gate-finger MOSFET is then $XY + XY/2 = 3/2\,XY$ and $3X + 6Y$. This simple calculation shows a 25% reduction in area and an X-length reduction in perimeter (since $Y \ll X$). Since the length of the gate material is reduced in the multiple gate finger MOSFET, the gate resistance R_G is also reduced by approximately $1/3N$, where N is the number of gate fingers and the factor

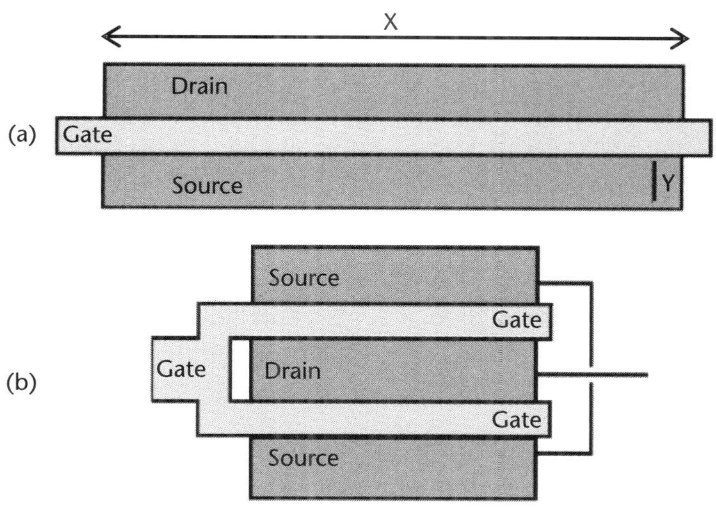

Figure 5.5 (a) MOSFET with single gate and large drain/source area and perimeter. (b) Two gate-finger MOSFET with a merged drain connection for capacitance reduction.

of 1/3 arises from the distributed nature of the resistance [22]. Modern MOSFET control devices will exhibit many more gate fingers than this simple example, with a corresponding reduction in source/drain diffusion capacitance and gate resistance that dramatically improves both frequency response and switching speed. Regardless of whether a single or multiple gate finger MOSFET is used, an RF equivalent circuit for the transistor useful in control circuit simulations is needed and is the focus of the next section.

5.4.3 RF Equivalent Circuit

The individual circuit elements described in the previous sections can be combined to create an RF equivalent circuit for the MOSFET control device (Figure 5.6(a)). There are a number of operational concepts that arise when studying this equivalent circuit. For low-frequency operation in the on-state, the capacitive reactances are large and the RF path through the drain (D) to the source (S) simply goes through the channel resistance. $R_C = R_{ON}$, governing the low-frequency insertion loss or isolation (depending on the overall circuit topology). As the frequency increases, the capacitive reactances start to connect the main D-S path to both the gate and bulk nodes, impacting the frequency response and power handling [5]. Note that if the gate resistance R_G is small, then the two gate capacitances, C_{GS} and C_{GD}, shunt RF straight to ground since the gate node is connected to a dc supply to create the switching action (Figure 5.2(b)). The high-frequency performance can be improved if R_G is relatively large (on the order of kΩ) to "float" the gate. This R_G for MOSFET control devices can either be the inherent resistance of a polysilicon gate or an added lumped element resistor (or a combination of the two) and acts to isolate the dc gate bias supply from the RF path. A gate bias circuit cutoff frequency is often defined as $F_G = 1/2\pi R_G(C_{GS} + C_{GD})$ (or its equivalent, a gate bias circuit time constant, $T_G = R_G(C_{GS} + C_{GD})$) and is on the order of a several hundred kilohertz into the megahertz range (microseconds) in high-quality MOSFET control devices. In the off-state, the inverting channel is removed and R_{ON} becomes essentially an open circuit; the low-loss RF path is therefore removed with drain-source transmission through the gate capacitance path (C_{GS}, C_{GD}) and the substrate path (C_{DB}, C_{SB}). Figure 5.6(b) shows insertion loss and isolation for the example nMOSFET RF switch using the equivalent circuit in Figure 5.6(a) for the MOSFET element (with $R_B = 0$) in the series connected switch in Figure 5.2(b). The on-state insertion loss is constant at low frequencies with monotonically increasing insertion loss as frequency increases and the impact of the MOS capacitances take hold. The off-state isolation monotonically decreases with increasing frequency as coupling from the drain to the source occurs through the device capacitances. This curve was computed using Matlab script 'fet_switch1b.m' available in the file repository.

5.4.4 RF Bulk Node Effects

The bulk node B shown in Figure 5.6(a) has an impact on control circuit performance in bulk CMOS technology. At low frequencies, the on-state resistance R_{ON} primarily governs the insertion loss in a series connected MOS RF switch. As the

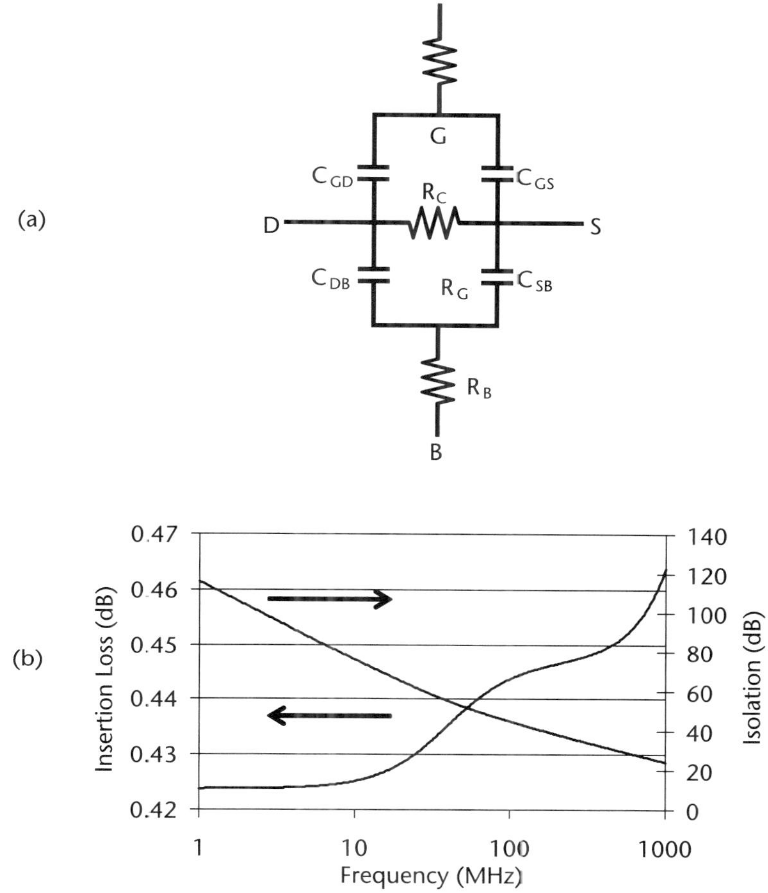

Figure 5.6 (a) RF equivalent circuit for MOSFET control element. (b) Insertion loss and isolation for the example MOS RF switch.

frequency increases, there is increased coupling through C_{DB} and C_{SB} to the substrate resistance, R_B, with R_B having an increasing impact at higher operating frequencies (Figure 5.7(a)). Huang and O found the insertion loss of the MOS RF switch was dependent on the substrate resistance and parasitic capacitances. They found an optimum value, $R_{B(MAX)}$, for which insertion loss was a maximum [23], as seen in Figure 5.7(a) and approximated by (5.7) [23, 24]:

$$R_{B(\mathrm{MAX})} = \frac{-\omega^2 C_T^2 Z_0\left(Z_0 + R_{\mathrm{ON}}\right) + \sqrt{\omega^4 C_T^4 Z_0^{\,2}\left(Z_0 + R_{\mathrm{ON}}\right)^2 + 4\omega^2 C_T^2\left(2Z_0 + R_{\mathrm{ON}}\right)^2}}{2\omega^2 C_T^2\left(2Z_0 + R_{\mathrm{ON}}\right)} \quad (5.7)$$

where $C_T = C_{DB} + C_{SB}$ and Z_0 is the system impedance (typically 50Ω). For minimal impact of the bulk resistance on the insertion loss, the bulk resistance should be either made as small as possible (direct RF to ground) or as large as possible (floating bulk node). Small R_B can be achieved by fully surrounding the transistors

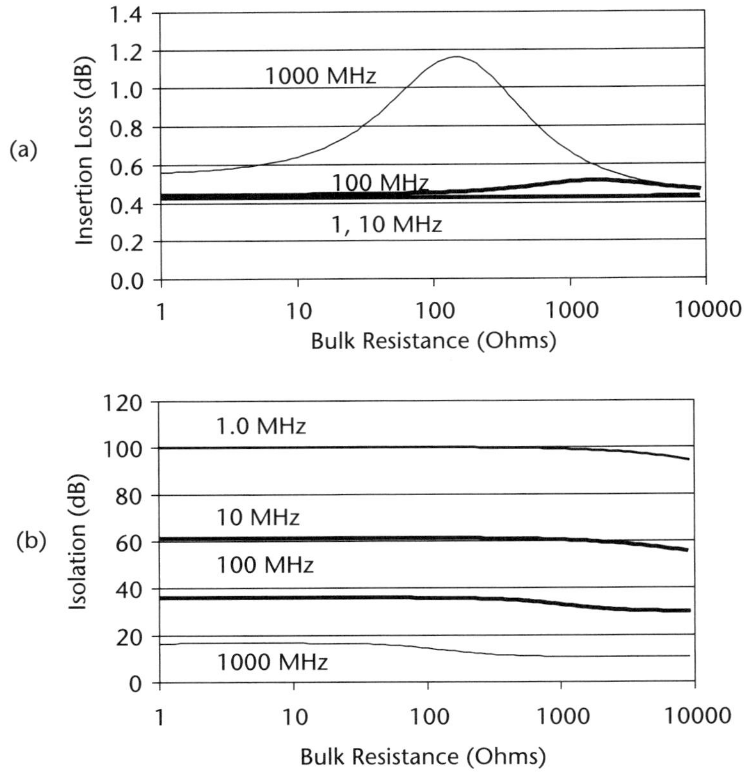

Figure 5.7 (a) On-state insertion loss as a function of bulk resistance, R_B with frequency as a parameter. (b) Off-state isolation as a function of bulk resistance, R_B with frequency as a parameter

with large area p⁺ substrate contacts surrounding the MOS device and liberal use of substrate contacts [23]. The use of p-implants close to the MOS transistors and placing the substrate connections far away can substantially increase the value of R_B [25, 26]. In the off-state, there is much less influence on the bulk resistance, with the isolation being primarily governed by the device capacitances and their coupling of RF energy around the off-channel region. The curves in Figure 5.7 were computed using the Matlab script 'fet_switch1.m'.

5.4.5 Silicon on Insulator (SOI)

The largest capacitance contributors in bulk CMOS are the source and drain diffused regions and their connections to the substrate. Since the substrate is typically tied to both RF and dc ground, these capacitances will shunt RF to ground through the relatively low resistivity substrate at one or both nodes (depending on circuit layout), thereby limiting the upper frequency limit of the device and control circuit. The impact of these two capacitances could be reduced by a high impedance path to ground, but unfortunately bulk CMOS technology requires this low resistivity substrate. SOI technology provides a solution to this capacitance issue by growing a very thin layer of silicon on an insulating layer (sapphire is widely used, which gives rise to the alternate definition of silicon on sapphire or SOS) grown over the usual lower resistivity bulk CMOS substrate [27–29]. This layer is sometime referred to as the buried oxide or BOX layer [30]. The layer is thick enough that all the key operational features of CMOS are retained but with reduced substrate coupling because of the highly insulating layer. Broadband switch time constants TC of a few hundred femtoseconds are possible [31].

5.4.6 Packaging Parasitics

Many of the mechanical characteristics of mounting the nMOSFET RF switch die are the same as for the case of two-terminal devices such as the PIN diode described in Section 2.1, except now there are potentially four terminals to deal with instead of the PIN diode's two. However, in the case of bulk MOS fabrication technology, the substrate is usually grounded to provide the lowest value of R_B to prevent the insertion loss degradation observed in Figure 5.7 [23]. Direct RF grounding of the substrate, however, impacts the power handling capability since RF voltages exceeding approximately 0.6V at the drain and/or source can forward bias the drain/source PN junctions. Using triple-well MOS technology (adding a deep heavily donor doped layer below the MOSFET layer, as shown in Figure 5.2(d)) and floating both the region under the gate and this new layer, an on-chip bulk resistance R_B can be added with subsequent improvements in insertion loss and power handling but with a slight loss in isolation [32, 33]. These circuit and fabrication choices tend to remove the need for a separate bulk node connection, leaving only a maximum of three connections needed. Since most MOS-based RF switch and switch modules use a variety of series and shunt-connected configurations (discussed in Chapters 7 and 8 on applications), internal connections between the various MOS devices reduce the external packaging requirements further so that the final connections between the die and package are primarily the RF inputs/outputs, the dc control inputs, and the RF ground (or possibly multiple ground connections).

The package parasitics between the RF switch module and the package are the bonding wire(s) inductance(s) and the parasitic package capacitances. The same general rule of thumb in computing the inductance of the bond wire is approximately 1 nH per millimeter of wire length, as presented in Section 2.1. Multiple bond wires may be employed to reduce the inductance on each connection with similar package styles used for both PIN diodes and MOS-based switches. Manufacturer or fabrication/packaging houses can provide estimates of these parasitics.

Finally, while not necessarily a package parasitic, other components are often added to the final MOS RF switch circuit to provide ESD (electrostatic discharge) protection. A common ESD protection technique is to lightly shunt the external RF nodes with an inductor because the inductor will present a low impedance path to ground for the low frequencies associated with static but will present a high impedance at RF frequencies [34, 35]. This added inductance serves to not only help with the ESD protection but can also be used to partially resonate out the parasitic capacitance at the same node, improving RF performance but narrowing the bandwidth [34, 35].

5.5 Detailed MOS Control Device Characteristics

5.5.1 High Field Effects in MOSFET Control Devices

As device geometries become smaller, the current-voltage expression for the MOSFET needs to be adjusted because of the high electric fields. These so-called high field effects primarily impact the electron mobility in the inversion channel. This field-dependent mobility was also seen back in Chapter 4 (4.35) in limiting the drift velocity of the carriers at high electric fields. Using a modified version of this mobility expression, (5.3) can be written to include this high field effect as [5, 36]

$$I_{DS} = \frac{KP}{1 + \dfrac{V_{DS}}{LE_c}} \frac{W}{L} V_{DS}\left(V_{GS} - V_T - 0.5V_{DS}\right) \tag{5.8}$$

where E_c is mobility critical electric field (approximately 1.5×10^6 V/m in silicon at 300 K [36] and dependent on doping density) and the other symbols have their usual meanings. The impact on this field-dependent mobility can be seen in Figure 5.8 where the $V_{GS} = 2.0$ I-V curve in Figure 5.3 is repeated using identical KP and V_T values but compared with the field dependent mobility I-V expression in (5.8). Note that at very low V_{DS}, the I-V curves and associated dynamic resistances are identical, but for larger electric fields occurring at larger V_{DS}, the I-V curve begins to saturate at significantly lower current and voltage, implying that nonlinear behavior will be observed in these small feature size devices at lower RF drives.

5.5.2 Gate Resistance

The gate resistance R_G, and the associated gate bias circuit cutoff frequency $F_G = 1/2\pi R_G(C_{GS} + C_{GD})$, has an important influence on the MOSFET control device operation. At low frequencies in the on-state, the gate capacitive reactances are much higher than the value of R_G, and so the gate is not a floating node. The gate-source voltage V_{GS} in this case is approximately equal to $-Z_0 I_{DS}$, implying that the gate-source voltage follows the RF signal drive level. This RF signal is superimposed

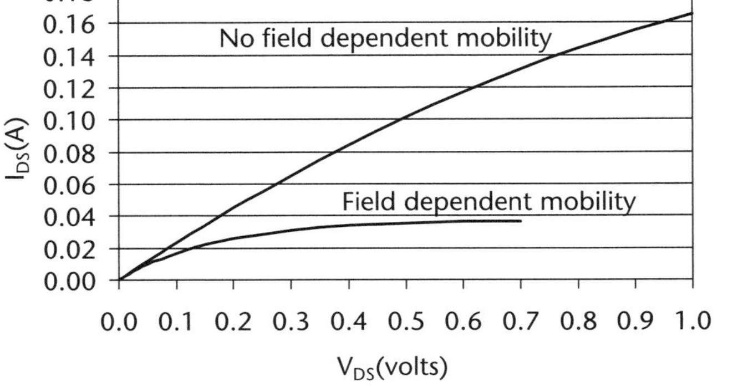

Figure 5.8 Comparison of I-V curves with field dependent and no field dependent electron mobility in an n-channel MOSFET in the triode region.

on the dc gate bias voltage V_{GS0} and could, at large RF levels, potentially drive the instantaneous gate voltage below the threshold voltage, V_T, effectively starting device turnoff. This low-frequency gate bias circuit effect therefore governs the power handling capacity of the MOSFET control device at low frequencies. Increasing the dc gate bias voltage will increase the power handling but only up to a point, since there is a maximum gate voltage that can be applied to the MOSFET before the gate oxide breakdown voltage is exceeded and device destruction occurs. As the frequency increases above F_G, the gate node becomes a floating node, since R_G is much greater than Z_0, and V_{GS} is half the voltage drop across the drain-source connection, V_{DS}, which is small since R_{ON} is much less than Z_0 as required for low insertion loss. In this case, the power limitation is governed by how much current the MOSFET can pass for a given V_{GS0} while still remaining in the triode/linear region of operation for that dc gate voltage.

In the off-state, the conducting channel disappears as R_{ON} goes toward an open circuit and high isolation occurs. At low frequencies, almost the entire applied RF voltage is dropped as the drain-source voltage V_{DS} of the MOSFET, with the drain node at the applied RF peak voltage, therefore making that node subject to any voltage breakdown effects. The source node voltage is very small because of the high isolation. As the frequency increases, RF coupling from the source to the load through the two gate capacitances occurs, causing the isolation to degrade. The gate-source voltage, V_{GS}, approaches $Z_0 I_{DS}$ and therefore tracks the applied RF voltage amplitude with similar impacts of the instantaneous RF swing approaching V_T as in the conducting state. Since the drain node is still at high RF potential throughout the high isolation state, breakdown at this junction could also occur if the instantaneous RF voltage exceeds the drain breakdown voltage.

At first glance, it appears that F_G can be made arbitrarily low to garner the better power handling capability at frequencies above F_G. R_G cannot be too high a value, however, because the gate bias circuit capacitances have to be charged or discharged to turn the device on or off, potentially slowing the switching speed of the device if R_G is set too high. In the case of MOSFETs, R_G is often a mix of a lumped element resistance combined with the inherent resistance of the polysilicon gate material and is typically in the range of 10 kΩ. Both should be considered when modeling the gate bias circuit. The portion of R_G consisting of the polysilicon gate material causes another issue for high-speed MOSFET switching, since this resistance is a distributed resistance over the entire gate width [37–41]. When the dc gate bias control voltage is applied to the distributed RC gate network to turn the MOSFET on, it takes a finite amount of time for the voltage pulse to travel down the line to each elemental MOSFET, making the large MOSFET control device turn on in a cascade-type effect (the inverse is true for the turn-off pulse). This turn-on time delay is a function of the MOSFET geometry (W, L) and gate construction (polysilicon layer sheet resistance, thin oxide thickness) and the number of gate fingers used. A study of the distributed nature of the polysilicon gate resistance showed that the optimum number of gate fingers for the MOSFET control device is in the range of 10 to 20 [37].

Some workers have looked at replacing the fixed portion of the gate resistance R_G with a series-connected MOSFET in the gate lead so that this device may be toggled on and off, depending on the desired switch state [30]. Improved off-state isolation (approximately 5 dB) with little change in on-state performance was noted up to Ku band (12–18 GHz) using this concept [30].

5.5.3 Nonlinear Operation in the On-State

The MOSFET RF control element will have its on-state channel resistance modulated by the applied RF signal, making the on-state resistance a time-varying quantity dependent on the applied RF signal characteristics (5.8) and therefore leading to nonlinear operation. Using the same technique as in Chapter 4's discussion of nonlinearities in PIN diodes, a Taylor series expansion can be used for defining the I_{DS}-V_{DS} characteristic for the device as a function of power of V_{DS}:

$$I_{DS} = \sum_{n=1}^{\infty} \alpha_n V_{DS}^n \tag{5.9}$$

where the α_n are the expansion coefficients determined by the series expansion of the I_{DS}-V_{DS} characteristic in (5.8). The gate bias circuit, composed of C_{GS}, C_{GD}, and R_G, links the RF voltages V_{GS} and V_{DS} and can be described by the ratio V_{GS}/V_{DS} and given the symbol γ [5]. Assuming a series connected FET-based switch element and a gate bias resistance much larger than the on-state resistance of the FET, the term γ can be written with the aid of the RF equivalent circuit in Figure 5.6(a) as [5]

$$\gamma = \frac{V_{GS}}{V_{DS}} = -\frac{\dfrac{Z_0}{R_{ON}} - j\omega R_G C_{DG}}{1 + j\omega R_G \left(C_{DG} + C_{GS}\right)} \tag{5.10}$$

where Z_0 is the source/load impedance (typically 50Ω). No bulk connections have been assumed in the derivation of (5.10). The α_n in (5.9) can be found through the Taylor series expansion of (5.8) as [5]

$$\begin{aligned}
\alpha_1 &= KP\frac{W}{L}\left[V_{GS0} - V_T\right] = 1/R_{ON}, \\
\alpha_2 &= KP\frac{W}{L}\left[\gamma - \frac{V_{GS0} - V_T}{LE_C}\right] = KP\frac{W}{L}\gamma - \frac{\alpha_1}{LE_C}, \\
\alpha_3 &= -KP\frac{W}{L}\frac{1}{LE_C}\left[\gamma - \frac{1}{LE_C}\left(V_{GS0} - V_T\right)\right] = -\frac{\alpha_2}{LE_C}
\end{aligned} \tag{5.11}$$

where V_{GS0} is the applied dc bias voltage on the gate. The term α_1 is the linear term, $1/R_{ON}$, and the remaining values are the nonlinear device coefficients. Only the first two nonlinear coefficients represent the most important nonlinear characteristics, the second and third order device nonlinearities.

The distortion produced by the MOSFET control element is dependent on both the electrical conductivity of the conducting channel under the gate in terms of magnitude and on the gate bias circuitry in terms of frequency dependence [42–44], as shown by the expansion coefficients α_n in (5.11). General expressions for load or output-referenced IP2 and IP3 can be written for the series reflective FET-based switch in terms of the expansion parameters in (5.11) by assuming a distortion current generator in parallel with R_C in Figure 5.6(a) as [5]

$$\begin{aligned} \text{IP2} &= \frac{Z_0}{2R_{ON}^4\alpha_2^2}\left(1+\frac{2Z_0}{R_{ON}}\right)^2 \\ \text{IP3} &= \frac{Z_0}{2R_{ON}^3\alpha_3}\left(1+\frac{2Z_0}{R_{ON}}\right) \end{aligned} \tag{5.12}$$

Figure 5.9(a) shows the results of calculations using this model in determining the on-state second and third order distortion intercept for a 180 nm CMOS RF switch element with example parameters discussed previously (Section 5.2) with a 10 kΩ gate bias resistor, R_G, and zero bulk resistance, R_B. Note that the intercept point is lower (higher level of distortion) at lower frequencies, but the intercept point increases as the frequency increases. The transition point at approximately 30 to 50 MHz is near the gate bias circuit cutoff frequency $2\pi F R_G(C_{GD} + C_{GS}) = 1$. The nearly constant IP3 performance for the silicon switch is primarily due to a

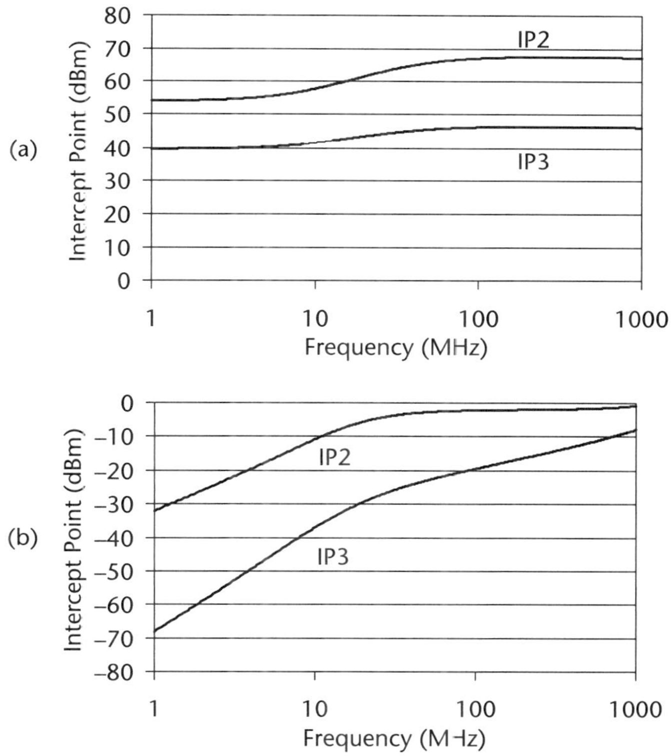

Figure 5.9 (a) On-state load-referenced distortion intercept points for MOSFET RF series-connected RF switch. (b) Off-state load-referenced distortion intercept points for MOSFET RF series-connected RF switch

reduction in surface mobility caused by short channel effects in these devices and the increased drain/source sidewall capacitances. Roll-off in high-frequency distortion of the control element is due to the switch's inherent high-frequency roll-off. This approach for on-state distortion modeling of MOS-based RF switches shows good agreement with experimental data on 1200 nm gate length switches [45].

5.5.4 Nonlinear Operation in the Off-State

In the off-state, the inverted channel disappears because the gate control signal and V_{GS} drop below the threshold voltage, making R_{ON} infinite. However, the drain and source diffused regions are still in the RF path, and the voltage variations on the drain and source nodes cause the associated capacitances to become time-varying and hence generate distortion. The distortion currents are also generated by these capacitances in the on-state but are much smaller than those due to distortion generated by time-varying channel resistance. In a series connected circuit in the off-state, the voltage on the drain side of the MOS switch is nearly the same as the generator voltage due to the high isolation of the switch and therefore will generate a higher level of distortion current than the source side because of the high isolation. However, this same high isolation of the fundamental drive signal also significantly isolates the drain side from the load, reducing the level of distortion currents at the load. A similar approach to distortion modeling was performed with PIN diodes in Chapter 4 and can be employed with MOSFET structures by looking at the voltage dependence of the drain and source node capacitances. Using a Taylor series expansion of (5.6), the first several terms of the expansion are

$$C_{\mathrm{DB(SB)}} = \sum_{k=0}^{\infty} c_k \left(V_{\mathrm{DB(SB)}}\right)^k = \left[(ab)CJ + 2(a+b)CJSW\right] \cdot \left[1 + \frac{1}{2}\frac{1}{V_{bi}} V_{\mathrm{DB(SB)}} - \frac{3}{8}\frac{1}{V_{bi}^2} V_{\mathrm{DB(SB)}}{}^2 - \frac{5}{16}\frac{1}{V_{bi}^3} V_{\mathrm{DB(SB)}}{}^3 \ldots\right] \quad (5.13)$$

for small $V_{\mathrm{DB(SB)}}/V_{bi}$. In (5.13), note that the voltage dependence is based on the drain or source potential with respect to the bulk node, implying that the level of distortion depends on the bulk node resistance. The linear ($k = 0$) and nonlinear ($k \geq 1$) currents can be written in terms of the c_k and the drain/source voltage using a similar form as (4.25)

$$I_{\mathrm{DB(SB)}}(t) = \frac{d}{dt}\sum_{k=0}^{\infty} c_k V_{\mathrm{DB(SB)}}(t)^{k+1} = \sum_{k=0}^{\infty} c_k \frac{dV_{\mathrm{DB(SB)}}(t)^{k+1}}{dt} \quad (5.13)$$

Figure 5.9(b) shows the results of calculations using this capacitance model in determining the load-referenced off-state second and third order distortion intercept for the example 180 nm series connected MOS RF switch element with parameters discussed previously and a 10 kΩ gate bias resistor, R_G, and zero bulk resistance, R_B. Note that the intercept point is lower (higher level of distortion) at lower frequencies, but the intercept point steadily increases as the frequency increases. This increase is due to both an increase in the distortion current level with increasing frequency and also more coupling of the drain distortion onto the load as the isolation falls off with increasing frequency.

5.5.5 MOS Stacking

To increase the power handling capability of the MOS control structure, "stacking" of MOS transistors is often used as shown in Figure 5.10 [46–48]. This stacking can be in both the series and shunt arms of any RF switch circuit. Here, the RF voltage is shared between a series of two or more MOS transistors. Discrete and multiple gate (not multiple gate finger) structures have been used to improve the power handling, but care must be taken that the devices are designed to keep the insertion loss within design limits and to also keep the additional nodal capacitances as small as possible for high-frequency operation. The insertion loss can be improved by increasing each MOSFET's individual aspect ratios, W/L, by reducing gate length, increasing gate width, or both, and increasing the applied gate control voltage. Increasing W, however, increases the capacitance and lowers the off-state performance and increases the nonlinearities of the circuit. In multiple gate MOS devices, some of the drain/source capacitance is shared by adjoining transistors in the stack, keeping the capacitance somewhat lower than it would be if individual devices were used [46].

5.5.6 Thermal Modeling

The ability to accurately model the thermal effects in MOS devices allows better prediction of operation over a wide range of temperatures. The major thermal effects fundamentally impact the channel mobility, $\phi_B = kT/q \ln(N_{\text{sub}}/n_i)$, and bulk PN junction built-in potential, V_{bi}, which then impact the threshold voltage, V_T, and the on-state resistance R_{ON}, which are defined in (5.2) through (5.4) and (5.6). Studies have shown that the on-state insertion loss degrades slightly with increasing temperature whereas the isolation is relatively constant over temperature [49]. This on-state insertion loss observation is consistent with (5.4) because the carrier mobility, which decreases with increasing temperature [19], also makes KP decrease

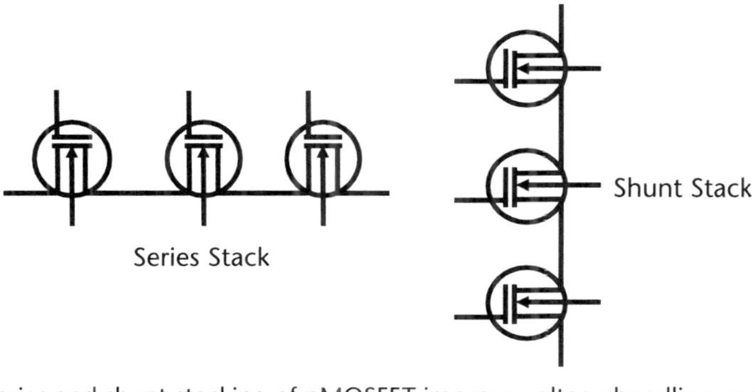

Figure 5.10 Series and shunt stacking of nMOSFET improve voltage handling capability but at the possible expense of increased insertion loss and reduced frequency response.

with increasing temperature, increasing the on-state resistance R_{ON} with increasing temperature. The capacitance, being less sensitive to variation in temperature as indicated in (5.6), remains relatively constant with increasing temperature, and since the isolation is primarily governed by the capacitance, the off-state isolation observation in [49] is also consistent with (5.6). This phenomenon is observed in both bulk and SOI-based MOSFET RF control devices.

5.6 SPICE/BSIM Models: SPICE Levels 1 through 3 and BSIM models

5.6.1 SPICE Level 3

There are many different simulators available for the RFIC designer, but a widely used "standard" simulator is SPICE [50]. In fact, almost all simulators (free, share, and proprietary) use the same set of parameters for MOSFET modeling as more generic SPICE simulators, and so in keeping with the idea of a generic simulator, this discussion will focus on the generic SPICE definitions.

Previous technology generations (gate lengths of approximately 1 μm and greater) used the SPICE Level 3 model (SPICE-3). IC fabricators would provide test MOSFET data with extracted SPICE-3 of previous runs to RFIC designers for use in future designs. SPICE-3 has approximately 25 parameters [51], but only a small subset is actually needed for first pass modeling. In addition, the SPICE-3 model is a physics-based model, which can be seen by looking at the design equation forms earlier in this chapter. A minimum set of SPICE-3 parameters for dc modeling would include the following:

Parameter	*Description*	*Units*
VTO	Threshold Voltage	Volts
UO	Surface Mobility	$cm^2/V\text{-}s$
KP	Intrinsic Transconductance	A/V^2
GAMMA	Body Effect Parameter	$V^{1/2}$
NSUB	Substrate or Well Doping	cm^{-3}

For RF modeling, additional capacitance terms are required for accurate modeling. SPICE-3 uses the following definitions for the model capacitances:

Parameter	*Description*	*Units*
CBD	Zero bias bulk drain capacitance	F
CBS	Zero bias bulk source capacitance	F
CGBO	Gate bulk overlap capacitance	F/m
CGDO	Gate drain overlap capacitance	F/m
CGSO	Gate source overlap capacitance	F/m
CJ	Bulk PN junction capacitance/area	F/m^2
CJSW	Bulk PN junction sidewall capacitance	F/m

The I_{DS}-V_{DS} equations in the SPICE-3 model are very similar to those developed in (5.3). The capacitance models are similar as well, which allows an easy one-to-one mapping of the MOSFET layout and associated lumped elements to simulator parameters using this model, increasing the utility of using SPICE-3 for RF control device modeling.

5.6.2 BSIM Parameters

As MOSFET device geometries entered the submicron realm, the SPICE-3 model was found to be less accurate in modeling MOSFET operation. The Berkeley Short-Channel IGFET Model for MOS Transistors (BSIM) was developed in the 1980s to provide for more accurate submicron MOSFET simulation [52]. Since then BSIM has gone through numerous versions and has branched out from the original BSIM3 version to BSIM6 (for sub-100 nanometer technologies), BSIMSOI (for silicon on insulator technology), and BSIMCMG/BSIMIMG (for multiple gate structures) at the time of this writing. The number of parameters describing the MOSFET has grown with increases in modeling accuracy, with the parameter number following a Moore's Law type of increase with the number of modeling parameters doubling every 10 years [51]. Full details and information can be found on the BSIM website [http://www-device.eecs.berkeley.edu/bsim/]. More recent BSIM models contain significantly more parameters than the original SPICEBSIM3 model in an effort to describe the more complex behavior of submicron MOSFET operation [53]. For example, BSIMSOI contains such enhancements as gate tunneling current (an issue for deep submicron MOSFETs), improved noise modeling (of interest to those working with LNAs, for example), better descriptions of bulk/substrate network (important to the RF designer), improved modeling in the weak and moderate inversion regions of operation, and better modeling of multiple-fingered MOSFETs [54].

One of the significant differences between SPICE-3 and early BSIM models was that while many (but not all) of the SPICE-3 parameters were physically based, some of the BSIM model parameters were not physically based but were rather determined from extensive measurements on fabricated MOSFET test structures. Curve fitting routines were then used with the measured data to determine the BSIM model parameters. This provided better accuracy but at the expense of some physical intuition, primarily because of the large number of parameters required in the BSIM model. Updated versions and releases of the various BSIM versions are available online [http://www-device.eecs.berkeley.edu/bsim/].

The most recent BSIM version, BSIM6.0, provides excellent accuracy for analog and RF modeling and is considered an industry standard model [55]. The listing of BSIM6.0 parameters fills more than 20 pages in the BSIM6.1.0 MOSFET *Compact Model Technical Manual* [53], and so the full set will not be repeated here. The *Technical Manual*, however, is readily available on the web[3] and contains in-depth discussions of the model equations and the measurement techniques used in extracting the model parameters. The SPICE simulator used in this book can use

3 http://www-device.eecs.berkeley.edu/bsim/?page=BSIM6.

Table 5.2 Example SPICE Netlist for Series-Connected MOSFET Switch

```
* SERIES-CONNECTED RF SWITCH SIMULATION EXAMPLE
MFET OUT GATE IN BULK M1
+ L=0.18U W=175U AD=75P AS=75P PD=321U PS=321U
RBULK BULK 0 .001
RGATE GATE0 GATE 10E3
RSOURCE IN VGEN 50
RLOAD OUT 0 50
VGATE GATE0 0 2
VGENERATOR VGEN 0 0 AC 1
.AC DEC 100 1E6 1000E6

.MODEL M1 NMOS LEVEL=3 KP=150E-6 VTO=0.45
+CGDO = 4.88E-10 CGSO = 4.88E-10
+CGBO = 1E-12 CJ = 8.386557E-4
+MJ = 0.5137012 CJSW = 2.196109E-10
+MJSW = 0.2172875

.END
```

up to BSIM4.6.1, and so the example parameters available from the UC Berkeley BSIM Research Group[4] will be used in the examples in this and later chapters. More recent versions of the example parameters can also be used for modeling if the SPICE simulator can accept the updated parameter list. These more recent BSIM parameters are available from the UC Berkeley website and may be available from the MOS device fabricator or vendor.

5.6.3 SPICE Simulation Example

This section will cover a simple SPICE example based on the 180 nm MOS RF switch that has been extensively discussed in this chapter. The SPICE netlist ('netlist5-1.net') for the series-connected MOS RF switch based on a SPICE Level 3 model is presented in Table 5.2 with the associated circuit layout and on-state insertion loss and off-state isolation shown in Figure 5.11. In Table 5.2, the SPICE circuit elements, connection nodes, values, and parameters are capitalized with no subscripts; this is the traditional structure found in the SPICE netlist. The .MODEL portion of the SPICE netlist provides the most important parameters for the MOSFET switch: KP, V_T and the various layer and diffusion capacitances per unit area or length. The various source, load, gate, and bulk resistances (set here to zero) are indicated. The bulk resistance (RBULK) is explicitly placed in the netlist as a placeholder for investigations of "floating bulk" operation. The description of the MOSFET

http://www-device.eecs.berkeley.edu/bsim/?page=BSIM4_Arc.

is devoted to the second and third lines of the netlist, with line 2 describing the device's connections to the remainder of the circuit and also calling the model M1 information from the .MODEL section. The third line of the MOSFET switch is also important because the drain/source area (AD/AS) and drain/source perimeter (PD/PS) are described. Computation of these area and perimeter values follows as described in Section 5.4.2. The '+' sign at the start of line 3 simply indicates this is a continuation of the MOSFET definition. These variables define the MOSFET layout geometry and therefore are used in computing the associated capacitances but are often overlooked by modeling engineers, at least on the first design pass or two. Computing the area and perimeter of the MOSFET layout has been described in Section 5.4.2; these parameters can be reduced if multiple gate finger MOSFET are used, but the modeler still has to compute the area and perimeter of each of the individual MOSFET layouts and then put the individual MOSFET in parallel (i.e., if two gate fingers are used, then two MOSFETs need to be placed in parallel for accurate modeling).

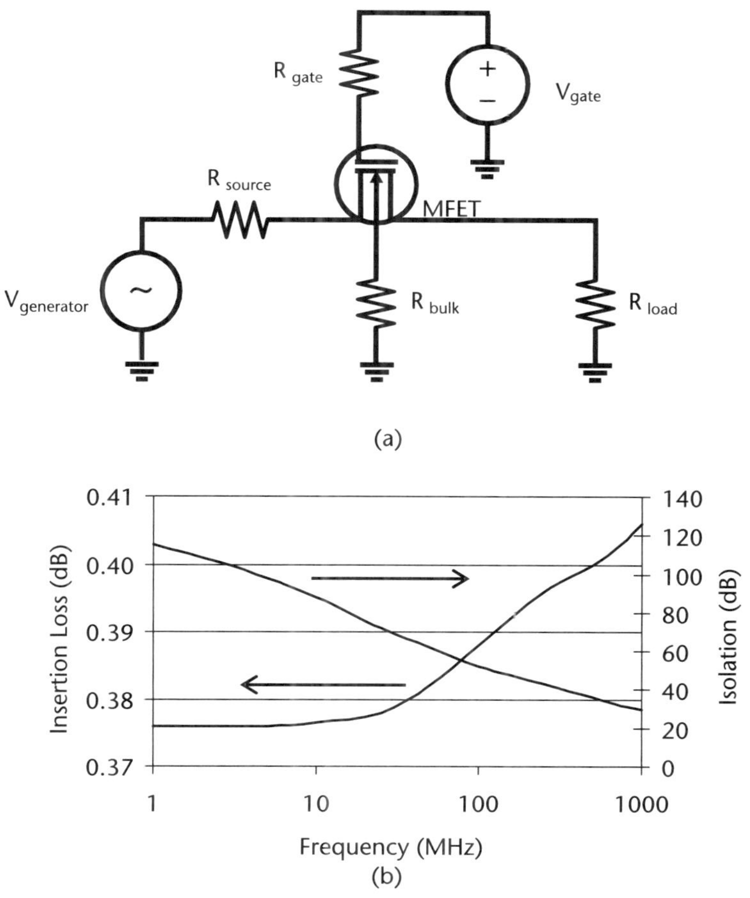

Figure 5.11 (a) SPICE simulator layout for series-connected MOS RF switch based on the 180 nm example. (b) Insertion loss and isolation for the MOS RF switch from 1.0 MHz to 1000 MHz (R_{bulk} = 0).

References

[1] Camilleri, N., et al., "Silicon MOSFETs, the Microwave Device Technology for the 1990s," *Proc. 1993 IEEE MTT-S Intl. Microwave Symp.*, Vol. 2, 1993, pp. 545–548.

[2] Abidi, A. A., "RF CMOS Come of Age," *IEEE Microwave Magazine*, Vol. 4, No. 4, 2003, pp. 47–60.

[3] L. Boglione, J. Lin, and W. Khalil, "Expanding RFIC Horizons," *IEEE Microwave Mag.*, Vol. 13, No. 1, 2012, pp. 10–12.

[4] Mak, P., and R. P. Martins, "Enhanced RFICs in Nanoscale CMOS," *IEEE Microwave Mag.*, Vol.13, No. 6, 2012, pp. 80–89.

[5] Caverly, R., "Linear and Nonlinear Characteristics of the Silicon CMOS Monolithic 50 Ohm Microwave and RF Control Element," *J. Solid-State Circuits*, Vol. 34, No. 1, 1999, pp. 124–126.

[6] Madihian, M., et al., "CMOS RF ICs for 900 MHz-2.4 GHz Band Wireless Communication Networks," *1999 IEEE Radio Frequency Integrated Circuits (RFIC) Symposium*, 1999, pp.13–16.

[7] Floating Ground SPNT MMIC Switch Driver Techniques, Application Note 17-134, v02.0402, Hittite Microwave Corp., 2004 (http://www.hittite.com/support/application-notes.html).

[8] Tsividis, Y. P., *Operation and Modeling of the MOS Transistor*, New York: McGraw-Hill, 1987.

[9] Enz, C., "An MOS Transistor Model for RF IC Design Valid in All Regions of Operation," *IEEE Trans. on Microwave Theory and Techniques*, Vol. 50, No. 1, 2002, pp. 342–359.

[10] Enz, C., and Y. Cheng, "MOS Transistor Modeling for RF IC Design," *IEEE Trans. on Solid-State Circuits*, Vol. 35, No. 2, 2000, pp. 186–201., 2000.

[11] Ou, J., et al., "CMOS RF Modeling for GHz Communication IC's," 1998, *VLSI98*.

[12] Das, M., "High Frequency Network Properties of MOS Transistors Including the Substrate Resistivity Effect," *IEEE Trans. on Electronic Devices*, Vol. ED-16, 1969, pp. 1049–1069.

[13] Reinhard, D., *Introduction to Integrated Circuit Engineering*, Houghton Mifflin, Boston, MA: Houghton Mifflin, 1987.

[14] Chau, R., et al. "High-κ/Metal-Gate Stack and Its MOSFET Characteristics," *IEEE Electron Device Letters*, Vol. 25, No. 6, 2004, pp. 408, 410.

[15] Bohr, M. T., et al., "The High-k Solution," *IEEE Spectrum*, Vol. 44, No. 10, October 2007, pp. 29–35.

[16] Banerjee, K., "Trends for ULSI Interconnections and Their Implications for Thermal, Reliability and Performance Issues," *Proc. 7th Intl. DCMIC*, 2001, pp. 38–50.

[17] Ducarouge, B., et al., "Power Capabilities of RF MEMS," *Proc. 24th Intl. Conf. Microelectronics (MIEL 2004)*, Vol. 1, 2004, pp. 65–70.

[18] Caverly, R., *CMOS RFIC Design Principles*, Norwood, MA: Artech House, 2007.

[19] Sze, S., *Physics of Semiconductor Devices*, New York: Prentice Hall, 1982.

[20] Hjelmgren, H., and A. Litwin, "Small-Signal Substrate Resistance Effect in RF CMOS Identified Through Device Simulations," *IEEE Trans. on Electronic Devices*, Vol.48, No. 2, 2001, pp. 397–399.

[21] Mead, C., and L. Conway, *Introduction to VLSI Systems*, Boston, MA: Addison-Wesley, 1979.

[22] Abou-Allam, E., and E. T. Manku, "A Small Signal MOSFET Model for Radio Frequency IC Applications," *IEEE. Trans. on Computer-Aided Design*, Vol. 16, No. 5, 1997, pp. 437–447.

[23] Huang, F.-J., and K. O, "A 0.5-μm CMOS T/R Switch for 900-MHz Wireless Applications," *IEEE J. of Solid-State Circuits*, Vol. 36, No. 3, March 2001, pp. 486–492.

[24] Li, X. J., and Y. P. Zhang, "Flipping the CMOS Switch," *IEEE Microwave Magazine*, Vol. 11, No. 1, February 2010, pp. 86–96.

[25] Kidwai, A. A., et al., "A Fully Integrated Ultra-Low Insertion Loss T/R switch for 802.11b/g/n Application in 90 nm CMOS Process," *IEEE J. of Solid-State Circuits*, Vol. 44, No. 5, 2009, pp. 1352–1360.

[26] Li, Z., et al., "5.8-GHz CMOS T/R Switches with High and Low Substrate Resistances in a 0.18-μm CMOS Process," *IEEE Microwave Wireless Compon. Lett.*, Vol. 13, No. 1, 2003, pp. 1–3.

[27] Botula, A., et al., "A Thin-Film SOI 180nm CMOS RF Switch Technology," *IEEE Topical Meeting on Silicon Monolithic Integrated Circuits in RF Systems, 2009. SiRF '09*, 2009, pp. 1–4.

[28] Tinella, C., et al., "A High-Performance CMOS-SOI Antenna Switch for the 2.5-5-GHz band," *IEEE J. of Solid-State Circuits*, Vol. 38, No. 7, 2003, pp. 1279–1283.

[29] Garcia, G. A., et al., "High-Quality CMOS in Thin (100 nm) Silicon on Sapphire," *IEEE Electron Device Letters*, Vol. 9, No. 1, 1988, pp. 32–34.

[30] Chen, C. L., et al., "Fully Depleted SOI RF Switch with Dynamic Biasing," *2007 IEEE Radio Frequency Integrated Circuits (RFIC) Symposium*, 2007, pp. 175–178.

[31] Peregrine Semiconductor Corp., http://www.psemi.com/ (accessed August 2015).

[32] Li, Q., and Y. P. Zhang, "CMOS T/R Switch Design: Towards Ultra-Wideband and Higher Frequency," *IEEE J. of Solid-State Circuits*, Vol. 42, No. 3, 2007, pp. 563–570. (See also Poh, A., and Yue Ping Zhang, "Design and Analysis of Transmit/Receive Switch in Triple-Well CMOS for MIMO Wireless Systems," *IEEE Trans. on Microwave Theory and Techniques*, Vol. 55, No. 3, 2007, pp. 458–466.)

[33] Kuo; C-S., et al., "A High-Isolation 60GHz CMOS Transmit/Receive Switch," *2011 IEEE Radio Frequency Integrated Circuits Symposium (RFIC)*, June 2011, pp. 1–4, 5–7.

[34] Wu; C-C., A. Yen, and Jen-Chung Chang, "A 0.13/spl mu/m CMOS T/R Switch Design for Ultrawideband Wireless Applications," *Proc. 2006 IEEE International Symposium on Circuits and Systems, ISCAS 2006*, 2006, pp. 1–4.

[35] Linten, D., et al., "A 5-GHz Fully Integrated ESD-Protected Low-Noise Amplifier in 90-nm RF CMOS," *IEEE J. of Solid-State Circuits*, Vol. 40, No. 7, 2005, pp. 1434–1441.

[36] Gray, P., and R. Meyer, *Analysis and Design of Analog Integrated Circuits*, 3rd ed. New York: Wiley, 1993.

[37] Caverly, R., and J. Manosca, "Transient Switching Behavior in Silicon MOSFET RF Switches," *IEEE Topical Meeting on Silicon Monolithic Integrated Circuits in RF Systems (SiRF 2008)*, 2008, pp. 179–182.

[38] Lin, H. C., et al., "Effect of Silicon Gate Resistance on the Frequency Response of MOS Transistors," *IEEE Trans. on Electronic Devices*, Vol. 22, No. 5, 1975, pp. 255–264.

[39] Allam, E., and T. Manku, "An Improved Transmission-Line Model for MOS Transistors," *IEEE Trans. on Circuits Systems II—Analog and Digital Signal Processing*, Vol. 46, No. 11, 1999, pp. 1380–1387.

[40] Martin, D., and N. C. Rumin, "Delay Prediction from Resistance-Capacitance Model in General MOS Circuits," *IEEE Trans. on CAD of Integrated Circuits and Systems*, Vol. 12, No. 7, 993, pp. 997–1003.

[41] Ding, L., and P. Mazumder, "On Optimal Tapering of FET Chains in High-Speed CMOS Circuits," *IEEE Trans. on Circuits Systems II—Analog and Digital Signal Processing*, Vol. 48, No. 12, 2001, pp. 1099–1109.

[42] Gutmann, R., and N. Jain, "Characterization of Linear and Nonlinear Properties of GaAs MESFETs for Broad-Band Control Applications," *IEEE Trans. on Microwave Theory and Techniques*, Vol. MTT-35, No. 5, 1987, pp. 516–520.

[43] Caverly, R., "Distortion in Broadband Gallium Arsenide MESFET Control and Switch Circuits," *IEEE Trans. on Microwave Theory and Techniques*, Vol. 39, No. 4, 1991, pp. 713–717.

[44] Caverly, R., "Distortion in GaAs MESFET Switches," *Microwave J.*, 1994, pp. 106–114.

[45] Caverly, R. H., "Nonlinear Properties of Gallium Arsenide and Silicon FET-Based RF and Microwave Switches," *Proceedings of the 1998 IEEE International Symposium on Circuits and Systems (ISCAS '98)*, Vol. 4, 1998, pp. 337–340.

[46] Ahn, M., C.-H. Lee, and J. Laskar, "CMOS High Power SPDT Switch Using Multigate Structure," *IEEE International Symposium on Circuits and Systems, 2007(ISCAS 2007)*, 2007, pp. 3283–3286.

[47] Ahn, M., et al., "A Novel Multi-Stack Device Structure and Its Analysis for High Power CMOS Switch Design," *Proc. IEEE 2007 International Microwave Symposium*, 2007, pp. 1393–1396.

[48] Levy, C. S., P. M. Asbeck, and J. F. Buckwalter, "A CMOS SOI Stacked Shunt Switch with Sub-500ps Time Constant and 19-Vpp Breakdown," *2013 IEEE Compound Semiconductor Integrated Circuit Symposium (CSICS)*, 2013, pp.1–4.

[49] Emam, M., D. Vanhoenacker-Janvier, and J. Raskin, "High Temperature DC and RF Behavior of Partially Depleted SOI versus Deep n-Well Protected Bulk MOSFETs," *IEEE Topical Meeting on Silicon Monolithic Integrated Circuits in RF Systems, 2009 (SiRF '09)*, 2009, pp.1–4.

[50] *MicroSim PSpice A/D Reference Manual*, MicroSim Corp., 1997 [see also http://bwrcs.eecs.berkeley.edu/Classes/IcBook/SPICE/].

[51] Foty, D., "MOSFET Modeling for Circuit Simulation," *IEEE Circuits and Devices Magazine*, Vol. 7, 1998, pp. 26–31.

[52] Sheu, B., et al., "BSIM: Berkeley Short-Channel IGFET Model for MOS Transistors," *IEEE J. Solid State Circuits*, Vol. SC-22(4), 1987, pp. 558–566.

[53] Agarwal, H., et al., *BSIM6.1.0 MOSFET Compact Model Technical Manual*, Regents of the Univ. Calif., 2014.

[54] BSIMSOIv4.5.0, officially released on Nov. 7, 2013, updated on Nov. 20, 2013, http://www-device.eecs.berkeley.edu/bsim/?page=BSIMSOI_LR, accessed August 2015.

[55] BSIM6, officially released on May 5, 2013, updated on July 30, 2015, http://www-device.eecs.berkeley.edu/bsim/?page=BSIM6, accessed August 2015.

[56] Wafer Electrical Test Data and SPICE Model Parameters, The MOSIS Service, https://www.mosis.com/requests/test-data, accessed August 2015.

CHAPTER 6

Modeling MESFET and HEMT Control Devices

6.1 Introduction

Up until the mid-1960s, most microwave circuits were still fabricated on thin dielectric substrates, which provided good performance but did not provide the capability for small footprints that integrated circuits would eventually provide. The first true microwave integrated circuit was actually a PIN diode-based switch operating at X-band and fabricated by Ertel [1, 2]. In 1979, Garver proposed using the then-emerging GaAs MESFET (metal semiconductor field effect transistor) technology as an RF control device [3], and Ayasli in 1980 developed a GaAs MESFET version of the X-band PIN diode switch [4]. Ayasli's low-power transmit-receive (TR) switch incorporated two quarter-wavelength 50Ω transmission lines with shunt-connected MESFETs controlling the flow of RF. Ayasli developed a higher power version of the TR switch several years later [5]. The development of these control applications was occurring concurrently with the rapid development of single-chip GaAs systems, and the control capability of GaAs MESFET switches aided the antenna duplexing capability of these new single-chip systems. GaAs was a vast improvement over silicon-based technology at the time, exhibiting a native fivefold increase in electron mobility, and hence switching speed and frequency response for a given device size. The current flow in GaAs is governed not by minority carrier flow as in the PN or PIN diode but rather is majority carrier based, another factor associated with its higher performance, because diffusion-based carrier flow is inherently slower than drift-based carrier flow. Since that time, GaAs MESFET control devices have become ubiquitous and are found throughout systems in both dedicated control-device packages as well as integrated with GaAs systems, and are sold by many different vendors. Of more recent interest for high-power RF and microwave control are devices fabricated from gallium nitride (GaN) [6]. GaN is of particular interest because of its high electron mobility and a wider energy band gap than GaAs, affording it a higher breakdown voltage and power handling capability. The first GaN MESFET was demonstrated in 1993 [7], with rapid development of the technology since [6]. Even more recent advances have used diamond and silicon carbide (SiC) as substrates, again for their high breakdown voltage and hence high-power handling features [8–14].

Since the introduction of GaAs technology, fabrication processes have further improved so that monolayers (single layers of atoms) can be deposited on substrates (heterostructures), creating a two-dimensional electron gas (2DEG) as a conducting

Table 6.1 Some Electronic Properties of Materials Used for MESFET-Based Microwave and RF Control [10, 15–18]

Property	*Silicon*	*GaAs*	*SiC*	*GaN*	*Diamond(C)*
ε_r	11.9	13	9.7	9	5.5
μ_n, m^2/V-s	0.15	0.85	0.05	0.125	0.18
E_G (eV)	1.12	1.42	3.0	3.36	5.47
$E_{\rm BR}$ (10^7 V/m)	3.7	4.0	30	20	70
$v_{\rm sat}$ (10^5 m/s)	1	2.0	2.0	1.1	2.7

Legend:
$E_{\rm BR}$—breakdown electric field
$v_{\rm sat}$—carrier saturation velocity
ε_r—relative permittivity
μ_n—low field electron mobility
E_G—semiconductor band gap (eV)

layer, further improving upon the already high electron mobility of GaAs. From quantum mechanics, the 2DEG theoretically exhibits very little scattering, hence, the higher electron mobility. These high electron mobility transistors (HEMTs, sometimes also called heterostructure FETs, HFETs or modulation doped FETs, MODFETs) operate at higher speeds and frequencies, further increasing the range of RF control devices based on MESFET-type technology. Popular heterostructures are GaAs/AlGaAs with slightly different lattice constants, and GaN/AlGaN, although other materials of different energy band gaps can be used. A modification of the HEMT, termed the pseudomorphic HEMT (pHEMT), allows the use of even larger differences in lattice constants/energy band gaps for improved device performance. The HEMT family of devices, however, is mathematically similar in operation to the standard MOSFET in that a gate voltage is used to modulate the conducting channel (2DEG instead of an inversion layer), changing the device from a conducting to a nonconducting state. These details will be covered in this chapter.

6.2 Review of Bulk MESFET Technology

The primary indicator of the MESFET's frequency response is the amount of time it takes a carrier to traverse the region under the controlling gate of length L. This transit time is dependent on the electron mobility μ_n and the applied electric field E and can be approximated as the distance to travel (L) divided by the fastest achievable velocity, the saturation velocity, $v_{\rm sat}$. Most of the semiconductors used in MESFET designs exhibit two or more conduction band valleys that, at high fields, cause the velocity-electric field profile to have a peak velocity at some critical electric field, with a decrease in velocity above this field. For the purposes of microwave and RF control modeling, operation is usually assumed to be well below this critical field, and hence operation is confined to this "low field regime." For GaAs, the saturation velocity is approximately 2×10^5 m/s [18], requiring reductions in gate length

L to reduce transit time with the associated increase in frequency response and hence switching speed. A quick calculation of frequency response would look at the inverse of the transit time (at least dimensionally) to determine the peak operating frequency per micron of MESFET gate length:

$$\tau = \frac{L}{\mu_n E} = \frac{L}{v_{\text{drift}}} \tag{6.1a}$$

$$f = \frac{1}{2\pi\tau_{\min}} = \frac{v_{\text{sat}}}{2\pi L} = 31 \text{ GHz}/\mu\text{m} \tag{6.1b}$$

Equation (6.1b) shows that significant improvements are garnered by reducing the gate length. Reducing L, however, also means varying many of the MESFET process and design parameters, such as channel thickness, channel doing, and terminal separation, a reason why the device modeler needs to know the very specific technology process parameters for accurate modeling.

6.2.1 Current-voltage (I-V) Characteristics of the Bulk MESFET RF Control Device

The standard bulk MESFET device structure starts with a high resistivity semi-insulating (SI) substrate material in which an n-type channel region (donor concentration N_D) is introduced of depth D, over which a layer of metal is deposited, creating a Schottky metal-semiconductor junction as the gate terminal (see Figure 6.1(a)). The width of this metal layer is termed the gate length L and can be less than 100 nm [19]. On either side of the channel region, regions of high n-type doping (N^+_D) are introduced, with metal layers deposited over these two regions, forming low-resistance ohmic contacts. These two connections comprise the MESFET drain and source connections. Some fabrication processes use a planar (Figure 6.1(a)) or recessed-gate (Figure 6.1(b)) structure to aid in reducing contact resistance [20].

Regardless of type of MESFET structure (planar or recessed), the gate material over the n-type channel creates a Schottky junction at the interface with a depletion region of depth h (Figure 6.1) that penetrates a short distance into the doped channel. With no dc bias applied on the gate terminal, any voltage applied between the drain and source terminals will cause a current to flow through the nondepleted portion of the channel with a vertical channel opening of $D - h$. The depletion region depth h is created by the metal-semiconductor Schottky junction and is therefore dependent on the channel doping concentration, the associated Schottky gate barrier potential V_{bi} at the interface, and the voltage variation down the channel $V(y)$ (Figure 6.2) [15]:

$$h = h\left(V_{\text{GS}}\right) = \sqrt{\frac{2\varepsilon_s}{qN_D}\left[V_{bi} - V_{\text{GS}} + V(y)\right]} \tag{6.2}$$

where ε_s is the permittivity of the semiconductor material. A negative voltage applied to the gate terminal V_{GS} will further increase the penetration of the Schottky depletion region into the channel, eventually causing the depleted region to fully extend

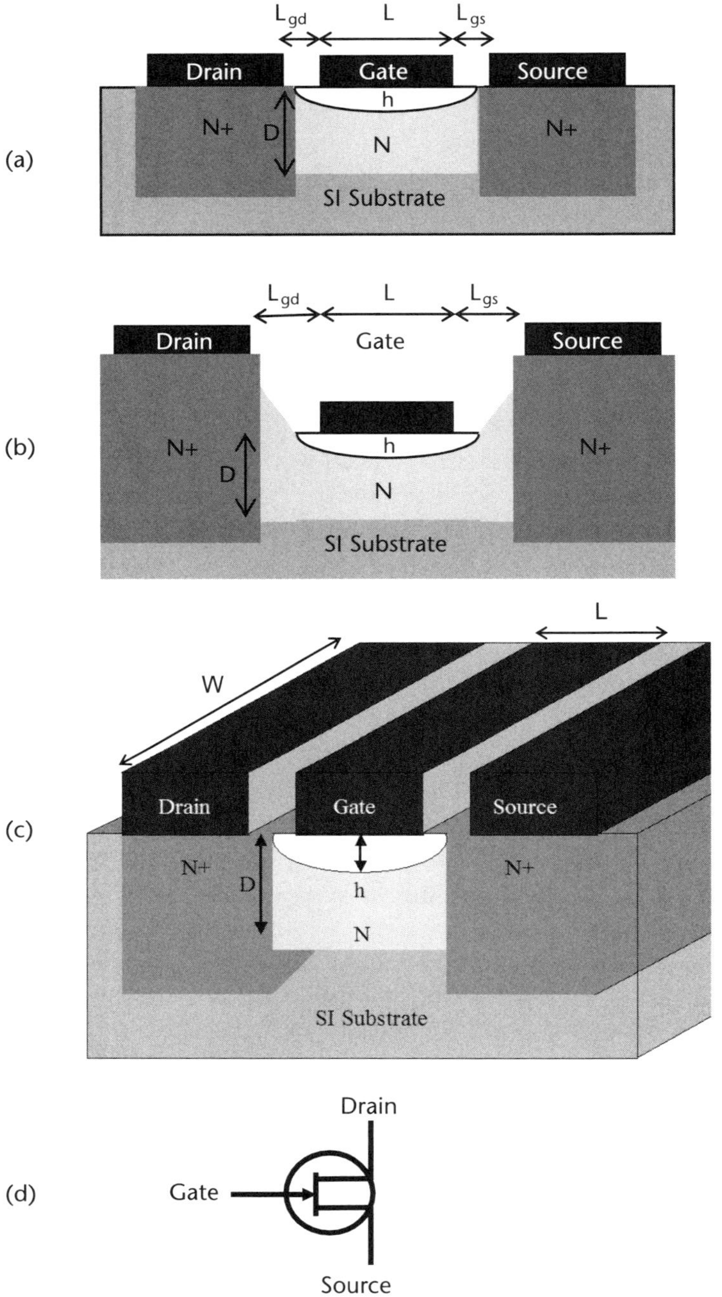

Figure 6.1 Cross section of (a) planar and (b) recessed-gate bulk MESFET structure, (c) three-dimensional view showing gate width W, and (d) general circuit symbol for the MESFET.

into the channel $D = h$, at which point the channel is termed "pinched off" and no current flows; the associated voltage is termed the pinchoff voltage, V_P, defined as

$$V_P = -\frac{qN_D}{2\varepsilon_s} D^2 \tag{6.3}$$

The pinchoff voltage is negative for n-channel depletion mode MESFETs (enhancement mode MESFETs do exist but are not widely used) [42–44].

The I-V characteristics of the MESFET can now be derived with this simple physical model. The following derivation is a straightforward one, assuming that the channel is at least partially open; that is, the gate voltage V_G is small enough to not exceed the pinchoff voltage, V_P. The research literature is quite rich with various MESFET I-V models that, over time, improved the accuracy of I-V modeling from the simple form to be presented here, but usually at the expense of developing intuition of the underlying physical operation. Some of the more famous models for MESFET modeling are the Statz, Curtice, Materka, and TOM (TriQuint's Own Model)[1] models [21–25], which are still used in microwave and RF simulators as well as general circuit simulators such as SPICE. These models, and others similar to them, provide extremely accurate modeling results that show excellent agreement with circuit and device measurements. The main applications that these models were developed for, however, was use of the MESFET in amplifier and other applications where the MESFET was operating in the saturation region, an area to be avoided in RF and microwave control applications. A short overview of these models will be mentioned later in this chapter.

The I-V characteristics of the channel can be developed in a similar fashion to that done by Sze [15]. In the conducting channel, the conductivity is directly related to the channel doping N_D and electron mobility, μ_n: $q\mu_n N_D$. The cross section of the channel looking from drain to source is the conducting channel depth $D - h(y)$ times the gate width W (Figure 6.2). The drain-source current, I_{DS}, can be written as the product of the resistance directly below the gate times the electric field in the channel ($E_{chan} = -dV/dy$):

$$I_{DS} = -q\mu_n N_D W[D - h(y)]\frac{dV}{dy} \tag{6.4a}$$

or

$$\int_0^L I_{DS}\, dy = I_{DS}L = -q\mu_n N_D WD \int_{V_{DS}}^{0} \left(1 - \sqrt{\frac{V_{bi} - V_{GS} + V(y)}{|V_P|}}\right) dV \tag{6.4b}$$

1 TriQuint was a semiconductor company heavily involved in GaAs MESFET technology. See Wikipedia for a more detailed history of the company [http://en.wikipedia.org/wiki/TriQuint_Semiconductor].

where the first integral is simply $I_{DS}L$, since I_{DS} can be assumed constant in the channel and the term $D - h(y)$ is replaced by $D\left[1 - \sqrt{[V_{bi} - V_G + V(y)]/|V_P|}\right]$. The resulting I-V expression for the drain current I_{DS} is found after taking the integral on the right-hand side of (6.4b) as

$$I_{DS} = I_{DSS}\left[\frac{V_{DS}}{|V_P|} - \frac{2}{3}\left(\frac{V_{bi} - V_{GS} + V_{DS}}{|V_P|}\right)^{3/2} + \frac{2}{3}\left(\frac{V_{bi} - V_{GS}}{|V_P|}\right)^{3/2}\right] \quad (6.5a)$$

where

$$I_{DSS} = \frac{q\mu_n N_D W D|V_P|}{L} = \frac{W}{L}\frac{q^2\mu_n^2 N_D^2 D^3}{\varepsilon_s} \quad (6.5b)$$

or, if the more detailed electron mobility expression under high fields is used,

$$I_{DSS} = \frac{W}{L}\frac{q^2\mu_{LF}^2 N_D^2 D^3}{\varepsilon_s}\left[\left(1 + \frac{\mu_{LF}V_{DS}}{Lv_{sat}}\right)\right]^{-2} \quad (6.5c)$$

with μ_{LF} being the low field mobility. I_{DSS} is often referred to as the open channel current; the open channel current definition is a good description of the parameter as (6.5b) shows that I_{DSS} is the current if there were no Schottky depletion region $h(y)$ reducing the current flow cross section. The I-V relationship derived in (6.5) only applies to the region directly under the gate. There are also two regions on either side of the gate region (L_{gs} and L_{gd}) that contribute a small amount of resistance to the circuit and can be included for more accurate modeling of the I-V characteristics at the drain and source terminals (Figure 6.1). These small resistances also influence the eventual on-state resistance of the MESFET and have a small impact on the on-state insertion loss [20].

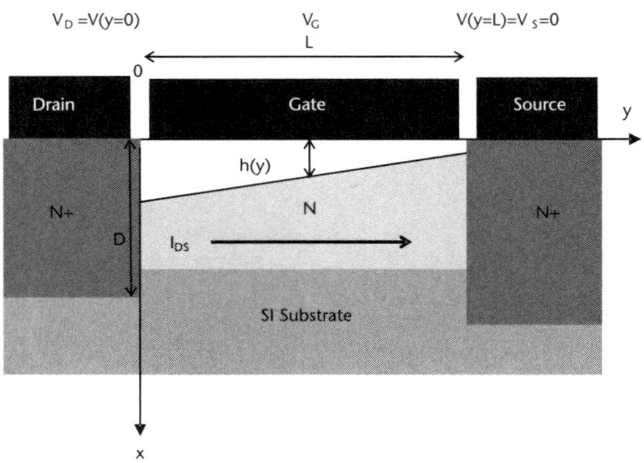

Figure 6.2 Details of the channel region under the gate in the MESFET structure with definitions for derivation of the I-V characteristics.

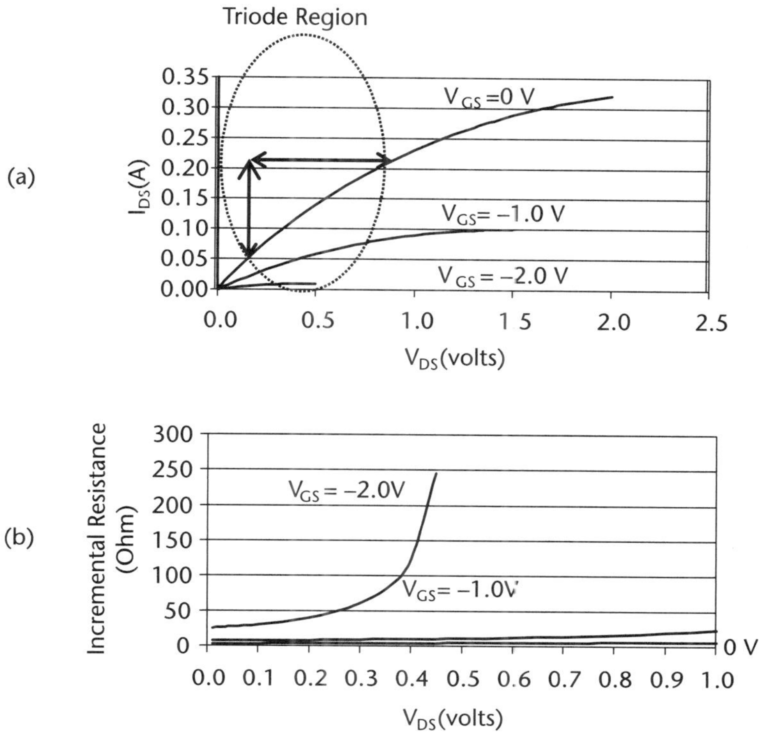

Figure 6.3 (a) Example I_{DS}-V_{DS} curves for various gate voltages. The triode region or linear region of operation is indicated. (b) Dynamic resistance for the three gate voltages.

6.2.2 RF On-State Resistance

In the usual textbook discussions of the I-V characteristics for a MESFET device, the discussion turns to a look at device behavior as V_{DS} begins to approach pinchoff. This so-called saturation region where the current is constant over V_{DS} for a fixed V_{GS}, while the major focus in active circuits such as amplifiers and oscillations, is not the region of interest for the engineer investigating microwave and RF control using MESFETs. In this case, the region of interest is confined to the so-called linear or triode[2] region for small V_{DS} well below the saturation point, with the MESET acting as a variable resistor controlled by the applied gate bias voltage. Figure 6.3(a) shows the I-V characteristics of a GaAs MESFET with an I_{DSS} of 1000 mA as a function of gate voltage V_{GS} using (6.5) and a −2.5V pinchoff voltage; the triode region is marked on the figure. The dynamic resistance of the device can be found by looking at the change in V_{DS} over I_{DS}, ($\Delta V_{DS}/\Delta I_{DS}$ or the slope of the I-V curve) at a fixed gate voltage $V_{GS} = 0$ as shown in Figure 6.3(a). For this condition, the change in current is approximately (0.2 − 0.05) = 0.15A with a corresponding change in

2 The term *triode region* is a holdover from vacuum tube terminology. In a vacuum tube, the output current (from the high-voltage plate) was controlled by a voltage applied to the two other terminals: the cathode (where the electrons originated) and the grid. The resulting curves were linear in nature, and since three terminals were used, the term triode was created.

voltage of (0.7 – 0.1) V = 0.6V, yielding a slope of approximately 0.6/0.15 V/A or 4Ω. Figure 6.3(b) shows the dynamic resistance over the three gate operating voltages shown in Figure 6.3(a); note that the dynamic resistance is relatively constant for small V_{DS} but increases as V_{DS} increases toward the onset of pinchoff. A slightly positive gate-source voltage V_{GS} can also be applied to the MESFET, which further reduces the dynamic resistance from the $V_{GS} = 0$ value, but with increased potential for forward biasing the Schottky gate and a resulting substantial gate current being drawn. The relatively high resistance occurring near the knee of the I-V curve could be used for attenuation applications but is usually avoided in practice since this is a highly nonlinear region of operation and can introduce spurious signals into the system, a topic for Section 6.5.1.

Using the same concept as the preceding example, a mathematical relationship for the channel resistance R_C yields the on-state RF resistance, $R_C = R_{ON}$, of the MESFET and can be derived by looking at the change of I_{DS} with respect to V_{DS} in (6.5). This computation shows the dependence of the various fabrication parameters on the MESFET resistance:

$$R_C = R_{ON} = \left(\frac{\partial I_{DS}}{\partial V_{DS}}\right)^{-1} = \left\{\frac{I_{DSS}}{V_P}\left[1 - \sqrt{\left(\frac{V_{bi} - V_{GS}}{V_P}\right)}\right]\right\}^{-1} \tag{6.6}$$

$$= \frac{V_P}{I_{DSS}} \frac{1}{\left[1 - \sqrt{\left(\frac{V_{bi} - V_{GS}}{V_P}\right)}\right]}$$

for small V_{DS}. Note as V_{GS} increases toward the pinchoff voltage, R_{ON} increases as shown in Figure 6.3(b). In RF switch applications, low on-state insertion loss using a series-connected MESFET (Figure 6.4) is needed, and so for an insertion loss of 0.4 dB in a 50Ω system, R_{ON} should be approximately 5Ω. This often requires a physically large perimeter MESFET with a large W/L aspect ratio (6.5). A MESFET with a 2.5V pinchoff voltage and V_{bi} of 0.8V is required to have an open channel current I_{DSS} of 1.15A for a 5.0Ω on-state resistance at zero gate-source voltage using (6.6). For a channel doping $N_D = 10^{17}$ cm^{-3}, the channel depth D is approximately 0.2 μm from (6.3). A MESFET aspect ratio of approximately 400 is required using (6.5b). For a 1.0 μm gate length L, the gate perimeter ($2W$) is nearly 0.8 mm.

A MESFET with this large aspect ratio often requires that the device be divided into a number of parallel elements with multiple gate fingers. This multiple gate finger layout of the device also has a positive impact on the overall capacitance of the device, a topic discussed in detail in MOSFETs in Chapter 5. The channel resistance is a function of V_{GS} as seen in (6.6). Since the source node is in the RF path, variation of V_{GS} by the applied RF voltage potentially causes a time-varying channel resistance. This time-varying channel resistance has ramifications in terms of nonlinear operation, a topic for Section 6.5.1.

Detailed study of Figure 6.1 shows that there are two other resistances in series with R_{ON} and hence in the RF path. These resistances are associated with the finite length between the gate-drain contact and the gate-source contact, L_{gd} and L_{gs}. These two resistances can be estimated from $R_{gX} = L_{gX}/q\mu_n N_D(D - h_s)$, where X is

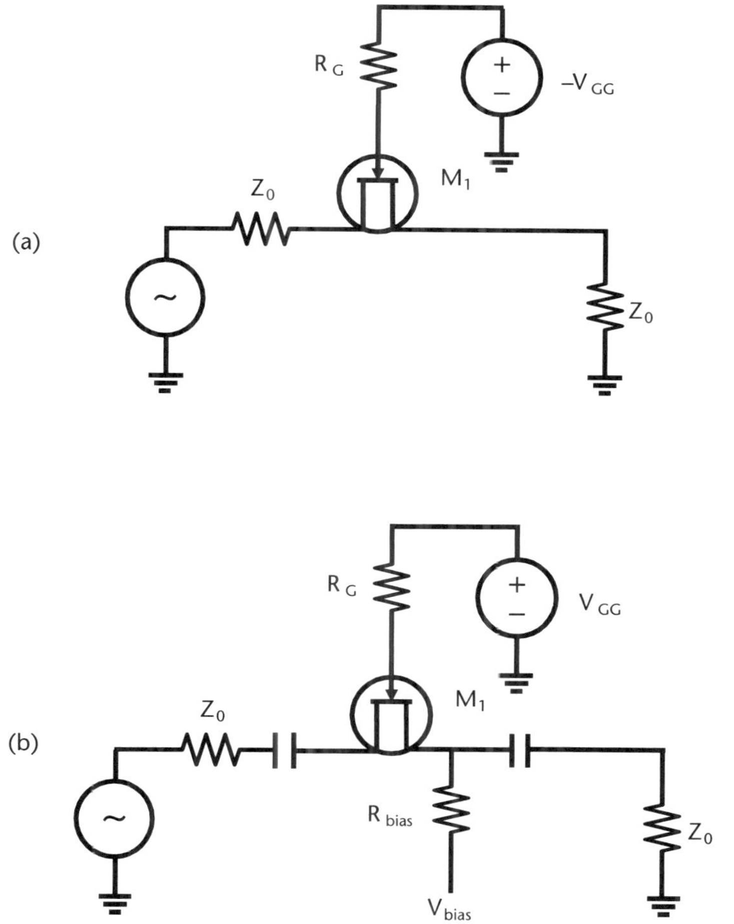

Figure 6.4 (a) Standard series-connected MESFET in a microwave and RF control application using a negative control voltage, V_{GG}. (b) Dc floated version of the series-connected MESFET using a positive control voltage, V_{GG} (substrate internally connected to ground for dc continuity). V_{bias} can be set externally for unipolar power supply operation. R_{bias} should be significantly larger than Z_0. The same circuit diagrams hold for HEMTs as well.

drain or source and h_s is the depletion thickness due to the surface potential (not shown) [20], which in many cases can be approximated as the pinchoff voltage, V_P [26]. In most MESFETs, these resistances are fractions of an Ohm.

6.2.3 RF Off-State Resistance

If the dc gate voltage, V_{GS}, exceeds the pinchoff voltage V_P, the Schottky depletion region completely extends through the channel, impeding current flow regardless of the polarity of the RF signal between drain and source, resulting in an effective dc open circuit between the two nodes. The value of the channel resistance R_C can now be considered infinite and the device is considered off. However, there are still RF paths around this channel due to device capacitances that can be lumped together into a single off-state capacitance value, C_{OFF}. A frequently cited figure of merit

(FOM) combines the on- and off-state characteristics into broadband switch cutoff frequency, $F_C = 1/2\pi R_{ON}C_{OFF}$ (or the associated broadband switch time constant, $T_C = R_{ON}C_{OFF}$). With typical on-state resistance of 5Ω and off-state capacitances of a few tenths of a picoFarad, F_C values for MESFETs can reach into hundreds of gigahertz range (with associated time constants in the femtosecond range). F_C is not, however, a measure of the operational upper frequency limit but simply a FOM that allows comparison between different devices. In this case, the higher FOM device will typically be usable at a higher frequency than the lower FOM device.

6.3 MESFET Capacitance Characteristics

The main factor limiting MESFET RF performance is the various capacitances that are an inherent part of the device structure (Figures 6.5 and 6.6). These MESFET capacitances scale with device gate width W, and since W is relatively large to keep the on-state resistance small, the capacitances can be significant. The main capacitances of interest in modeling the MESFET for RF control are the gate overlap (C_{GXO}) capacitances common in both on- and off-state as well as the capacitance associated with the Schottky depletion region. Capacitance modeling of the MESFET has been the focus of considerable work, with many different models proposed for the various regions of operation, with these references [20–27] being just a few. The focus in this chapter is on capacitance models developed primarily for MESFET and HEMT RF control device modeling, where operation is limited to applied dc bias only on the gate, with the drain and source seeing no applied dc bias (this is in contrast to the use of these devices in active operation such as amplifiers or oscillators, where at least the gate and drain are dc biased). This focus dramatically reduces the number of potential models available in the literature to discuss.

6.3.1 Intrinsic Device Capacitance Origin

In the on-state, C_i is the Schottky depletion region capacitance between the gate and the channel, and in the off-state, C_{GD} and C_{GS} are the gate-drain and gate-source capacitances associated with the depletion region voltages. In the off-state, C_{GD} and C_{GS} are functions of their respective terminal voltages (V_{GD} and V_{GS}, respectively), as indicated in the expressions for these capacitances [27].

$$
\begin{aligned}
C_i &= \frac{\varepsilon_s WL}{D - h\left(V_{GS} = 0\right)} \\
C_{GXO} &= \varepsilon_s W \frac{K\left(1 - k^2\right)}{K(k)} \\
C_{GX} &= \varepsilon_s W \tan^{-1}\left(\sqrt{\frac{\left|V_P\right|}{V_{bi} + \left|V_P\right| - V_{GX}}}\right)
\end{aligned}
\tag{6.7}
$$

where the X is either drain (D) or source (S), $k = \sqrt{L_{gX}/(L_{gX} + L)}$ (L less than L_{gX}) and K is the complete elliptic integral of the first kind [20]. For the example device presented in Section 6.2.2 and assuming a GaAs MESFET (Table 6.1) with gate-drain and gate-source separations of 3 μm (k = 0.866), and a zero bias depletion depth of 0.2 μm, C_i and C_{GXO} from (6.7) are approximately 0.2 pF and 0.04 pF, respectively. For small RF voltages, C_{GX} is approximately 0.03 pF.

6.3.2 RF Equivalent Circuit

The individual circuit elements described in the previous sections can be combined to create an RF equivalent circuit for the MESFET control device (Figures 6.5(a) and 6.6(a)). There are a number of operational concepts that arise when studying this equivalent circuit. For low-frequency operation in the on-state, the capacitive reactances are large and the path that the RF current follows through the drain (D) to the source (S) goes through the channel resistance $R_C = R_{ON}$, ultimately governing the low-frequency insertion loss or isolation (depending on the overall circuit topology). As the frequency increases, the capacitive reactances start to connect the main D-S path to the gate node, impacting the frequency response. If the gate is directly connected to a dc control source to govern the switching action, then the two gate capacitances (C_{GS} and C_{GD}) shunt RF straight to ground since the gate node is at RF ground (Figures 6.5(a) and 6.6(a)). In the off-state, the inverting channel is removed and R_{ON} becomes essentially an open circuit; the low-loss RF path is therefore removed with drain-source transmission through the gate capacitance

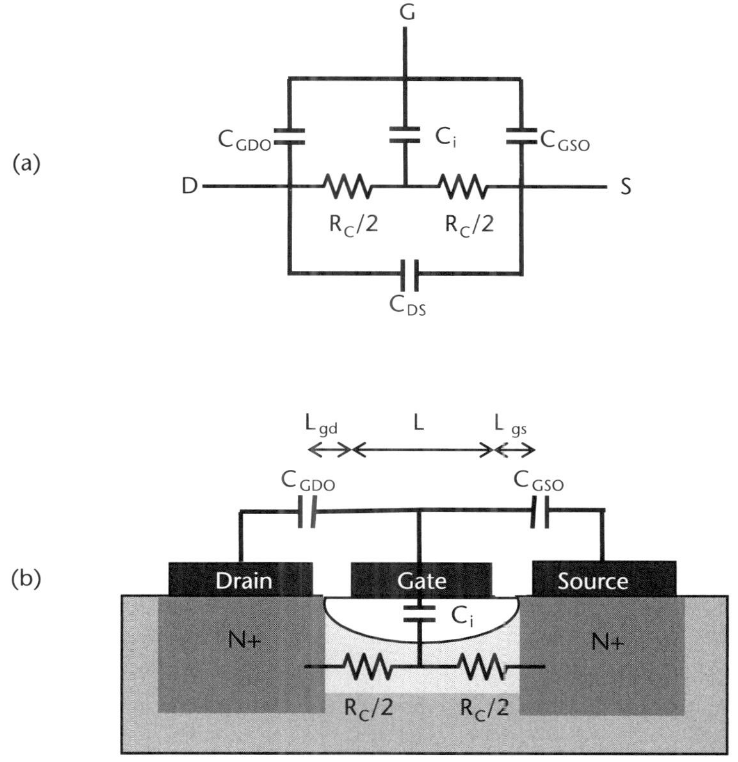

Figure 6.5 (a) RF equivalent circuit for the on-state MESFET control element. (b) Origin of the on-state equivalent circuit elements.

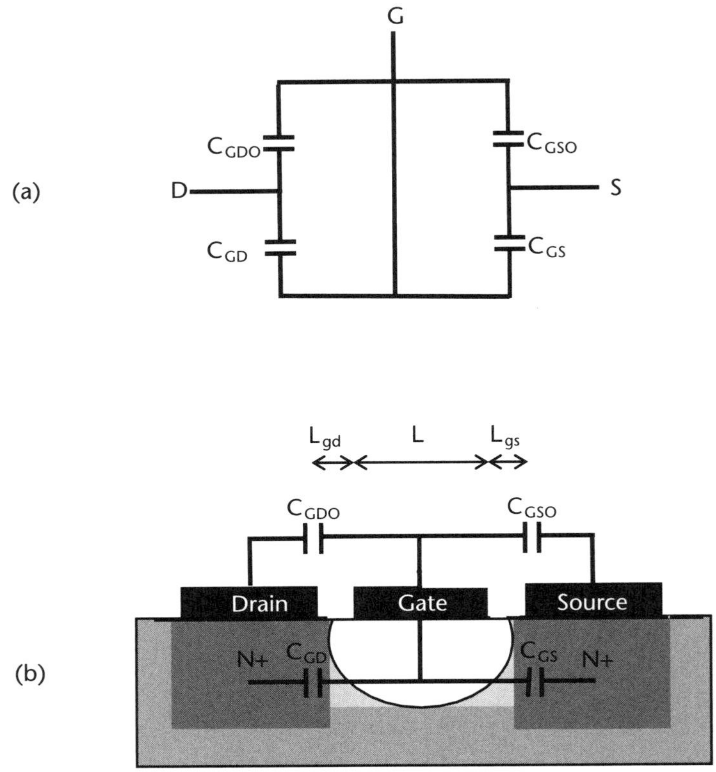

Figure 6.6 (a) RF equivalent circuit for the off-state MESFET control element. (b) Origin of the off-state equivalent circuit elements.

path (C_{GS}, C_{GD}). The high-frequency performance can be improved if an external resistor, R_G, is added in series with the dc gate voltage to "float" the gate. This R_G for MESFET/HEMT control devices is usually an on-chip lumped element acting to isolate the dc gate bias supply from the RF path. A gate bias circuit cutoff frequency can then be defined as $F_G = 1/2\pi R_G(C_{GS} + C_{GD})$ or its equivalent, a gate bias circuit time constant, $T_G = R_G(C_{GS} + C_{GD})$ and is on the order of a several hundred kilohertz into the megahertz range (microseconds) in high-quality MESFET/HEMT control devices. Details of the impact of R_G on RF performance are discussed in the next section.

One important feature of the MESFET-based RF control device is the need for a negative gate voltage to pinch off the channel, as opposed to a positive control voltage to control the inversion layer in the MOSFET (Chapter 5). Many systems, especially those that are battery powered, have unipolar supplies (of usually positive voltage) to reduce system complexity and cost. The use of a positive unipolar power supply requires that the MESFET-based RF control device be dc "floated" with capacitors so that a positive voltage can be used to switch the MESFET (Figure 6.4(b)). A high-value resistance R_{bias} can connect the source node to an external bias supply for use with positive gate control voltages. These capacitors need to exhibit very good high-frequency response as well as be of sufficient value to exhibit low reactance at the lowest desired frequency of operation. DC floating the RF module does, however, prevent the device being used down to very low frequencies or even dc. It should be noted that some commercial MESFET/HEMT RF control modules have a power supply inverter than converts the positive unipolar supply to the required negative control potential.

6.3.3 Packaging Considerations

Just as in the case of PIN diodes and MOSFETs (covered in Chapters 4 and 5), there are many different types of packages used in MESFET/HEMT control device technology with a wide variety of abbreviations. For surface mounting, there are small outline packages (SOICs such as SOP and SOT, for example) and a variety of other definitions used by various manufacturers, each with their own advantages in terms of RF and thermal performance. Plastic and ceramic packages are available, with the plastic package frequently chosen for low cost and high volume usage; ceramic packages are used for very high frequencies because of their lower loss. The package parasitics between the RF switch module and the package are the bonding wire(s) inductance(s) and the parasitic package capacitances. The same general rule of thumb in computing the inductance of the bond wire is approximately 1 nH per millimeter of wire length, as presented in Chapter 2. Multiple bond wires may be employed to reduce the inductance on each connection. Since similar package styles are used for PIN diodes and MOS switches, typical values of the bond wire inductance range from 0.1 to 0.3 nH, with package capacitances ranging from 0.1 to 0.2 pF. Manufacturer or fabrication/packaging houses can provide estimates of these parasitics.

6.3.4 Gate Resistance, R_G

The gate resistance R_G, and the associated gate bias circuit cutoff frequency $F_G = 1/2\pi R_G(C_{GS} + C_{GD})$, has an important influence on the MESFET/HEMT control device operation. At low frequencies in the on-state, the gate capacitive reactances are much higher than the value of R_G and so the gate is not a floating node. The gate-source voltage V_{GS} in this case is approximately equal to $-Z_0 I_{DS}$, implying that the gate-source voltage follows the RF signal drive level. This RF signal is superimposed on the dc gate bias voltage V_{GS0} and could, at large RF levels, potentially drive the instantaneous gate voltage into device turnoff. This low-frequency gate bias circuit effect therefore governs the power handling capacity of the MESFET/HEMT control device at low frequencies. Increasing the dc gate bias voltage will increase the power handling but only up to a point, since there is a maximum gate voltage that can be applied to the MESFET/HEMT before the Schottky diode goes into forward conduction for the bulk MESFET or the parasitic MESFET is activated in the HEMT. As the frequency increases above F_G, the gate node becomes a floating node, since R_G is much greater than Z_0, and V_{GS} is half the voltage drop across the drain-source connection, V_{DS}, which is small since R_{ON} is much less than Z_0 as required for low insertion loss. In this case, the power limitation is governed by how much current the MESFET/HEMT can pass for a given V_{GS0} while still remaining in the triode/linear region of operation for that dc gate voltage.

In the off-state, the conducting channel disappears (R_{ON} ideally becomes infinite) and high isolation occurs. At low frequencies, the entire applied RF voltage is essentially dropped as the drain-source voltage V_{DS} of the MESFET/HEMT, with the drain node at the applied RF peak voltage, therefore making that node subject to any voltage breakdown effects. The source voltage is very small because of the high isolation. As the frequency increases, RF coupling from the source to the load through the two gate capacitances occurs, causing the isolation to degrade. The gate-source voltage, V_{GS}, approaches $Z_0 I_{DS}$ and so tracks the applied RF voltage amplitude with similar impacts of the instantaneous RF swing approaching V_T as in the conducting state. Since the drain node is still at high RF potential throughout the high isolation state, breakdown at this junction could also occur if the instantaneous RF voltage exceeds the drain breakdown voltage.

At first glance, it appears that F_G can be made as arbitrarily low as possible to garner the better power handling capability at frequencies above F_G. R_G cannot be too high a value, however, because the gate bias circuit capacitances have to be charged or discharged to turn the device on or off, potentially slowing the switching speed of the device if R_G is set too high. Added gate bias resistances R_G are typically in the range of 2 to 10 kΩ.

6.3.5 Equivalent Circuit Simulation

The previous discussion outlined an approach to estimate the important resistance and capacitances in the MESFET equivalent circuit that can be used to model the RF transfer characteristics, insertion loss, and isolation in an RF and microwave control application. In this section, these values, summarized in Table 6.2, are programmed into a Matlab script, 'mesfet_switch.m', and the RF transfer characteristics swept over the frequency range 1.0 to 1000 MHz using the equivalent circuit diagrams shown in Figure 6.7(a). The results of this simulation, shown in Figure 6.7(b), show that the insertion loss is constant at both low and high frequencies with a transition region in the vicinity of the gate bias circuit cutoff frequency, F_G (approximately 30 MHz in this example). There will be insertion loss degradation at frequencies well above 1000 MHz as other parasitic capacitances, primarily between drain/source and ground, come into play; these parasitics are not part of the basic MESFET RF equivalent circuit but are part of the Matlab script. The isolation monotonically decreases from greater than 140 dB at 1.0 MHz down to below 35 dB at 1000 MHz as the various capacitances in the RF equivalent circuit shunt RF around the channel.

Table 6.2 Summary of MESFET Simulation Parameters

Symbol	*Parameter*	*Value*
V_P	pinchoff voltage	–2.5V
I_{DSS}	open channel current	1000 mA
C_i	Schottky gate capacitance	0.1 pF
C_{GS}, C_{GD}	gate-source, gate-drain capacitance	0.05 pF

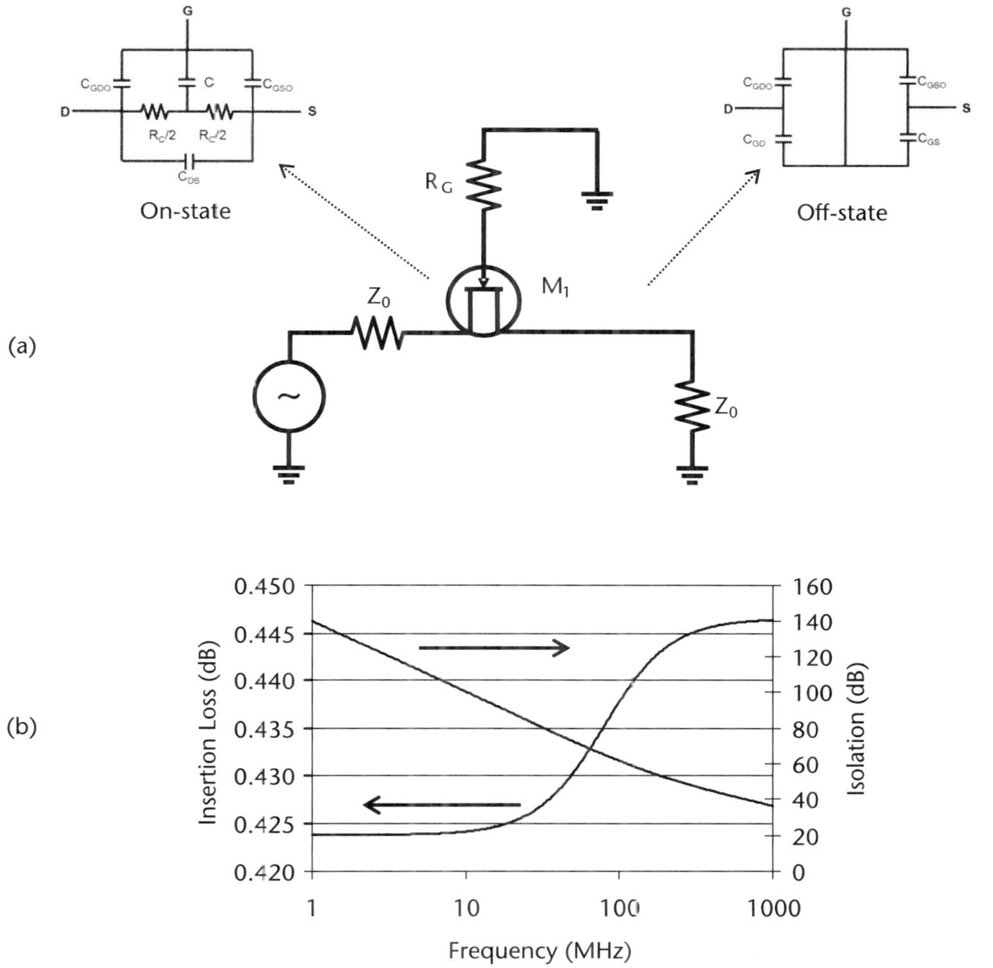

Figure 6.7 (a) The ac equivalent circuit for the MESFET RF switch simulation. (b) Insertion loss and isolation of the example switch circuit.

6.4 HEMT Technologies

The high electron mobility transistor (HEMT) has been the subject of intense research and commercial interest since the early 1980s. Just by its very name, the HEMT exhibits a higher electron mobility than found in bulk channel technologies. The fabrication of these high-frequency devices, however, requires tight fabrication control since the conducting region is only achieved by creating a heterostructure between materials with different energy band gaps, such as the AlGaAs/GaAs and the AlGaN/GaN systems. These heterostructures create an interface two-dimensional electron gas (2DEG) at the metallurgical junction (Figure 6.8). The electrons in this 2DEG are of very high mobility due to reduced scattering in the undoped layer, and when coupled with a high carrier density in the 2DEG, a very thin but highly conducting layer is created. Both enhancement and depletion mode HEMTs can be created, primarily based on the thickness of the two AlGaAs layers, $d + d_i$ [28]. If

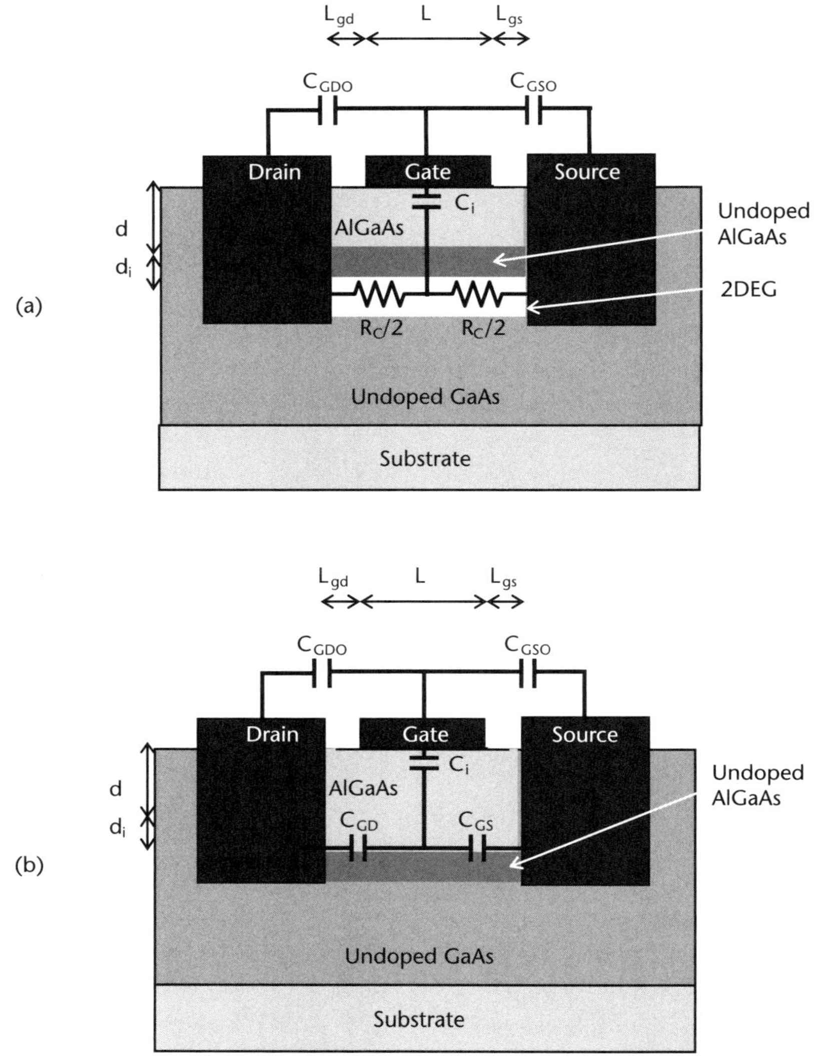

Figure 6.8 Simple AlGaAs/GaAs HEMT cross section and equivalent circuit parameters: (a) on-state (with 2DEG), (b) off-state (with 2DEG suppressed). Replace AlGaAs/GaAs layers with AlGaN/GaN for GaN-based technology.

$d + d_i$ is thin, the depletion region associated with the Schottky gate junction will extend through both the AlGaAs layer and the 2DEG, interrupting current flow, in which case a positive V_{GS} is required to reduce the depletion region to allow current flow. In this case, the HEMT device is said to be operating in enhancement mode. If $d + d_i$ is thick, the depletion region does not extend to the 2DEG and current can then flow through the 2DEG; the device is then said to be operating in depletion mode. The carrier density in this 2DEG layer during depletion mode operation is modulated by the application of a negative gate voltage V_{GS}, in turn, modulating I_{DS}. The analysis in this section assumes depletion mode operation. Another interesting structural phenomenon in HEMTs can be seen by further study of the

Table 6.3 Table of Low Field Mobility and Saturation Velocity for the 2DEG in GaAs [31] and GaN [32] Heterostructure Systems

Technology	*AlGaAs/GaAs [31]*	*AlGaN/GaN [32]*
μ_{LF} (m^2/V-sec)	0.68	0.15
v_{sat} (m/sec)	2×10^5	2.5×10^5

heterostructure in Figure 6.8. Although there is the 2DEG channel layer between the drain and source, there is also a MESFET-like structure in the heavily doped region under the gate that can be activated with a high enough gate voltage to open the otherwise fully depleted channel under normal operation [29–30]. The discussion in this section assumes this parasitic MESFET is not activated and that HEMT operation is purely related to the conducting characteristics of the 2DEG. Table 6.3 shows some characteristics of two popular heterostructure systems that will be used in the derivation of the I-V characteristics [31, 32].

The I-V characteristic for the HEMT is derived in a similar fashion as the MOSFET (outlined in detail in Chapter 5); the carrier density in the 2DEG above threshold is a function of the capacitance per unit area of the gate-2DEG structure and the gate-source voltage V_{GS} [29]:

$$n_{2DEG} = \frac{\varepsilon_i}{\left(d + d_i\right)}\left(V_{GS} - V_{Off}\right) \tag{6.8}$$

where V_{Off} (a negative value) depends on the fabrication technology and d and d_i are the thicknesses of the various layers shown in Figure 6.8. The I-V characteristics of the HEMT are also strongly influenced by the field dependent mobility in the 2DEG. High electric fields will occur in the conducting channel layer during control of high-power signals because of narrow gate lengths, with a corresponding reduction in the channel mobility with increasing voltage drop across the drain-source. A simple model for the velocity-electric field profile is assumed, and the expression for the mobility is derived from the derivative of this profile [15]. In the 2DEG, a simple approximation for the drain-source 2DEG channel electric field is V_{DS}/L [33], yielding a 2DEG channel mobility versus channel voltage relation of form

$$\mu_n(V) = \frac{\mu_{LF}}{\left(1 + \dfrac{\mu_{LF} V_{DS}}{L v_{sat}}\right)} \tag{6.9}$$

where μ_{LF} is the low field mobility value from Table 6.3. By assuming a linear variation of voltage in the channel (similar to the derivation of the MOSFET I-V characteristics in Chapter 5) and the mobility relation in (6.9), the nonlinear drain-source current can be written as [30, 34]

$$I_{DS} = \frac{W}{L}\frac{\varepsilon_i}{d + d_i}\frac{\mu_{LF}}{\left(1 + \dfrac{\mu_{LF} V_{DS}}{L v_{sat}}\right)} V_{DS}\left(V_{GS} - V_{Off} - 0.5V_{DS}\right) \tag{6.10a}$$

or, for small V_{DS} (low channel field V_{DS}/L),

$$I_{DS} = \frac{W}{L}\frac{\mu_{LF}\varepsilon_i}{d + d_i} V_{DS}\left(V_{GS} - V_{Off}\right) \tag{6.10b}$$

where V_{Off} and V_{GS} are both negative. Equation (6.10) shows a form that is different from the bulk MEFET I-V characteristics in (6.5), indicative of the operation of the two devices being quite different. For bulk MESFETs, the gate voltage V_{GS} is used to vary the channel cross section via the Schottky depletion region to modulate the current flow through the device, whereas in the depletion mode HEMT, the gate potential V_{GS} varies the 2DEG charge density, thereby modulating the drain-source current flow.

6.4.1 HEMT On-State Resistance

When there is drain-source current flow, the HEMT is considered in the on-state, and the on-state channel resistance can be derived in a similar manner as (6.6):

$$R_C = R_{ON} = \left(\frac{\partial I_{DS}}{\partial V_{DS}}\right)^{-1} = \frac{1}{\left[\frac{W}{L}\frac{\mu_{LF}\varepsilon_i}{d + d_i}\left(V_{GS} - V_{Off}\right)\right]} \tag{6.11}$$

assuming low level RF fields. Equation (6.11) shows that, just as in the case of the bulk MESFET, a large aspect ratio W/L device operated far from the 2DEG suppression voltage provides a low on-state resistance; larger capacitances also are created with the larger aspect ratio devices. However, the high mobility inherent to HEMT operation reduces the need for extremely large aspect ratios, thereby keeping the capacitance low and providing for operation at higher frequencies.

6.4.2 HEMT Capacitance Characteristics

The main factor limiting HEMT RF performance is the various capacitances that are an inherent part of the device structure (Figure 6.8) and scale with device gate width W. The main capacitances of interest in modeling the HEMT for RF control in the on-state are the Schottky gate capacitance due to the fully depleted layer under the gate (C_i) and the gate overlap capacitances that make up the gate source and capacitances, C_{GS} and C_{GD}, as shown in (6.12):

$$C_i = \frac{\varepsilon_s}{d + d_i}\cdot WL \tag{6.12}$$

$$C_{GXO} = \varepsilon_i W \frac{K\left(1 - k^2\right)}{K(k)}$$

where the X is either drain or source and K and k were defined as part of (6.7) [20, 35].

In the off-state, the gate overlap capacitances C_{GXO} are still present, but the 2DEG has now been suppressed, so that there is now capacitive coupling between the gate and the drain/source nodes, similar to the bulk MESFET capacitance, since this region is still fully depleted. These two capacitances are given by the C_{GX} terms, here repeated from (6.7):

$$C_{GX} = \varepsilon_s W \tan^{-1}\left(\sqrt{\frac{|V_{off}|}{V_{GX} - V_{Off}}}\right) + C_{GXO} \tag{6.13}$$

where the X is either drain or source and $|V_{GX}|$ is assumed to be greater than $|V_{Off}|$. The same MESFET RF equivalent circuits for the on- and off-state as shown in Figures 6.5 and 6.6 can be used for the HEMT control device with a simple change in the capacitance values, using (6.12) and (6.13) instead of (6.7).

6.5 Detailed MESFET/HEMT Control Device Characteristics

In the design of circuits using GaAs MESFETs, a key design specification is the level of distortion introduced into the system. The GaAs MESFET can introduce distortion into a circuit in both switch states, with the primary distortion caused by resistive nonlinearities in the on-state and capacitance nonlinearities in the off-state. This section will cover both mechanisms.

6.5.1 Nonlinear Operation in the On-State MESFET/HEMT

Expressions for the full I-V characteristics of MESFETs and HEMTs, (6.5b) and (6.10a), show that the applied RF signal will modulate the channel resistance through the time-varying nature of V_{GS} and V_{DS}, making R_{ON} a function of time and therefore potentially generating spurious signals at the load. Using the same technique as in Chapter 4's discussion of nonlinearities in PIN diodes or MOSFETs in Chapter 5, a Taylor series expansion can be used for defining the I_{DS}-V_{DS} characteristic for the device as powers of V_{DS}:

$$I_{DS} = \sum_{n=1}^{\infty} \alpha_n V_{DS}^n \tag{6.14}$$

where the α_n are the expansion coefficients determined by the series expansion of the I_{DS}-V_{DS} characteristic. The gate bias circuit, composed of C_{GS}, C_{GD}, and R_G, links the RF voltages V_{GS} and V_{DS} and can be described by the ratio V_{GS}/V_{DS} and given the symbol γ [36]. The only dc signal in the circuit is assumed to be the gate control signal, V_{GS0}. Assuming a series connected MESFET/HEMT switch element and a gate bias resistance much larger than the on-state resistance of the MESFET/HEMT, the term γ can be written with the aid of the RF equivalent circuit in Figure 6.5(a) as [36]

$$\gamma = \frac{V_{GS}}{V_{DS}} = -\frac{\frac{Z_0}{R_{ON}} - j\omega R_G C_{DG}}{1 + j\omega R_G\left(C_{DG} + C_{GS}\right)} \tag{6.15}$$

where Z_0 is the source/load impedance and is typically 50Ω. The I_{DS}-V_{DS} relationships between the MESFET and the HEMT are different, and so their associated series expansion coefficients will be different as well. For the MESFET, the α_n in (6.15) can be found through the Taylor series expansion of (6.5a) as [37]

$$\begin{aligned}
\alpha_1 &= \frac{I_{DSS}}{|V_P|}\left(1 - \sqrt{\frac{V_{bi} - V_{GS0}}{|V_P|}}\right) = 1/R_{ON} \\
\alpha_2 &= -\frac{1-2\gamma}{2}\frac{I_{DSS}}{|V_P|^2}\sqrt{\frac{|V_P|}{\left(V_{bi} - V_{GS0}\right)}} \\
\alpha_3 &= \frac{1}{4}\frac{I_{DSS}}{|V_P|^3}\left(\frac{|V_P|}{\left(V_{bi} - V_{GS0}\right)}\right)^{1.5}\left(1 - 3\gamma + 3\gamma^2\right)
\end{aligned} \tag{6.16}$$

and for the HEMT using (6.10a) as

$$\begin{aligned}
\alpha_1 &= \frac{W}{L}\frac{\mu_{LF}\varepsilon_i}{\left(d + d_i\right)}\left(V_{GS0} - V_{Off}\right) = 1/R_{ON} \\
\alpha_2 &= -\frac{W}{L}\frac{\mu_{LF}\varepsilon_i}{\left(d + d_i\right)}\frac{1+2\gamma}{2} - \frac{W}{L}\frac{\mu_{LF}\varepsilon_i}{\left(d + d_i\right)}\frac{\mu_{LF}}{v_{sat}L}\left(V_{GS0} - V_{Off}\right) \\
&= -\frac{W}{L}\frac{\mu_{LF}\varepsilon_i}{\left(d + d_i\right)}\frac{1+2\gamma}{2} - \alpha_1\frac{\mu_{LF}}{v_{sat}L} \\
\alpha_3 &= \frac{W}{L}\frac{\mu_{LF}\varepsilon_i}{\left(d + d_i\right)}\frac{1+2\gamma}{2}\frac{\mu_{LF}}{v_{sat}L}
\end{aligned} \tag{6.17}$$

In both cases, the term α_1 is the linear term, $1/R_{ON}$, while the remaining values are the two nonlinear device coefficients that give rise to the second and third order device nonlinearities.

The distortion produced by the MESFET/HEMT control element is dependent on both the conducting properties of the channel under the gate in terms of magnitude and on the gate bias circuitry in terms of frequency dependence [34, 37–38], as shown by the expansion coefficients in (6.16) and (6.17). The distortion circuit, however, is similar to that of the MOSFET control element, and so the same general expressions for load or output-referenced IP2 and IP3 can be used for the series reflective FET-based switch in terms of the α_n expansion parameters [34, 37]:

$$\mathrm{IP2} = \frac{Z_0}{2R_S^4\alpha_2^2}\left(1 + \frac{2Z_0}{R_S}\right)^2 \tag{6.18}$$

$$\mathrm{IP3} = \frac{Z_0}{2R_S^3\alpha_3}\left(1 + \frac{2Z_0}{R_S}\right)$$

Figure 6.9(a) shows the results of calculations using this model in determining the on-state second and third order distortion intercept point (IP2 and IP3, respectively) for the example MESFET with I_{DSS} of 1000 mA and a 10 kΩ gate bias resistor being used as a series-connected RF switch element (Section 6.2). Note that the intercept point is lower (higher level of distortion) at lower frequencies, but the intercept point increases (distortion decreases) as the frequency increases. The transition point occurs near the gate bias circuit cutoff frequency $\omega R_G(C_{GD} + C_{GS}) = 1$. The intercept points for the MESFET and MOSFET are nearly identical, although the MESFET does exhibit slightly improved distortion performance. It should be noted in the MESFET calculation that the only capacitances are in the gate bias circuit; there are no capacitances from drain or source to ground. These gate-ground and drain-ground capacitances will influence the high-frequency roll-off, and a degradation in both insertion loss and distortion performance will occur. This approach for on-state distortion modeling of MESFET-based RF switches shows good agreement with experimental results [37].

Distortion modeling of HEMT-based RF switches can be performed by simply replacing the α_n in (6.16) with those values in (6.17). For a HEMT with similar on-state resistance as the MESFET (and hence similar insertion loss) in the preceding

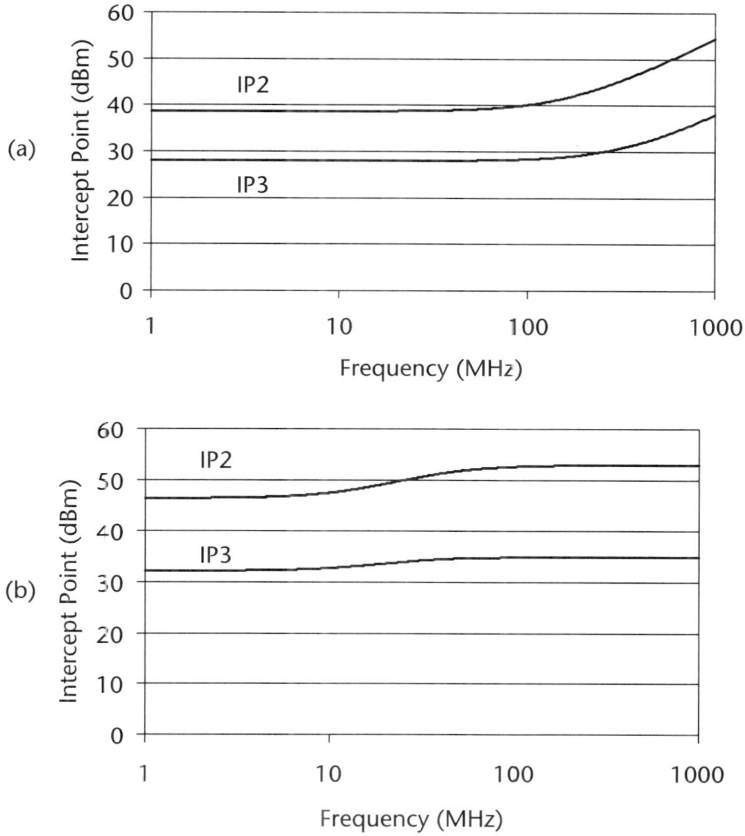

Figure 6.9 (a) On-state load-referenced distortion intercept points for MESFET RF series-connected RF switch. (b) On-state load-referenced distortion intercept points for HEMT RF series-connected RF switch.

example, the distortion intercept point exhibits significantly different behavior, as shown in Figure 6.9(b). For the HEMT, the distortion intercept points do not dramatically differ about the gate bias circuit cutoff frequency as observed in MESFETs. This phenomenon can be explained by noting that the gate-drain and gate-source circuit couples RF voltage to the gate, causing the 2DEG carrier density (6.8) itself to vary with the applied RF signal [34]. This effect is in direct contrast to the MESFET switch, where the Schottky depletion layer varies with the ac signal and the channel carrier density remains constant. The increase in intercept point associated with the gate bias circuit for MESFETs is still there, but at the same time, the 2DEG is being suppressed by the RF signal, partially compensating for the IP increase and yielding the relatively frequency-independent intercept point [34]. The MESFET exhibits substantially improved distortion performance compared with the HEMT well above the gate bias circuit cutoff frequency; this improvement will be noted until the natural switch roll-off frequency begins to degrade the distortion performance in both devices.

6.5.2 Nonlinear Operation in the Off-State

In the off-state, the Schottky depletion region extends through the channel into the semi-insulating substrate in MESFETs, or suppresses the HEMT 2DEG charge density in the channel; both cause the conducting channel to disappear, making R_{ON} infinite. However, the depleted regions in both devices are still in the RF path, and the voltage variations on the drain and source nodes cause the associated capacitances to become time-varying and hence generate distortion. Distortion currents are also generated by these capacitances in the on-state but are much smaller than those due to distortion generated by time-varying channel resistance. In a series connected circuit in the off-state, the voltage on the drain side of the MOS switch is nearly the same as the generator voltage due to the high isolation of the switch and therefore will generate a higher level of distortion current than the source side because of the high isolation. However, the high isolation afforded by the off-state FET also isolates the drain side generated distortion currents from the load, reducing the overall level of load distortion current. A similar approach to distortion modeling as performed with PIN diodes in Chapter 4 and MOSFETs in Chapter 5 can be done by looking at the voltage dependence of the drain and source node capacitances. Using a Taylor series expansion of the gate-drain or gate-source capacitance from (6.7) or (6.13), the first three terms of the expansion of the time-varying capacitance are [39]

$$
\begin{aligned}
C_{\mathrm{GX}}(t) &= \varepsilon_s W \tan^{-1}\left(\sqrt{\frac{V_P}{V_{bi} + V_P - V_{\mathrm{GX}}(t)}}\right) + C_{\mathrm{GXO}} = \sum_{k=0}^{\infty} c_k V_{\mathrm{GD(GS)}}(t)^k \\
c_0 &= C_{\mathrm{GXO}} \\
c_1 &= \frac{\varepsilon_s W}{2} \frac{1}{V_{bi} + V_P - V_{\mathrm{GX}}} \sqrt{\frac{V_P}{V_{bi} + V_P - V_{\mathrm{GX}}}} \\
c_2 &= \frac{3}{2}\frac{\varepsilon_s W}{2}\left(\frac{1}{\left(V_{bi} + V_P - V_{\mathrm{GX}}\right)^2}\right)\sqrt{\frac{V_P}{V_{bi} + V_P - V_{\mathrm{GX}}}}
\end{aligned}
\tag{6.19a}
$$

for the MESFET and

$$C_{GX}(t) = \varepsilon_s W \tan^{-1}\left(\sqrt{\frac{V_{Off}}{V_{GX} - V_{Off}}}\right) + C_{GXO} = \sum_{k=0}^{\infty} c_k V_{GX}(t)^k$$

$$c_C = C_{GXO}$$

$$c_1 = \frac{\varepsilon_s W}{2}\frac{1}{V_{GX} - V_{Off}}\sqrt{\frac{V_{Off}}{V_{GX} - V_{Off}}} \tag{6.19b}$$

$$c_2 = \frac{3}{2}\frac{\varepsilon_s W}{2}\left(\frac{1}{\left(V_{GX} - V_{Off}\right)^2}\right)\sqrt{\frac{V_{Off}}{V_{GX} - V_{Off}}}$$

for the HEMT. Equation (6.19) shows that harmonic currents generated in the as-fabricated MESFET/HEMT can be reduced only by increasing the reverse bias on the gate electrode, V_{GS0}. The lower limit of V_{GS0} is the pinchoff voltage value to actually turn the MESFET off, and the gate breakdown voltage at the upper limit [40]. An appropriate bias point for both distortion and power handling would lie approximately midway between these two voltages [20, 40].

The linear ($k = 0$) and nonlinear ($k \geq 1$) gate-drain and gate-source currents can be written in terms of the c_k and the drain/source voltage using a similar form as (4.25)

$$I_{GD(GS)}(t) = \frac{d}{dt}\sum_{k=0}^{\infty} c_k V_{GD(GS)}(t)^{k+1} = \sum_{k=0}^{\infty} c_k \frac{dV_{GD(GS)}(t)^{k+1}}{dt} \tag{6.20}$$

Figure 6.10 shows the results of calculations using this capacitance model in determining the input-referenced off-state second and third order distortion intercept for the previously used example series-connected MESFET RF switch element and a 10 kΩ gate bias resistor [39]. Note that the input-referenced intercept point is higher (lower level of distortion) at lower frequencies, but the intercept point steadily decreases as the frequency increases. This increase is due to both an increase in the distortion current level with increasing frequency and also more coupling of the drain distortion onto the load as the isolation falls off with increasing frequency. The change in slope near 30 MHz is in the vicinity of the gate bias circuit cutoff frequency, F_G. The relatively constant intercept point at frequencies above F_G can be explained using (6.19). For example, on the positive swing of the source voltage, there is an instantaneous decrease in the value of C_{GD} and a corresponding instantaneous increase in C_{GS}. The resulting distortion currents partially cancel each other and track with frequency since V_{DG} and V_{GS} are comparable in magnitude above F_G. Conversely, at frequencies below F_G, the drain distortion current is effectively isolated from the load by the high switch isolation; this same isolation lowers the RF voltage at the source and hence its overall distortion at the load. The lower intercept points at frequencies above F_G occur because of the increased frequency and hence increased distortion current magnitude (6.20) [39]. The HEMT will exhibit similar behavior but at a slightly different IP2/IP3 level due to the different capacitance-voltage characteristic as shown in (6.19). Figure 6.10 shows input-referenced

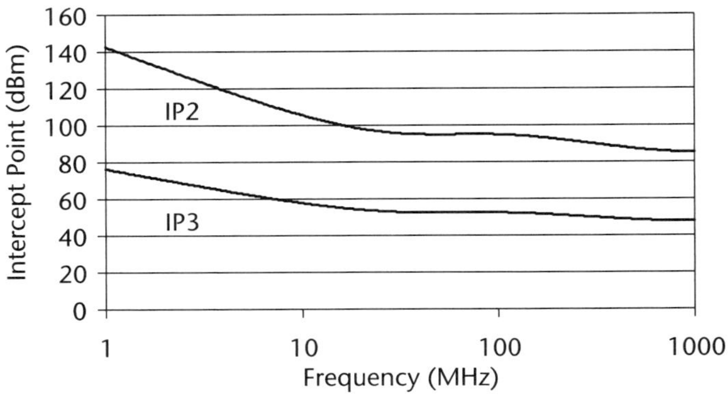

Figure 6.10 Input referenced IP2 and IP3 versus frequency for the example RF control MESFET biased at twice the pinchoff voltage. Note that at frequencies above approximately 30 MHz, the intercept points are relatively constant with frequency.

intercept points; Figure 5.8(b) for the MOSFET RF control device shows output or load-referenced intercept points. The two different intercept point references were displayed to show the differences between the two. These two intercept points are related by the gain (or in this case, the isolation) of the control device, which was presented in Chapter 2.

6.6 SPICE Modeling

There are many different simulators available for the RFIC designer, but a widely used "standard" simulator is SPICE [41]. In fact, almost all simulators (free, share, and proprietary) use a very similar set of parameters for modeling MESFETs and HEMTs as more generic SPICE simulators, and so in keeping with the idea of a generic simulator, this discussion will focus on the generic SPICE definitions.

6.6.1 SPICE MESFET (Statz) Model

One of the early problems with MESFET simulation using SPICE involved agreement with mathematical approaches such as outlined by Sze [15] based on the junction FET (JFET) derivations and measurements of MESFETs through the linear and saturation regions [21]. Curtice proposed using a hyperbolic tangent (tanh) form in SPICE with adjustable parameters for the I_{DS}-V_{DS} relationship because tanh is a smoothly varying function from zero to its asymptotic maximum and can be used to model a smooth transition from linear to saturation [21]:

$$I_{DS}\left(V_{GS}, V_{DS}\right) = \beta\left(V_{GS} + V_T\right)^2\left(1 + \lambda V_{DS}\right)\tanh\left(\alpha V_{DS}\right) \tag{6.21}$$

where α, β, and λ are constants to be fit to the measured data, with λ usually associated with the output drain conductance. Statz and colleagues [22] found that the Curtice I-V model in (6.21) provided a good fit to measured data for a given V_{GS}, but that (6.21) did not perform as well if the MESFET V_P was large. In addition, Statz and colleagues found that the tanh function consumed considerable computing resources. Using the Curtice model as a starting point, they proposed an updated current-voltage expression for better data fit and faster SPICE computation [22] for low and high V_{DS}:

$$I(V_{GS}, V_{DS}) = \beta \frac{(V_{GS} - V_T)^2}{1 + b(V_{GS} - V_T)}(1 + \lambda V_{DS})\left[1 - \left(1 - \frac{1}{3}\alpha V_{DS}\right)^3\right] \quad (6.22a)$$

$$0 < V_{DS} < \frac{3}{\alpha}$$

$$I(V_{GS}, V_{DS}) = \beta \frac{(V_{GS} - V_T)^2}{1 + b(V_{GS} - V_T)}(1 + \lambda V_{DS}) \quad V_{DS} > \frac{3}{\alpha} \quad (6.22b)$$

The device capacitances were also presented based on a model of the total gate charge as a function of V_{GS} and V_{GD}. The Statz model is the standard model in SPICE [41], with the n-channel MESFET using the model keyword "NMF" in SPICE. An abbreviated list of SPICE3 model parameters based on the Statz model is shown in Table 6.4.

For the RF control circuit designer, it is useful to see how to link the I_{DS}-V_{DS} relationships derived earlier and shown in (6.6) with the Statz I_{DS}-V_{DS} model in (6.22). Using the Statz model [22] in (6.22a), since that form is for small V_{DS}, the dynamic resistance for small V_{DS} can be derived using the same approach as previously in the chapter:

$$\frac{1}{R_{ON}} = \frac{\partial I_{DS}}{\partial V_{DS}} \approx \alpha\beta \frac{(V_{GS} - V_T)^2}{1 + b(V_{GS} - V_T)} \quad (6.23)$$

which, with the additional assumption that $b = 0$, shows $1/R_{ON}$ to be approximately $\alpha\beta(V_{GS} - V_T)^2$. Equating this expression to (6.6) shows $\alpha\beta(-V_T)^2$ to be approximately equivalent to I_{DSS}/V_P at zero gate voltage. From Figure 6.3 for $V_{GS} = 0$, linear operation of the MESFET is observed for V_{DS} less than approximately 1.0V,

Table 6.4 Abbreviated Table of SPICE Statz Model Parameters for Modeling the MESFET.

Parameter Description	*Units*
VTO pinchoff voltage	V
BETA transconductance parameter	A/V^2
RD drain ohmic resistance	Ω
RS source ohmic resistance	Ω
CGS zero-bias G-S junction capacitance	F
CGD zero-bias G-D junction capacitance	F

and using the range of V_{DS} in (6.22a) implies that the fit parameter $\alpha = 3$. For the example MESFET described earlier ($V_P = 2.5$V and I_{DSS} of 1000 mA) yields at $V_{GS} = 0$ the translation expression $\beta = I_{DSS}/\alpha V^2{}_P = 0.0213$ A/V^3 for $|V_T| = |V_P|$. These parameters are used in the SPICE simulation example in the next section.

6.6.2 SPICE Simulation Example

This section covers a simple SPICE simulation example based on the series-connected MESFET RF switch that has been discussed earlier in this chapter with $V_P = -2.5$V and I_{DSS} of 1000 mA). The SPICE netlist for the series-connected MESFET RF switch based on a SPICE model is presented in Table 6.5, with the associated circuit layout shown in Figure 6.11a. In Table 6.5, the SPICE circuit elements, connection nodes, values, and parameters are capitalized with no subscripts; this is the traditional structure found in the SPICE netlist. The .MODEL portion of the SPICE netlist provides the most important parameters for the MESFET switch; the parameters in the previous section have been used as input to the device model ($b = \lambda = 0.001$ to minimize their impact on the simulation but to also indicate their location in the SPICE netlist). The MESFET, element name Z1, is defined in the second line and describes the device's connections to the remainder of the circuit and also calls

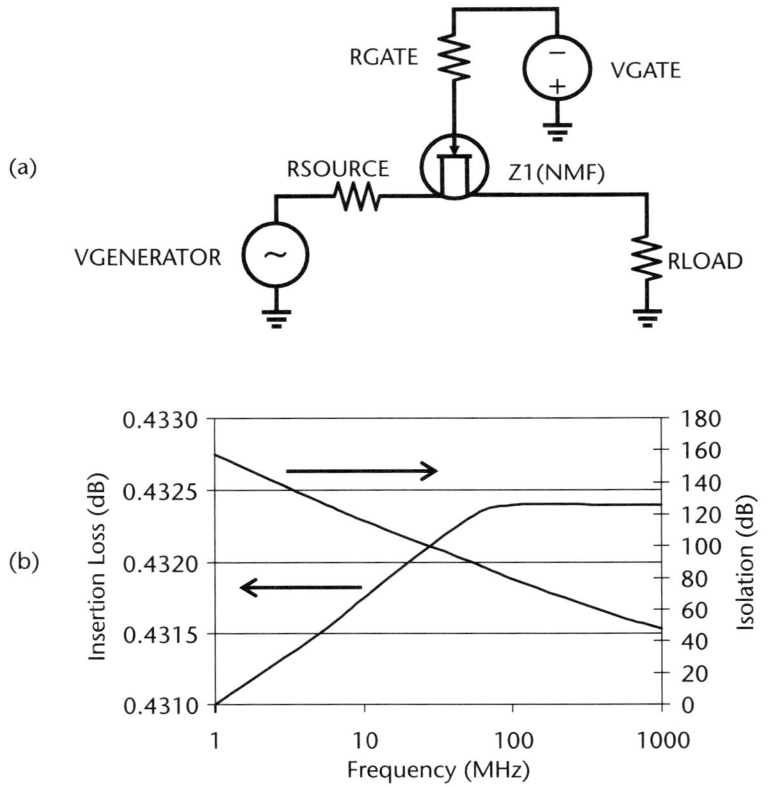

Figure 6.11 (a) SPICE simulator layout for series-connected MESFET RF switch based on the MESFET example in the text. (b) Insertion loss and isolation for the MESFET RF switch from 1.0 MHz to 1,000 MHz.

Table 6.5 Example SPICE Netlist for Series-Connected MESFET Switch

```
* SERIES-CONNECTED RF SWITCH SIMULATION EXAMPLE
Z1 IN GATE OUT NMF
RGATE GATE0 GATE 10E3
RSOURCE IN VGEN 50
RLOAD OUT 0 50
VGATE GATE0 0 0
VGENERATOR VGEN 0 0 AC 2
.MODEL NMF NMF BETA=0.0213 B=0. 0.001 VTO=-2.5 LAMBDA=0. 0.001
+ CGS=0.05PF CGD=0.05PF ALPHA=3
.AC DEC 100 1E6 1000E6
.END
```

the model NMF information from the .MODEL section. The remaining lines (with elements starting with R and V) describe the resistances and voltage sources in the circuit. The results of a frequency sweep from 1.0 to 1000 MHz are shown in Figure 6.11(b) with VGATE=0 (on-state) and VGATE=−3V (off-state) showing the insertion loss and isolation afforded by this circuit. The SPICE simulation results show good agreement with the simulations based on the physical equations outlined earlier in the chapter and shown in Figure 6.7.

References

[1] Ertel, A., "Monolithic IC Techniques Produce First All-Silicon Xband Switch," *Electronics*, 1967, pp. 76–81.

[2] Pucel, R.A., "Looking Back at Monolithic Microwave Integrated Circuits," *IEEE Microwave Magazine*, Vol. 13, No. 4, 2012, pp. 62–76.

[3] Garver, R. V., "Microwave Semiconductor Control Devices," *IEEE Trans. on Microwave Theory and Techniques*, Vol. MTT-27(5), 1979, pp. 523–529.

[4] Ayasli, Y., et al., "A Monolithic X-Band Single-Pole, Double-Throw Bi-Directional GaAs FET Switch," *GaAs IC Symp. Res. Abstracts*, 1980, Paper No. 26.

[5] Ayasli, Y., "Microwave Switching with GaAs FETs," *Microwave J.*, Vol. 25, 1982, pp. 61–74.

[6] Campbell, C. F., et al., "GaN Takes the Lead," *IEEE Microwave Magazine*, Vol. 13, No. 6, 2012, pp. 44–53.

[7] Khan, A., et al., "Metal Semiconductor Field Effect Transistor Based on Single Crystal GaN." *Applied Physics Letters*, Vol. 62, No. 15, 1993, p. 1786.

[8] Ueda, K., et al., "Diamond FET Using High-Quality Polycrystalline Diamond with fT of 45 GHz and fmax of 120 GHz," *IEEE Electron Dev. Letters*, Vol. 27, No. 7, 2006, pp. 570–572.

[9] Umezawa, H., T. Matsumoto, and S.-I. Shikata, "Diamond Metal–Semiconductor Field-Effect Transistor with Breakdown Voltage over 1.5 kV," *IEEE Electron Device Letters*, Vol. 35, No. 11, 2014, pp. 1112–1114.

[10] Camarchia, V., et al., "An Overview on Recent Developments in RF and Microwave Power H-Terminated Diamond MESFET Technology," *2014 International Workshop on*

Integrated Nonlinear Microwave and Millimetre-wave Circuits (INMMiC), 2014, pp. 1–6, 2–4.

[11] Weitzel, C. E., "Comparison of SiC, GaAs, and Si RF MESFET Power Densities," *IEEE Electron Device Letters*, Vol. 16, No. 10, 1995, pp. 451–453.

[12] Clarke, R. C., and J. W. Palmour, "SiC Microwave Power Technologies," *IEEE Proc.*, Vol. 90, No. 6, 2002, pp. 987–992.

[13] Kameche, M., and N. Drozdovski, "High-Temperature Distortion and Power-Handling Properties of GaAs-4H-SiC-GaN-MESFET's Switches," *IEEE Mediterranean Electrotechnical Conf., 2006 (MELECON 2006)*, 2006, pp. 169–172.

[14] Alexandru, M., et al., "SiC Integrated Circuit Control Electronics for High-Temperature Operation," *IEEE Trans. on Industrial Electronics*, Vol. 62, No. 5, 2015, pp. 3182–3191.

[15] Sze, S., *Physics of Semiconductor Devices*, New York: Prentice Hall, 1982.

[16] Takata, I., "A Simple Mobility Model for Electrons and Holes," *Proc. 11th International Symposium on Power Semiconductor Devices and ICs (ISPSD '99)*, 1999, pp. 269–272.

[17] Oxley, C. H., and M. J. Uren, "Measurements of Unity Gain Cutoff Frequency and Saturation Velocity of a GaN HEMT Transistor," *IEEE Trans. on Electron Devices*, Vol. 52, No. 2, 2005, pp. 165–169.

[18] Shenai, K., R. S. Scott, and B. Jayant Baliga, "Optimum Semiconductors for High-Power Electronics," *IEEE Trans. on Electron Devices*, Vol. 36, No. 9, 1989, pp. 1811–1823.

[19] Kimura, S., et al., "Artificial-Line-Division Distributed ICs with 0.1-μm-Gate-Length GaAs MESFET and Three-Dimensional Transmission Lines," *IEEE Trans. on Microwave Theory and Techniques*, Vol. 50, No. 6, 2002, pp. 1603–1608.

[20] Jain, N., and R. J. Gutmann, "Modeling and Design of GaAs MESFET Control Devices for Broad-Band Applications," *IEEE Trans. on Microwave Theory and Techniques*, Vol. 38, No. 2, 1990, pp. 109–117.

[21] Curtice, W. R., "A MESFET Model for Use in the Design of GaAs Integrated Circuits," *IEEE Trans. on Microwave Theory and Techniques*, Vol. 28, No. 5, May 1980, pp. 448–456 (see also W. R. Curtice and M. Ettenberg, "A Nonlinear GaAsFET Model for Use in the Design of Output Circuits for Power Amplifiers," *IEEE Trans. on Microwave Theory and Techniques*, Vol. MTT-33 (12), 1985, pp. 1383–1394).

[22] Statz, H., et al., "GaAs FET Device and Circuit Simulation in SPICE," *IEEE Trans. on Electron Devices*, Vol. 34, No. 2, 1987, pp. 160–169.

[23] McCant, A. J., G. D. McCormack, and D. H. Smith, "An Improved GaAs FET Model for SPICE," *IEEE Trans. on Microwave Theory and Techniques*, Vol. MTT-38(6), June 1990, pp. 822–824. (see also R. B. Hallgren and D. S. Smith, "TOM3 Equations," *Triquint Internal Report* (unnumbered), December 2, 1999; and R. B. Hallgren and P. H. Litzenberg, "TOM3 Capacitance Model: Linking Large- and Small-Signal MESFET Models in SPICE," *IEEE Trans. on Microwave Theory and Techniques*, Vol. 47, No. 5, 1999, pp. 556–561).

[24] Parker, A. E., and D. J. Skellern, "Improved MESFET Characterization for Analog Circuit Design and Analysis," *Technical Digest 1992, 14th Annual IEEE Gallium Arsenide Integrated Circuit (GaAs IC) Symposium, 1992*, October 1997, pp. 225–228.

[25] Kacprzak, T., and A. Materka, "Compact DC Model of GaAs FETs for Large-Signal Computer Calculation," *IEEE Journal of Solid-State Circuits*, Vol. 18, No. 2, 1983, pp. 211–213.

[26] Gopinath, A., and B. Rankin, "GaAs FET RF Switches," *IEEE Trans. on Electron Dev.*, Vol. ED-32(7), 1985, pp. 1272–1278.

[27] Takada, T., et al., "An MESFET Variable Capacitance Model for Gallium Arsenide Integrated Circuit Simulation," *IEEE Trans. on Microwave Theory and Techniques*, Vol. MTT-30 (5), 1982, pp. 719–723.

[28] Vendelin, G. D., A. M. Pavio, and U. L. Rohde, *Microwave Circuit Design Using Linear and Nonlinear Techniques*, New York: John Wiley, 1990.

[29] Wang, G-W., and L. F. Eastman, "An Analytical Model for I-V and Small-Signal Characteristics of Planar-Doped HEMTs," *IEEE Trans. on Microwave Theory and Techniques*, Vol. 37, No. 9, 1989, pp. 1395–1400.

[30] Wang, G-W., and W. Ku, "An Analytical and Computer-Aided Model of the AlGaAs/GaAs High Electron Mobility Transistor," *IEEE Trans. on Electron Devices*, Vol. 33, No. 5, 1986, pp. 657–663.

[31] Lee, K., et al., "Current–Voltage and Capacitance-Voltage Characteristics of Modulation-Doped Field Effect Transistors," *IEEE Trans. on Electron Devices*, Vol. ED-30 (3), 1983, pp. 207–212.

[32] Mishra, U., et al., "GaN Microwave Electronics," *IEEE Trans. on Microwave Theory and Techniques*, Vol. 46, No. 6, 1998, pp. 756–761 (see also Mishra, U. K., P. Parikh, and Yi-Feng Wu, "AlGaN/GaN HEMTs-an Overview of Device Operation and Applications," *IEEE Proc.*, Vol. 90, No. 6, 2002, pp. 1022–1031).

[33] Singh, R., and C. Snowden, "Small-Signal Characterization of Microwave and Millimeter-Wave HEMTs," *IEEE Trans. on Microwave Theory and Techniques*, Vol. 44, No. 1, 1996, pp. 114–121.

[34] Caverly, R. H., and K. J. Heissler, "On-State Distortion in High Electron Mobility Transistor Microwave and RF Switch Control Circuits," *IEEE Trans. on Microwave Theory and Techniques*, Vol. 48, No. 1, 2000, pp. 98–103.

[35] Drozdovski, N. V., and R. H. Caverly, "GaN-Based High Electron-Mobility Transistors for Microwave and RF Control Applications,"*IEEE Trans. on Microwave Theory and Techniques*, Vol. 50, No. 1, 2002, pp. 4–8.

[36] Caverly, R. H., "Linear and Nonlinear Characteristics of the Silicon CMOS Monolithic 50 Ohm Microwave and RF Control Element," *J. Solid-State Circuits*, Vol. 34, No. 1, 1999, pp. 124–126.

[37] Caverly, R. H., "Distortion in Broadband Gallium Arsenide MESFET Control and Switch Circuits," *IEEE Trans. on Microwave Theory and Techniques*, Vol. 39, No. 4, 1991, pp. 713–717.

[38] Caverly, R. H., "Nonlinear Properties of Gallium Arsenide and Silicon FET-Based RF and Microwave Switches," *Proc. 1998 IEEE International Symposium on Circuits and Systems (ISCAS '98)*, Vol. 4, 1998, pp. 337–340.

[39] R. H. Caverly, "Distortion in Off-State Arsenide MESFET Switches," *IEEE Trans. on Microwave Theory and Techniques*, Vol. 41, No. 8, 1993, pp. 1323–1328.

[40] Gutmann, R., and D. Fryklund, "Characterization of Linear and Nonlinear Properties of GSAS MESFETS for Broad-Band Control Applications," *IEEE Trans. on Microwave Theory and Techniques*, Vol. MTT-35 (4), 1987, pp. 516–520.

[41] *MicroSim PSpice A/D Reference Manual*, MicroSim Corp., 1997 (see also http://bwrcs.eecs.berkeley.edu/Classes/IcBook/SPICE/).

[42] Wilson, D. O., et al., "A Comparison of High-Speed Enhancement and Depletion Mode GaAs MESFET's," *Proc. 1978 International Electron Devices Meeting*, Vol. 24, 1978, pp. 600–602.

[43] Gusenkova, A. V., et al., "Integration of Depletion-and Enhancement-Mode AlGaAs/GaAs MESFET for High-Speed MMIC Application," *Proc. 16th International Crimean Conference Microwave and Telecommunication Technology, 2006 (CriMiCO '06)*, Vol. 1, 2006, pp. 193–194.

[44] Sussman-Fort, S. E., J. C. Hantgan, and F. L. Huang, "A Spice Model for Enhancement- and Depletion-Mode GaAs FET's," *IEEE Trans. on Microwave Theory and Techniques*, Vol. 34, No. 11, 1986, pp. 1115–1119.

CHAPTER 7
Switch and Switched Circuit Applications

The fundamental applications of solid-state control elements involve their role in directing, routing, or switching RF and microwave energy. One area in which solid-state control elements are found is in communication systems where one or more transmitters and one or more receivers share a common antenna. Another area is toggling or switching passive elements into a circuit to change the resonant frequency, filter bandwidth, or to help improve impedance matching. Each of these applications will be discussed in turn in this chapter. Because this book is about modeling semiconductor control elements, the use of SPICE netlists and Matlab scripts for circuit simulation and analysis based on the previously discussed device models will be emphasized. Earlier chapters of this book describe details of various SPICE models for the PIN diode and MESFET and MOSFET-based switching elements. The reader may need to iteratively move back and forth between the next two chapters on applications and the relevant preceding chapters during the development of the circuit model that fits their needs.

It is assumed that the reader has some familiarity with basic SPICE circuit entry and simulation techniques; this book is not meant to be a primer on how to perform circuit simulations using SPICE or writing scripts in Matlab but rather to use them as tools to aid in understanding of circuit operation. For SPICE in particular, most of the widely use commercial microwave and RF CAD tools provide some means of introducing SPICE models into their work environment. Many of the netlists and scripts used in the simulation examples are found at the repository website listed in Chapter 1. The files are fully editable, allowing the reader to explore the impact of various changes to circuit elements on control circuit performance. The LT-SPICE version of the generic SPICE simulator has been used in all simulations in Chapter 8 [1].

7.1 Transmit/Receive (TR) Switches

One of the fundamental theorems in antenna work is the so-called reciprocity theorem, which states that the transmit and receive properties of an antenna are identical, whether discussing impedance, bandwidth, or directionality, assuming these properties do not vary with power level. Because of reciprocity, the first, and often optimal, design solution for the RF engineer is to use the same antenna for both transmit and receive.[1] The design problem then becomes how to provide a

1 Some RF systems designs may require separate transmit and receive antennas. These are very interesting design problems with many design issues, but the design of these systems is beyond the scope of this book and so will not be covered further.

low insertion loss connection for transmit and receive and to also well-isolate the receiver during high-power transmit. This function is performed using a transmit/receive or TR switch.

The following sections of Chapter 7 will focus on a number of TR switch circuits with PIN diodes and FETs as the control elements. The TR switch circuits discussed are by no means meant to be an exhaustive list of such circuit topologies, but rather to show simulation and modeling procedures that can be used regardless of TR switch circuit topology. The performance of these switch circuits will be discussed based on SPICE simulations of small-signal frequency response, switching time, and distortion; the latter two simulations will be performed in the time domain. An example set of specifications will be assumed so that the performance of various switching circuits can be best compared. The first time a concept is introduced in this chapter, a detailed discussion will be presented. Later discussion of the same concept in different circuits will be limited to significant differences compared with the earlier discussion. However, the reader is encouraged to go back to the relevant prior discussion for details.

7.1.1 Introduction

Early TR switches often used mechanical switches to toggle the antenna between the transmitter (XMTR) and receiver (RCVR) as shown in Figure 7.1 [2–4]. Although effective because they exhibited good insertion loss and isolation, they were bulky and slow since they depended on human control and had reliability problems that impacted the entire system's reliability. Early electronic TR high-power switches used gas discharge tubes [5], and significant advances in TR switches using these devices in coaxial and waveguide structures were developed and well described in volume eight of the MIT Radiation Laboratory Series. This volume discussed in detail many control circuits using microwave duplexers as TR switches; current standard designs often use circuit structures based on those described in this volume [6].

As mentioned in Chapter 1, semiconductor TR switches first started to appear in the 1950s, soon after the invention of the transistor. The first solid-state TR switches used PN junction devices but the introduction of the PIN diode supplanted early PN junction diodes for TR switches due to its much higher power handling

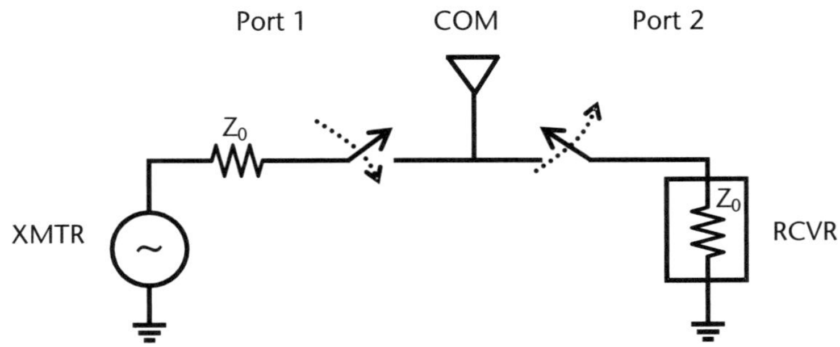

Figure 7.1 Schematic of TR switch in a Z_0 system: when the left switch is closed and the right switch open, the transmitter (XMTR) at Port 1 is connected to the common port (COM) and the receiver (RCVR) at Port 2 is isolated from COM. The reverse occurs when Port 2 (RCVR) is connected to COM.

capability. MESFET-based TR switches were introduced in the 1970s and 1980s, with MOSFET-based TR switches seeing increased use in the 1990s; both FET-based TR switches are still in use today and should continue in the far future. Chapter 1 presented a detailed look at this evolution of solid-state switches.

7.1.2 Basic Switching Structures

The simplest switching structure is the single pole, single throw (SPST) switch; operation of this switch was the basis for PIN diode, MESFET, and MOSFET modeling examples in their respective chapters. The SPST switch operates essentially as an on-off switch, connecting two terminals together. Though useful for illustrative purposes, the SPST switch does not exhibit enough degrees of freedom to provide the changeover requirements of a TR switch. In this case, the single pole, double throw (SPDT) switch provides this flexibility, allowing a single common (COM) terminal or port (usually the antenna in a standard transceiver TR switch) to be switched between components (transmitter and receiver) connected to the other two terminals (Port 1 and Port 2).

Several classic SPDT circuit topologies have evolved to create the necessary TR switch changeover. The two-device TR switch is the simplest of the SPDT topologies (Figure 7.2(a)), and has the potential for low insertion loss operation. Each control device is toggled separately, and the two devices should not be on (that is, in their low impedance state) simultaneously. Isolation, especially at the higher frequencies, can be poor because the single series off-state element can exhibit a significant amount to capacitive coupling. An SPDT switch with low insertion loss and improved isolation uses four control devices, two in series and two in shunt (Figure 7.2(b)). Each control line is connected to an opposing series-shunt pair; the opposing on-state shunt device provides increased isolation by reflecting any RF energy that leaks through the off-state series device.

Each of the semiconductor control elements that make up the TR switch will have their operational state determined by a control signal, usually generated external to the switch. This control signal must be decoupled from the RF and microwave path because the control signal is usually a very slowly changing signal compared with that of the RF being controlled and is often considered at a dc potential. The series connected devices in Figure 7.2 require dc return paths that may limit bandwidth or switching time. Shunt-connected control elements, however, require fewer decoupling elements, thereby reducing circuit complexity (Figure 7.2(c)). The problem with shunt elements, however, occurs because the low impedance on-state puts a nearly direct RF short across the RF path. By introducing $\lambda/4$ transmission lines, the low impedance shunt devices appear as high impedances $\lambda/4$ away, minimizing circuit loading. These tuned TR switches have simpler dc decoupling networks but exhibit reduced bandwidths because of the low bandwidth nature of the $\lambda/4$ transmission lines. Lumped inductor/capacitor networks can be used in place of the $\lambda/4$ transmission lines using (1.40). Other tuned switches include those based on quad hybrids.

Many other TR switch circuit topologies have been developed, and others in the future will undoubtedly be created, but the next two chapters look at standard control structures and cover techniques for simulation and interpretation of the simulation results using the models presented in the previous chapters.

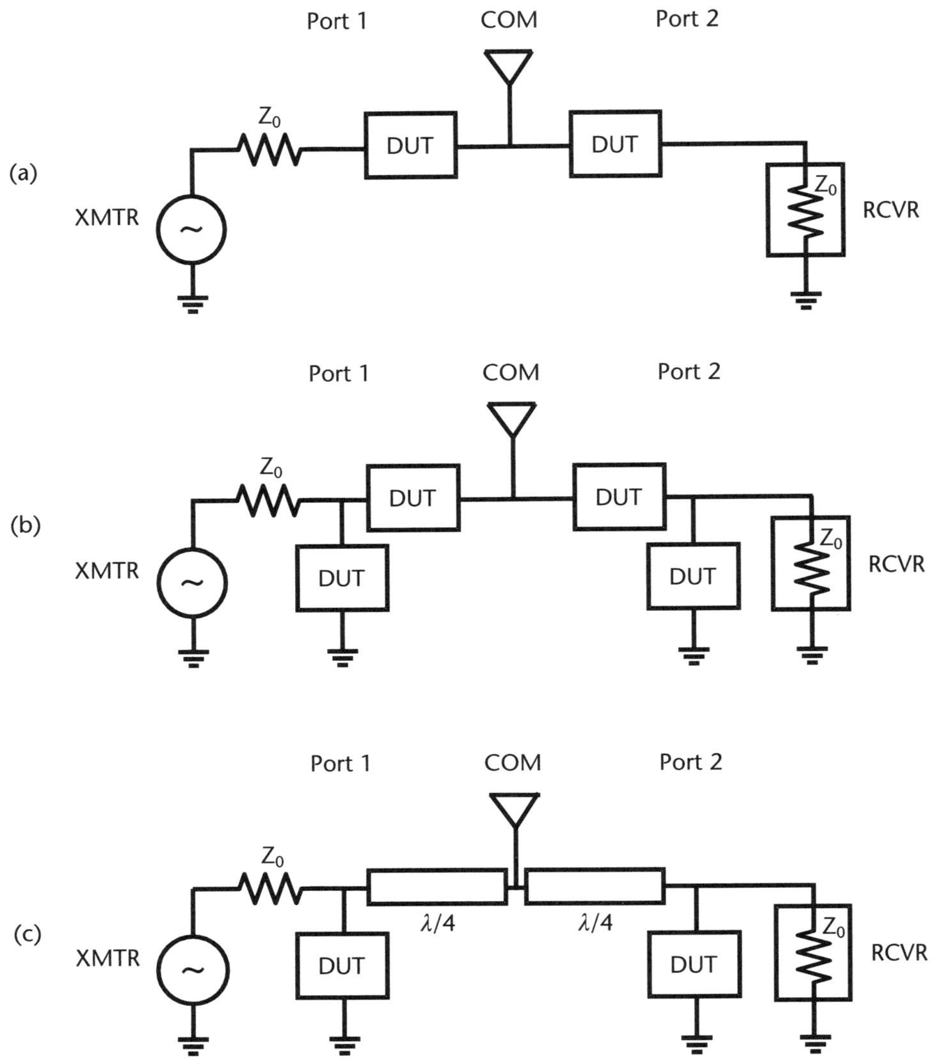

Figure 7.2 General TR switch configurations: (a) series two-control component circuit topology, (b) series-shunt control component circuit topology, and (c) tuned shunt control component circuit topology using $\lambda/4$ transmission lines.

7.2 Specific TR Switches

The models for the various semiconductor control devices covered in Chapters 3 through 6 will be discussed in this section for TR switches using two and four control devices in various series and shunt configurations, as well as using external components such as quad or 90° hybrids.

7.2.1 Two-Device SPDT TR Switch

The simplest SPDT TR switch uses two series control devices (Figure 7.2(a)). The control devices are controlled by complementary signals so that neither device is

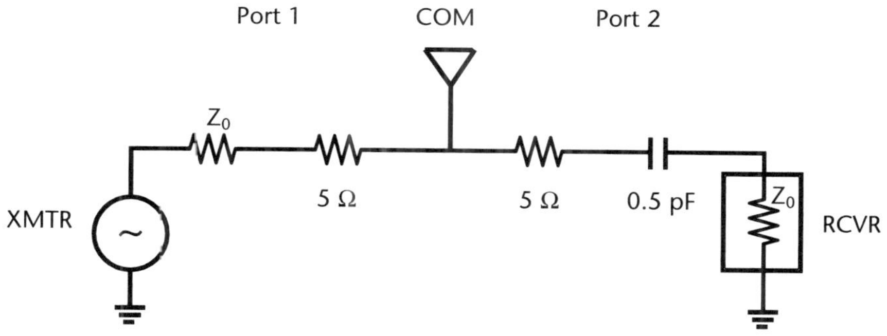

Figure 7.3 Two-device illustrative example TR switch using lumped control elements.

on at the same time. The SPDT TR switch is simply two SPST switches connected to a common port.

Some fundamental features of simulations of the SPDT TR switch can be discussed by replacing the generalized control elements with ideal switch elements shown in Figure 1.1. For illustrative purposes, an on-state impedance of 5Ω (Port 1) and off-state impedance of 5Ω in series with a 0.5 pF capacitor (Port 2) will be used (Figure 7.3). This combination of control elements exhibits a broadband cutoff frequency of 63.7 GHz using (1.4). From (1.18), the ideal insertion loss between Port 1 and common port COM is 0.42 dB with an ideal insertion loss between Port 2 and COM of over 3.0 dB at 1.0 GHz. The corresponding Switch Q at 1.0 GHz is 63.7 from (1.4b). This set of specifications will be seen to be a good balance between TR switch performance and the eventual PIN diode and FET control component requirements.

Figure 7.4 shows insertion loss and isolation results of a SPICE frequency sweep analysis using the SPICE netlist file 'netlist7-1.net'. This frequency sweep shows a nearly constant insertion loss up to approximately 200 MHz, above which the reactance of the 0.5 pF capacitor is becoming low enough to allow additional loading on the COM port with a corresponding increase in insertion loss. The isolation is high at low frequencies but monotonically decreases as frequency increases due to decreasing capacitive reactance. The isolation is approximately 15 dB at 1.0 GHz.

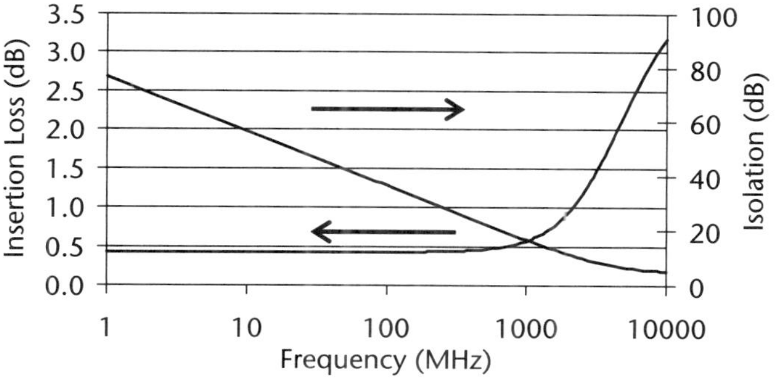

Figure 7.4 Insertion loss and isolation for the example two-device TR switch using lumped control elements.

Table 7.1 Example Control Element Package Parasitic Element Values

Element	*Value*
Bond wire inductance	0.65 nH
Shunt capacitance	0.14 pF
Coupling capacitance	0.05 pF

The above simulation assumes an ideal control component with no bond wire inductances or shunt parasitic capacitances. These parasitic elements, as explained in detail in Chapter 2, will introduce additional RF paths that will lead to degradation of both insertion loss and isolation with increasing frequency. The impact of these parasitic elements can be seen if a packaged control element is used with single bond wire inductance, shunt capacitances at input and output, and a package coupling capacitor are used. Similar values for the SOT-23 package from Chapter 2, Figure 2.5, are used and summarized in Table 7.1.

Figure 7.5(a) shows insertion loss and isolation SPICE simulation results of this packaged control element over the same frequency range (SPICE netlist file 'netlist7-2.net'). Figure 7.5(b) shows a comparison of the insertion loss values of the two-device SPDT TR switch using both packaged and unpackaged devices and illustrates that at frequencies below approximately 1000 MHz, the impact of the parasitic elements on the insertion loss is negligible. However, at frequencies above

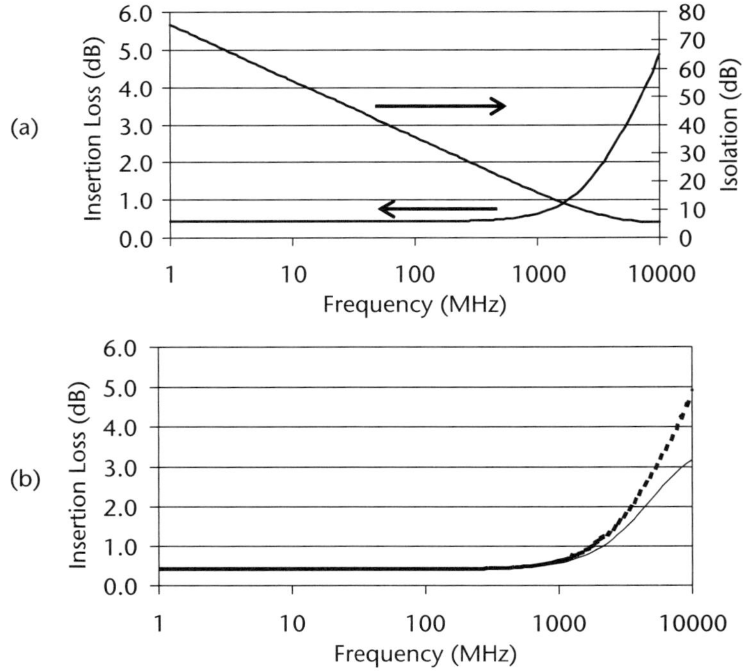

Figure 7.5 (a) Insertion loss and isolation for the example two-device TR switch using lumped control elements and including package parasitics shown in Table 7.1. (b) Comparison of insertion loss for the packaged (dotted line) and unpackaged (solid line) lumped control elements.

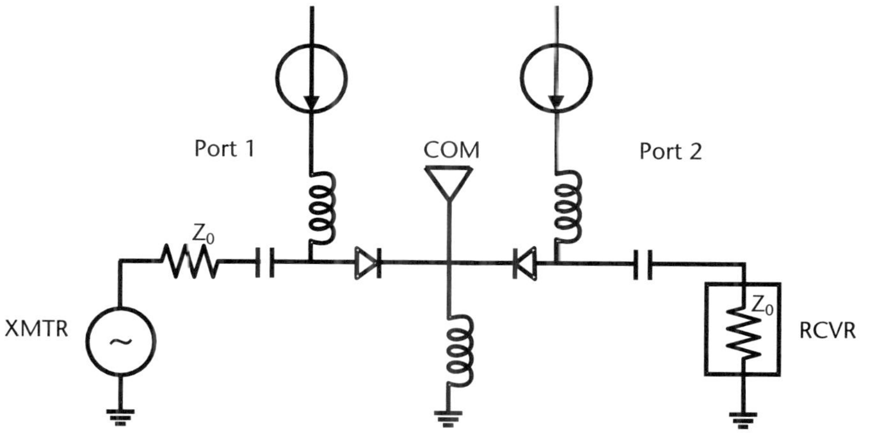

Figure 7.6 Two PIN diode SPDT TR switch showing dc decoupling network and dc bias current sources.

1000 MHz, the insertion loss dramatically increases so that at 10 GHz, the insertion loss is approximately 2.0 dB greater for the packaged control element compared with the ideal control element. Isolation is not as severely impacted by the package as is the insertion loss.

This analysis using ideal components forms a baseline to compare with results using PIN diode and FET-based control elements, details of which are in the next section.

7.2.1.1 PIN Diodes as Switch Elements

The ideal control elements in the two-device SPDT TR switch discussed in the previous section are now substituted by PIN diodes as shown in Figure 7.6. Additional inductances and dc current sources are added; the PIN diodes require dc bias to place them into the appropriate control state with the high-value inductances needed for dc decoupling. The choice of inductance for this dc decoupling requires a number of tradeoffs between TR switch usable bandwidth and switching time due to inductance-related time constants. The reactance of the inductors should be at four to five times that of the system impedance (200 to 250Ω reactive in a 50Ω system) at the lowest desired frequency of operation.[2] DC blocking capacitors may be added at Ports 1 and 2 and COM if needed; these blocking capacitors should exhibit low reactances ($Z_0/20$ or $Z_0/25$) at the lowest frequency of operation and block dc from the transmitter, receiver, and the common terminal.

Two main PIN diode models suitable for use in SPICE were discussed in Chapter 4 and will be discussed in the next two sections.

2 For wide bandwidth systems, a large value of inductance may be required; this inductor then has to be usable over the entire frequency band of operation, since it is directly attached to the RF path. Care must be taken in selection of the inductor to ensure that the highest desired frequency of operation is below the inductor self-resonant frequency (SRF). In some cases, a series combination of inductors can be used, with the higher(est) quality inductor (with highest SRF) close to the RF path point of attachment, and the larger value inductors with decreasing SRFs placed in series toward the dc control point.

Walston PIN Diode Model Example

The Walston PIN diode model [7] was discussed in detail in Chapter 3 and uses a series combination of a voltage-controlled current source (GPIN in the SPICE subcircuit for the Walston model shown in Table 7.2) as the current-dependent resistance and the internal SPICE model for the PN junction diode. There are four other parameters that are required to fully implement the SPICE model (italic values in Table 7.2): RSERIES (series contact and other device resistance), RSHUNT (shunt resistance modeling leakage around the intrinsic device), TT (I-region transit time; the I-region carrier lifetime is used for this model element), and CSHUNT (PIN diode reverse bias capacitance). These three elements can be determined from measurements or manufacturer's data sheets. The gain of the GPIN controlled current source, GPINGAIN is dependent on the PIN diode's resistance R_S at current I_{dc}, I-region width W, mobility, and I-region carrier lifetime τ and computed using (3.11) and (3.12):

$$\text{GPINGAIN} = \frac{W^2}{(\mu_n + \mu_p) R_S \tau} \tag{7.1}$$

An example SPICE simulation of the SPDT TR switch using the Walston PIN diode model for the control elements can be compared with the ideal TR switch described in the previous section by careful selection of the five parameters in the model. The most important parameter in the forward bias case is GPINGAIN. To achieve a 5Ω on-state resistance at 10 mA dc forward current (consistent with the ideal switch simulation), a 25 μm I-region width, 150 nS carrier lifetime PIN diode is assumed. The model factor value of GPINGAIN is computed with the aid (7.1) and is 0.00416A. The remaining SPICE subcircuit values for the simulation are shown in Table 7.3, and the SPICE netlist file for this simulation is part of the file repository under file name 'netlist7-3.net'. The following assumptions have been used in the simulation: the I-region resistance varies as $1/I_{dc}$, the dc decoupling inductors are the same 40 μH value (250Ω at 1.0 MHz), and no package parasitics.

Figure 7.7 shows the results of the SPICE simulation for the two-PIN diode SPDT TR switch using the Walston PIN diode model and Table 7.3 model parameters. The results are in good agreement with the ideal lumped element simulation

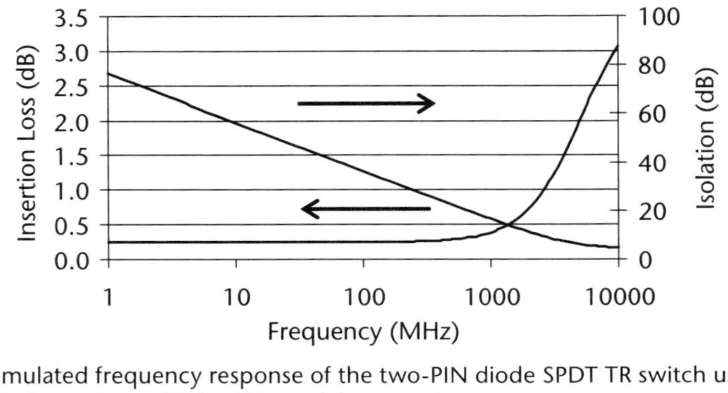

Figure 7.7 Simulated frequency response of the two-PIN diode SPDT TR switch using the Walston PIN diode model and Table 7.3 model parameters.

Table 7.2 SPICE Subcircuit for Walston PIN Diode [7]

```
.SUBCKT PINT 1 2
RSERIES 1 3 RSERIES
RSHUNT 3 4 RSHUNT
GPIN 3 4 POLY(2) 3 4 9 0 0 0 0 0 0 GPINGAIN
CSHUNT 3 2 CSHUNT
* MODEL COMPUTATIONS - SEE [7] FOR DETAILS
D1 4 2 PINDIO1
G2 0 5 4 2 1.0
D3 0 5 PINDIO3
V1 5 6 0.0
D4 6 0 PINDIO3
H1 7 0 V1 1.0
V2 7 8 0.0
D2 8 0 PINDIO2
F1 0 9 V2 10.0
R3 9 0 1000.0
* DIODE MODELS FOR ABOVE
.MODEL PINDIO1 D IS=1E-9 N=2 TT=TAU
.MODEL PINDIO2 D IS=1E-9 N=2
.MODEL PINDIO3 D IS=1E-20
.ENDS
```

shown in Figure 7.5, showing the utility of the Walston model, and explains why this model is part of the element catalog in many commercial microwave and RF CAD simulation packages.

Physical PIN Diode Model Example

The Physical PIN diode model was discussed in detail in Chapter 4, and, similar to the Walston model, uses a combination of a linked controlled sources and passive elements. Unlike the Walston PIN diode model, the physical model is based on semiconductor transport phenomenon directly and so represents a more accurate (both numerical and physical) approach in modeling the I-region conductivity modulation instead of using a simpler controlled resistance in the Walston model. The tradeoff between the two is that the physical model has more parameters to define, although many of these parameters are based on the geometry and electrical properties of the PIN diode. Instead of having to calculate parameters (such as GPINGAIN), the

Table 7.3 Table of Values and Definitions for Walston SPICE Model

RSERIES	0.5 Ohm	series contact and other device resistance
RSHUNT	1e9 Ohm	shunt leakage resistance
GPINGAIN	0.00416A	gain of controlled current source
CSHUNT	0.5 pF	PIN diode reverse bias capacitance
TT	150 nS	transit time (carrier lifetime)
IDC	0.01A	dc forward current

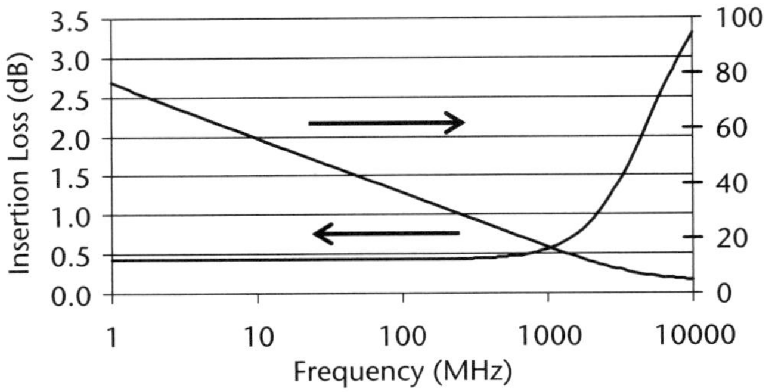

Figure 7.8 Simulated frequency response of the two-PIN diode SPDT TR switch using the physical PIN diode model and Table 7.4 model parameters.

physical model only requires entry of parameters. A full list of parameters is shown in Table 7.4; the full SPICE subcircuit implementation was discussed previously and can be found in Table 4.3 and SPICE netlist file 'netlist7-4.net'.

Figure 7.8 shows the results of the SPICE simulation for the two-PIN diode SPDT TR switch using the physical PIN diode model and Table 7.4 model parameters. The results are in good agreement with the ideal and Walston PIN diode models shown in Figures 7.5 and 7.7. The physical model may also be found in the element catalog of many commercial microwave and RF CAD simulation packages.

7.2.1.2 Time Domain Simulations

One of the SPICE circuit simulator's main features is the ability to perform time domain simulations, with both of the previous PIN diode SPICE models being capable of time domain simulations. These time domain simulations are useful because the waveforms computed can be analyzed for signal impacts such as distortion introduced into the circuit by the device. Many SPICE programs have the capability of performing a Fast Fourier Transform (FFT) on the time domain waveform as a means to determine the frequency spectrum of the waveform. In addition, the computed waveforms provide insight into circuit behavior during switching when the devices change state from on to off (or vice versa).

Table 7.4 Table of Values and Definitions for Physical SPICE Model

RMIN	0.5 Ohm	series resistance
RMAX	1E9 Ohm	shunt leakage resistance
W	25E-6 m	I-region thickness
TAU0	150 nS	carrier lifetime
CP	0.5 pF	PIN diode reverse bias capacitance
N	2	diode ideality factor

One of the important characteristics of a TR switch is the speed with which the control devices change state. There are a number of factors that control this switching speed, including the electrical properties of the device itself, the passive elements comprising the dc decoupling circuit, and the circuit controlling the switching (the so-called driver circuit). If the dc decoupling inductors and capacitors are too large, switching speeds will be slower due to LC time constant effects, and so minimizing the decoupling component values can improve switching speed. The tradeoff, however, is that the required TR switch bandwidth defines the component values at the lower end of the desired switch bandwidth since the decoupling network must work at the lowest frequency of operation. For PIN diodes, the charge stored in the I-region must be removed to turn the device off. The impact of the LC dc decoupling network can be somewhat mitigated by having the driver circuit remove current rather than rely on carrier recombination to dissipate the I-region stored charge; this can reduce the switching time. I-region stored charge removal by use of reverse switching current was discussed earlier in Chapter 3 when relating the I-region carrier lifetime τ to the switching time delay T_D and quantified by (3.36). PIN diode driver circuits are available as commercial modules [8], or examples of discrete circuits are contained in the literature [9, 10]. Switching rise and fall times for TR switches are defined using the usual 10%–90% rule: the fall time is the time duration to drop from 90% to 10% of its initial value (the rise time is the opposite) [11].

A series of SPICE simulations is performed to observe the impact of reverse current on the turn-off speed of the SPDT PIN diode TR switch using the Physical PIN diode model (SPICE netlist file ‘netlist7-5.net’). In the simulation, an ideal dc current source is set up to provide an initial dc forward current of 10 mA and then switched into the opposite state after 30 μS with dc reverse currents of zero, 1.0 mA and 10 mA. A 1.0 MHz signal is being controlled. Figure 7.9 shows the three time domain waveforms over a 0.1 mS time period. The series of figures show the dramatic improvement in switching speed by providing a reverse dc current to aid in removing the I-region stored charge. For a 10 mA reverse switching current, the TR switch turns off in approximately 4 μS. Additional simulations at 100 mA reverse switching current (not shown in Figure 7.9) provide sub-microsecond turn-off time. In Figures 7.9(a) and 7.9(b), the noticeable voltage modulation occurring during the turn-off transition is due to the LC time constant associated with the circuit's dc decoupling elements; the modulation frequency is approximately $1/2\pi\sqrt{LC}$.

Switch turn-on time delay can be simulated as well using SPICE. Figure 7.10 shows simulation results of the two-device SPDT TR switch using the physical PIN diode model. In the simulation, an ideal dc source is set up to provide an initial dc reverse voltage of 5 volts and is then switched into the opposite state after 30 μS with a dc forward current of 10 mA. Again, a 1.0 MHz signal is being controlled. The simulations show an approximate 4 μS transition delay to 90% full output; this long time delay is primarily due to the LC dc decoupling network. Further simulations (not shown) indicate that reducing this LC time constant by an order of magnitude shortens this transition delay to less than 1 μS.

Time domain simulations also provide insight into the amount of distortion introduced into the circuit by the control elements. In multicomponent TR switches such as the two-device switch, distortion is introduced by both elements even though

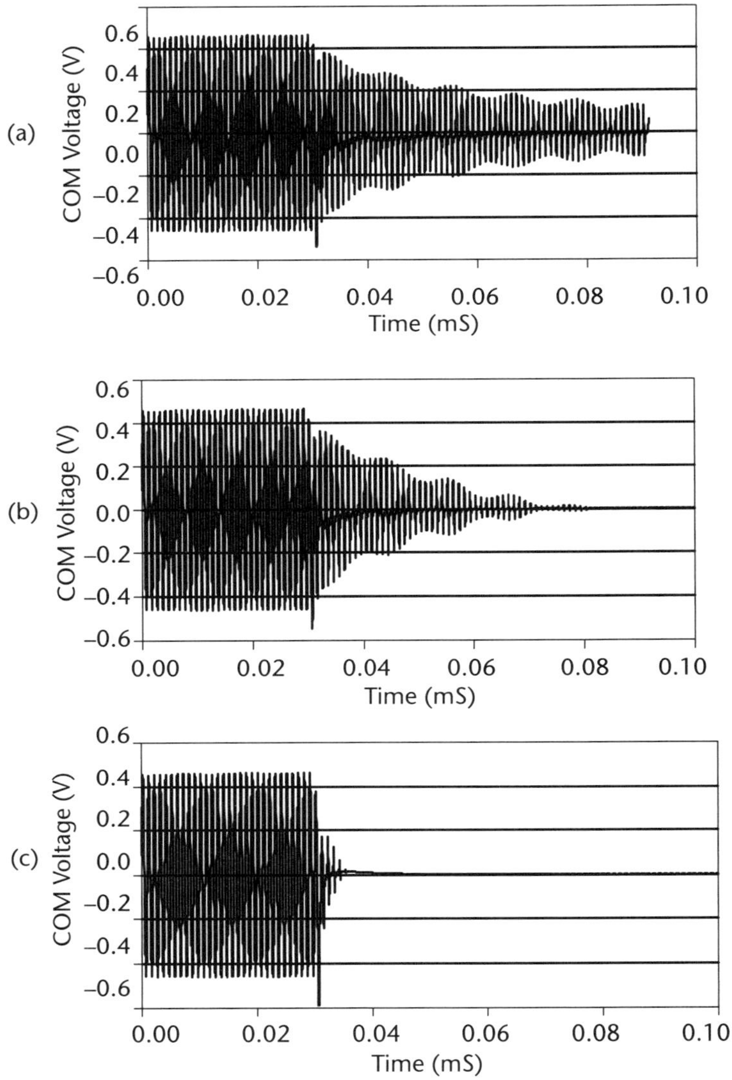

Figure 7.9 SPICE-generated time domain waveforms showing switching speed for the SPDT TR switch using the physical PIN diode model for (a) zero reverse switching current, (b) 1.0 mA reverse switching current, and (c) 10 mA reverse switching current.

they are in two different states (see detailed discussion in Chapter 4). The dominant distortion mechanism in TR switches is the distortion introduced by the on-state device, which is especially problematic because this will be the path of the higher power transmitter as it connects to the COM port (or antenna).

Many versions of SPICE are capable of performing an FFT on the time waveforms to calculate the frequency content in the signal. For determining the level of intermodulation distortion, the two-tone test is traditionally used. This test uses two RF sources at two slightly different frequencies but at the same power level (see Chapter 4 for details). Following this tradition, a SPICE netlist ('netlist7-6.

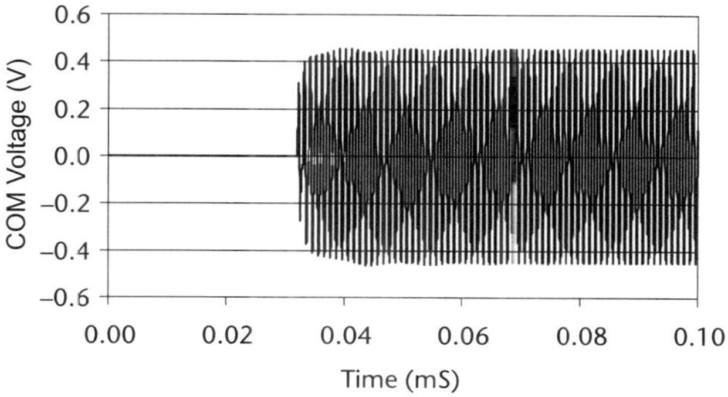

Figure 7.10 SPICE-generated time domain waveforms showing switch turn-on of the SPDT TR switch using the physical PIN diode model when switching between −5V dc reverse voltage and +10 mA dc forward current.

net') constructed using two 1.0V peak 50Ω RF sources (power at the COM port of approximately +1.0 dBm) at 1.0 and 1.1 MHz can be used. Low frequencies are used in the SPICE netlist to illustrate the distortion simulation because distortion is higher at lower frequencies in PIN diode circuits. The physical PIN diode model, a 25 μm and 150 nS I-region width and carrier lifetime biased at 10 mA dc forward current, is used in this example. The corresponding QF/R ratio defined in (4.21) can then be computed as $QF/R = I_{dc}\tau F/R = 3 \cdot 10^{-4}$ C · Hz/Ω. The corresponding theoretical load-referenced OIP2 and OIP3 using (4.21) will be 26 and 15 dBm, respectively. The SPICE time domain simulation results and subsequent FFT provide the frequency spectrum at the COM (antenna) as shown in Figure 7.11(a). The corresponding load intermodulation distortion powers are computed from the simulated frequency spectrum results as −26 and −34 dBm for the second and third order intermodulation power, respectively. The corresponding OIP2 and OIP3 can then be computed from (2.17a) as +24 and +15.5 dBm, in good agreement with the theoretical values in (4.21). Distortion power at the load decreases with increasing frequency as seen in (4.19); SPICE simulations confirm this phenomenon. Figure 7.11(b) shows the identical SPDT TR switch with the physical PIN diode model with excitation frequencies an order of magnitude higher, 10 and 11 MHz. The corresponding 20 and 30 dB decrease in distortion voltage at the COM port is observed in the SPICE simulation results of Figure 7.11.

From the discussion in Chapter 4, distortion in PIN diode switches was shown to be dependent on the ratio $QF/R = I_{dc}\tau F/R$, and the designer can vary a number of PIN diode parameters to reduce the distortion introduced by the switching circuit. For a given PIN diode and frequency of operation, increasing the stored charge Q by increasing the dc forward current will improve this ratio but at the expense of increased power consumption. Alternatively, a device with a longer carrier lifetime can also increase the stored charge at a given current; longer carrier lifetimes, however, tend to be seen in wider I-region devices with correspondingly higher resistances at a given current. These are the types of tradeoffs the design engineer faces when choosing the right PIN diode; however, as previous discussion has shown, there are a variety of PIN diode models that agree well with theory that will help the designer in verifying the PIN diode choice for a particular application.

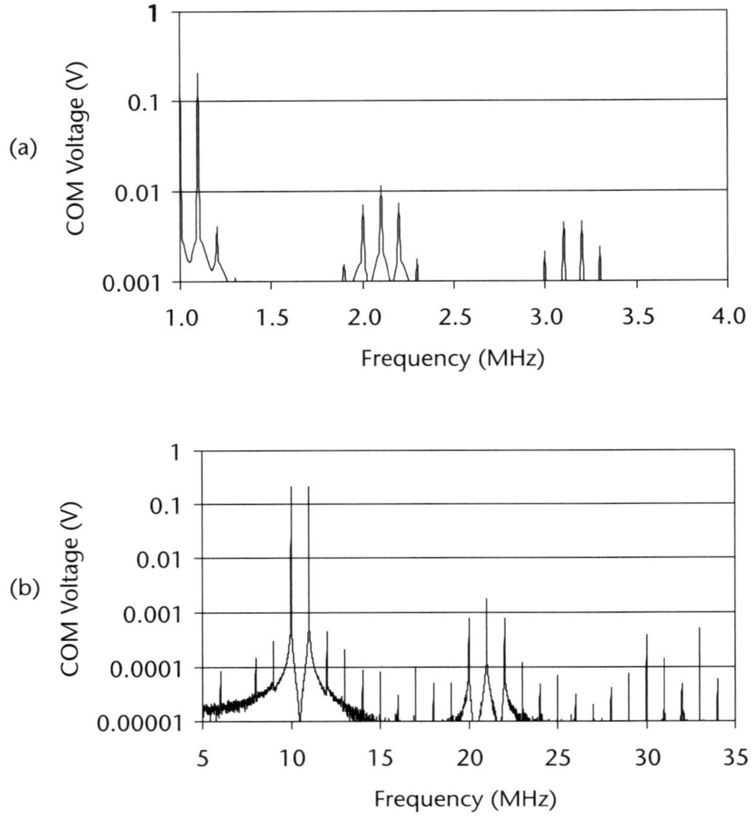

Figure 7.11 Two-device TR switch using the physical PIN diode model parameters shown in Table 7.4 at 10 mA dc forward current with RF excitation at (a) 1.0 and 1.1 MHz and (b) 10 and 11 MHz.

7.2.1.3 FET-Based Switch Elements

The ideal control elements in the two-device SPDT TR switch may also have FET-based control components used instead of PIN diodes. The use of FET-based control elements greatly simplifies the dc decoupling network required of PIN diodes since the control port (the FET gate) is only capacitively coupled to the RF path (drain to source) through the gate-drain and gate-source capacitances. As discussed in Chapters 5 and 6 on MOSFET and MESFET control device modeling, a single gate resistor R_G of value in the 5 to 10 kΩ range is often sufficient to provide the dc control source isolation. The use of R_G not only simplifies the dc decoupling network but also makes the TR switch broadband. Limitations on bandwidth at the low end are primarily an issue with MESFETs because these components typically require negative control voltages and so capacitive "floating" is required for use in unipolar power supply systems. In general, MOSFET-based TR switches do not require the dc float because the control voltages are positive as is the device threshold voltage.

This section discusses the simulation of the two-device TR switch using FET-based control elements. Both MOSFET and MESFET-based TR switches will be discussed using the same MOSFET and MESFET model parameters as the illustrative

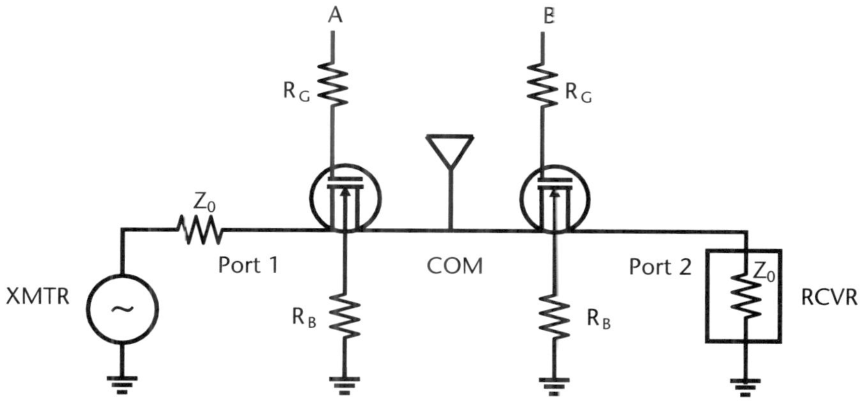

Figure 7.12 SPDT TR switch using two series-connected 180 nm MOSFETs. The bulk resistance R_B is used for floating bulk studies.

examples of these control components in Chapters 5 and 6, respectively. Both SPICE Level 3 and BSIM parameters for a 180 nm CMOS process will be used for the MOSFET discussion; the Statz model will be used in the MESFET discussion.

MOSFET Model Example

Two categories of SPICE MOSFET models were discussed in detail in Chapter 5. The simple Level 3 model is closely based on the theoretical current-voltage characteristics and so provides insight into the physical operation of the device. However, this model has been found to be inaccurate for sub-micron MOSFETs, and so BSIM parameters are used at these smaller geometries [27–29]. The BSIM parameters are determined by extensive measurements of test MOSFETs postfabrication, with the additional device parameters increasing the simulation accuracy over the Level 3 model. The tradeoff is that the BSIM model does not provide the same level of physical insight into MOSFET operation as the Level 3 model, but does provide better simulation accuracy for submicron MOSFETs.

This first part of this section is a study of TR switch performance using both Level 3 and BSIM models for a 180 nm MOSFET-based switch with the bulk node grounded (SPICE netlist file 'netlist7-7.net' available in the online repository). The Level 3 parameters are listed in Table 5.3, and the BSIM parameters are BSIM4.6.1 example parameters available from the UC Berkeley BSIM Research Group[3] (BSIM4.6.1 parameters are the latest version that the SPICE simulator version used in this book can use—more recent versions of the BSIM parameters are available from UC Berkeley and may be available from the MOS device fabricator or vendor; updated simulator versions may also be available). The floating bulk will be modeled using a fixed resistance for R_B (Figure 7.12). A 10 kΩ fixed resistance for gate bias resistance R_G will also be assumed.

Figure 7.13 shows the results of the SPICE simulation for the two-MOSFET SPDT TR switch using the SPICE Level 3 (Figure 7.13(a)) and BSIM4.6.1 (Figure 7.13(b))

3 http://www-device.eecs.berkeley.edu/bsim/?page=BSIM4_Arc

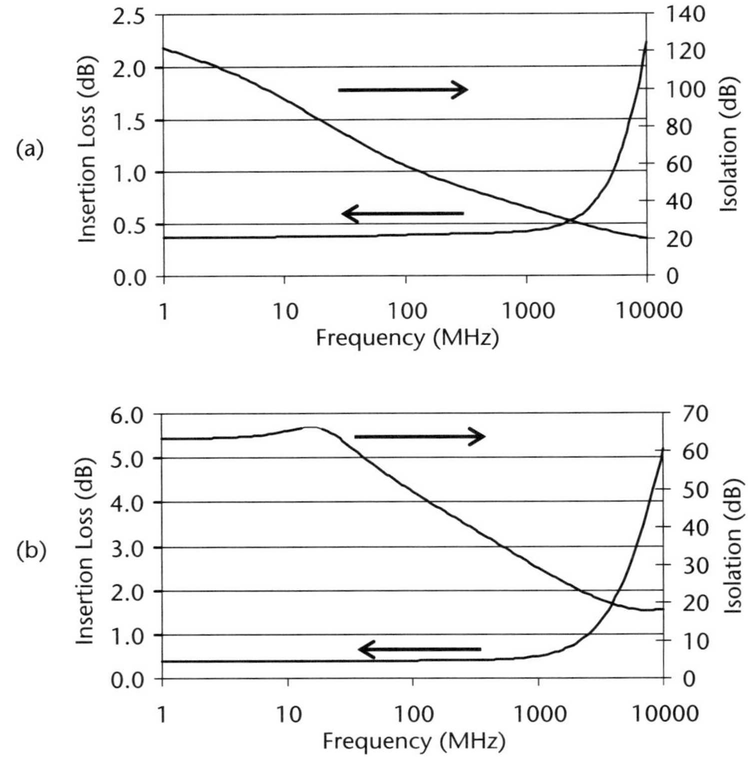

Figure 7.13 Two-MOSFET SPDT TR switch frequency response using (a) SPICE Level 3 and (b) BSIM4.6.1 parameters.

test parameters for a grounded bulk node ($R_B = 0$). There are significant differences between the two model results that can be seen by comparing Figure 7.13(a) with 7.13(b). Though the insertion loss is comparable at low frequencies, including the degradation beginning at approximately 1000 MHz, the BSIM parameter simulation shows a much larger degradation in insertion loss as frequency increases. This difference is due to the improved capacitance modeling in the BSIM model. Significant differences are also observed in the isolation response. Though the Level 3 model simulation shows monotonically decreasing isolation throughout the frequency range of the simulation, the BSIM parameter simulation shows a constant low-frequency isolation that monotonically decreases above approximately 30 MHz. The approximate 70 dB low-frequency isolation is consistent with commercial MOSFET-based switches, indicating that the BSIM parameter-based simulation provides the better simulation. The slight increase in isolation corresponds to operation near the gate bias circuit frequency given by $F_G = 1/2\pi R_G(C_{GS} + C_{GD})$ and described in Chapter 5.

As discussed in Chapter 5, the MOSFET bulk connection can influence the switch insertion loss depending on the value of bulk resistance R_B to the RF ground. Either high or low values of bulk resistance are required to maintain a low insertion loss over a wide bandwidth, and so careful simulation of the impact of R_B on performance is required. To illustrate the impact of R_B on TR switch performance, the results of a BSIM4.6.1 parameter simulation (SPICE netlist file 'netlist7-8.net' available in the online repository) with the bulk node resistance R_B varying between 0 and 10 kΩ is shown in Figure 7.14. The results of this simulation show that either a low or high value of R_B provides lower insertion loss over the switch bandwidth, with the floating bulk (R_B = 10 kΩ) providing lower insertion loss over a wider bandwidth than the ground bulk node. Bulk node resistance on the order of 100Ω yields the poorest insertion loss performance. This relatively low value of bulk node resistance can easily occur in MOSFET switch design if steps are not taken during the fabrication design process to ensure that the bulk to RF ground node is controlled; some of these steps are discussed in Chapter 5. The bulk node resistance also impacts isolation, with the floated bulk showing degraded isolation performance compared with lower values of R_B.

The small-signal simulations just discussed do not provide information on important specifications such as the power handling and its impact on insertion loss, isolation, and potential survivability in the MOSFET-based TR switch. Time domain simulations, however, can provide some of this needed insight into the power handling by providing detailed voltage and current waveforms at all nodes in the MOSFET TR switch. Some of these issues were mentioned in Chapter 5. RF voltage swings exceeding a few tenths of a volt, especially on the drain node since this node is at a higher potential, can cause the drain-bulk junction to become forward biased with a resulting high current being drawn. In addition, the RF voltage swing on the gate node could be of sufficient value to cause the device to turn off or to exceed the gate breakdown voltage during part of the RF cycle. Signal distortion can also be found by computing the spectral content of the signals, similar to the PIN diode case.

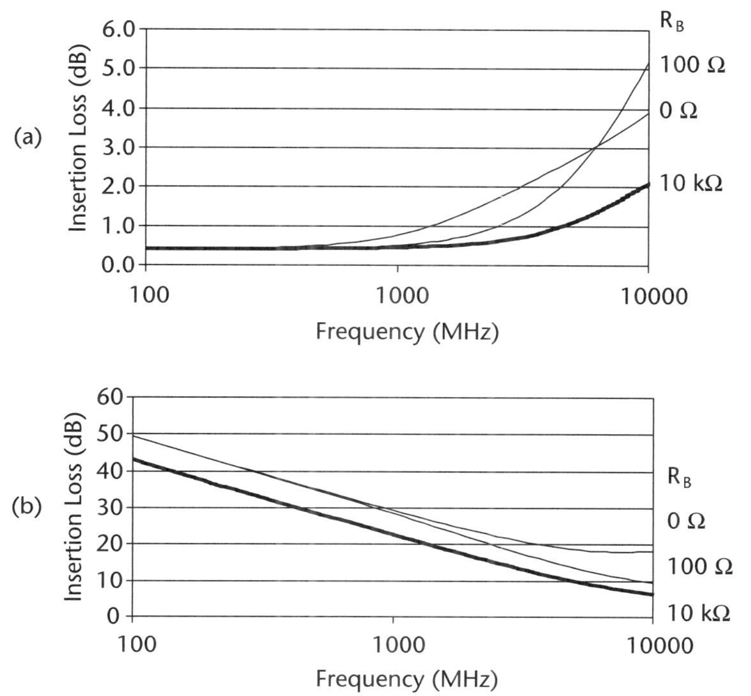

Figure 7.14 Impact of bulk resistance R_B on the two-MOSFET TR switch (a) insertion loss and (b) isolation performance.

As discussed in Chapter 5, the switching speed for FET-based control elements is primarily determined by the RC time constant associated with the gate bias control circuitry. For a given insertion loss/isolation response, the gate bias circuit time constant, $R_G(C_{GS} + C_{GD})$, can only be changed by changing R_G. Larger values of R_G provide better dc decoupling but at the expense of switching speed. To observe simulations of this TR switch switching speed, a SPICE netlist ('netlist7-9.net') of the two-MOSFET TR switch with a pulsed control voltage on one arm can be used. The time constant for a 10 kΩ gate bias resistance and a nominal gate capacitance of 0.5 pF is 5 nS. The time domain simulation for this circuit, shown in Figure 7.15 for a 1000 MHz input signal with control pulse applied at 20 nS, shows that it takes approximately 3.0 nS for the RF signal to rise to within 10% of its final on-state value. This switching time is consistent with the calculated gate bias control circuit time constant.

As an illustrative example of the distortion information that can be found from time-domain simulation, a SPICE netlist ('netlist7-10.net') of a two-MOSFET TR switch operating a 100 MHz can be used. The dc control voltage for the on-state MOSFET is set at 2.0V and for the off-state MOSFET at 0.0V, and 10 kΩ gate bias resistors are used with the bulk nodes tied to RF ground. The results of two simulations representing small (0 dBm) and large (20 dBm) signals are shown in Figure 7.16. The 0 dBm signal (Figure 7.16(a)) shows the low insertion loss predicted in the small-signal frequency response (Figure 7.14) at the COM port, with the RF swing on the gates (Figure 7.16(b)) of both MOSFETs low enough so that both devices remain in their respective state. Increasing the signal to +20 dBm shows changes throughout the TR switch beyond just a voltage magnitude increase. The insertion loss (Figure 7.16(c)) is significantly degraded because the voltage swing at the COM port is lower and the waveform is significantly perturbed, which will cause significant frequency content in the waveform and hence distortion. The negative clamping of the output waveform at approximately −0.6V is indicative of the RF voltage switch causing conduction of the drain-bulk and source-bulk PN junctions. The gate voltage curves (Figure 7.16(d)) show that the MOSFETs are changing state during the negative portion of the RF cycle, adding another cause to the distorted output waveform. A spectral analysis of this waveform using the SPICE FFT function yields

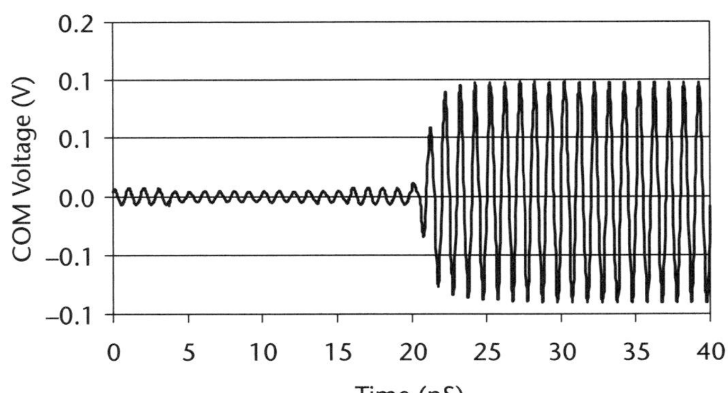

Figure 7.15 SPICE-generated time domain waveforms showing switch turn-on of the MOSFET-based SPDT TR switch when switching from 0 and 3 volts to change the device state.

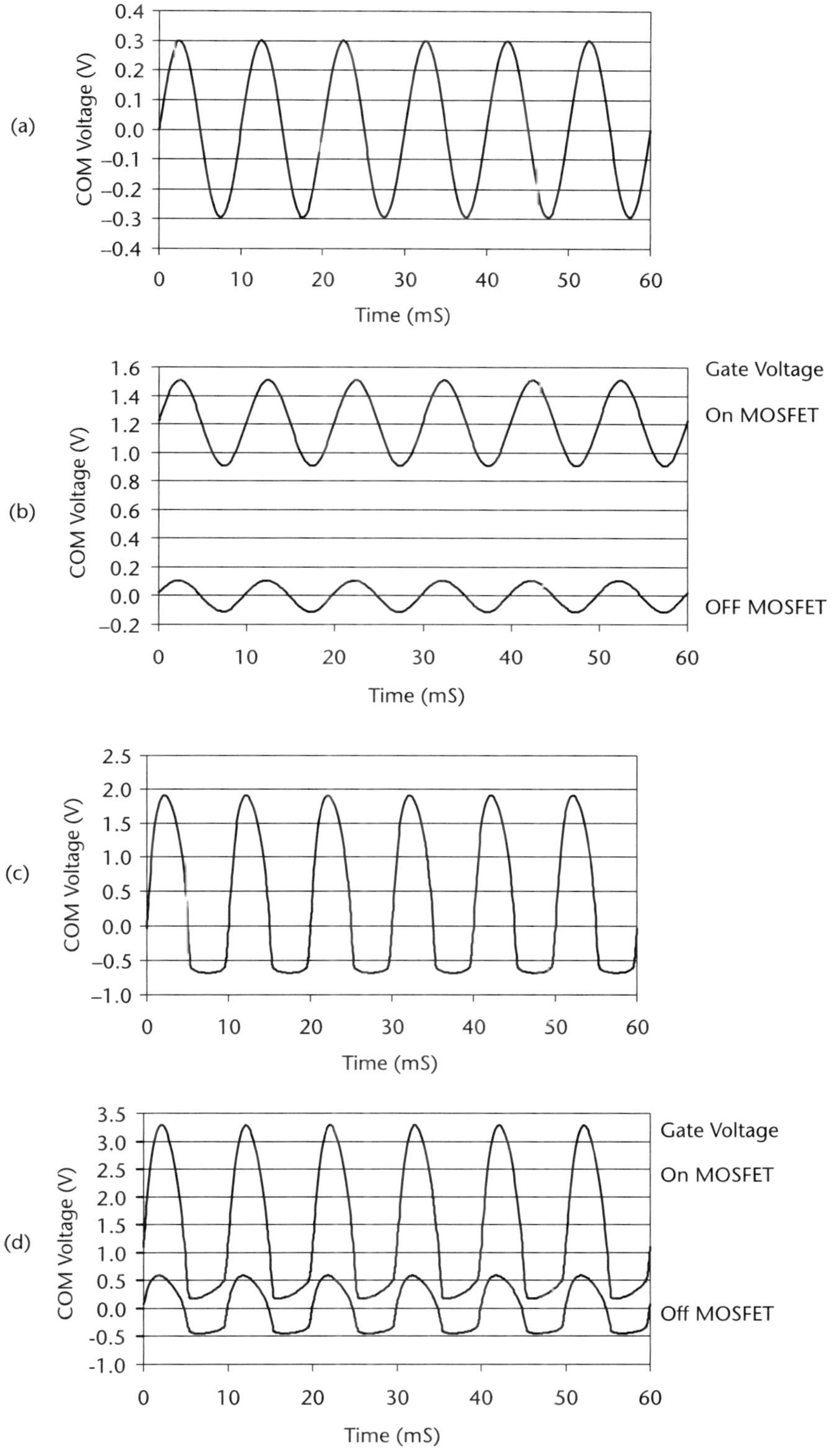

Figure 7.16 (a) COM and (b) gate terminal voltages for 0 dBm operation of the two-MOSFET TR switch; (c) COM and (d) gate terminal voltages for +20 dBm operation of the two-MOSFET TR switch.

OIP2 and OIP3 values of +42 and +32 dBm, respectively. These values are consistent with the theoretical values for MOSFETs outlined in Chapter 5 and illustrated in Figure 5.8. If such power levels need to be controlled with a MOSFET-based TR switch, a number of series-connected MOSFETs may be the solution, with each MOSFET operating more in its small-signal regime due to RF voltage sharing. This solution, however, can impact the overall insertion loss or bandwidth of the TR switch. To keep the same insertion loss, larger MOSFETs may be required because each device must have lower RF resistance; this will be accompanied by an increase in the overall device capacitance and with a corresponding reduction in bandwidth. Smaller MOSFETs may be used to keep the capacitance low and the bandwidth high, although an increase in RF resistance will degrade the insertion loss.

MESFET Model Example

The MESFET can also be used in the TR switch in lieu of either the PIN diode or the MOSFET (Figure 7.17). Like the MOSFET, the control decoupling circuitry is simpler than the PIN diode, and similar value gate bias resistors are used for dc decoupling of the control source (Chapter 6). Unlike the MOSFET, bulk effects are usually not as much of a concern, as the semi-insulating substrate under the MESFET channel produces a naturally high bulk resistance to RF ground. The MESFET often requires negative control voltages to toggle between the on and off states. Many systems use only unipolar (and usually positive) voltage sources, and so the MESFET switch module must be dc floated at all nodes using high-value capacitors. These capacitors limit the low-frequency response due to their reactance, but, if the values of capacitance are high enough, the self-resonant frequency may limit the upper frequency range as well.

Figure 7.18 shows the results of the SPICE simulation ('netlist7-11.net') for the two-MESFET SPDT TR switch using the Statz model with Table 7.5 simulation parameters. To ensure good insertion loss performance in the low-frequency range of 1.0 MHz, the TR switch is dc floated at all nodes using 0.1 μF capacitors. Control voltages of −3.0 and zero volts are used with 10 kΩ gate dc isolation decoupling

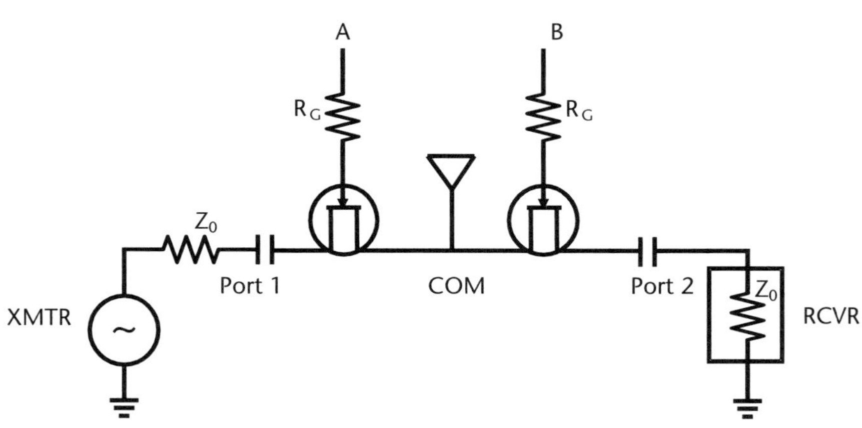

Figure 7.17 SPDT TR switch using two series-connected MESFETs. The SPICE Statz model is used for simulation.

Table 7.5 Table of Values and Definitions for Statz MESFET SPICE Model

BETA	*0.011 A/V²*
B	0.001 V^{-1}
VTO	−2.5 V
LAMBDA	0.001
CGS	0.25 pf
CGD	0.25 pf
ALPHA	3 V^{-1}

resistors. The simulated insertion loss and isolation characteristics of the MESFET TR switch resemble those of the MOSFET TR switch simulation using the SPICE Level 3 model. The change of slope of the isolation curve in the vicinity of 30 MHz is due to operation near the gate bias circuit cutoff frequency, $F_G = 1/2\pi R_G(C_{GS} + C_{GD})$. The improved insertion loss at high frequencies is a consequence of the MESFET's inherent lower capacitance structure, primarily the reduced drain and source capacitance to ground when compared with the MOSFET.

Time domain simulations can also be performed in a similar manner as for MOSFET-based switches. The time domain simulation files in the file repository are 'netlist7-12.net' and 'netlist7-13.net' but will not be covered because the results are similar to those previously discussed for the MOSFET. However, the reader is encouraged to perform simulations on these files to observe the similarities and differences between the two technologies.

7.2.2 Four-Device SPDT TR Switch with Improved Isolation

The simple two-device TR switches covered in Section 7.2.1 all exhibit good isolation at low frequencies, but this isolation degrades as the frequency increases because of increased capacitive leakage through the off-state control element. For the examples cited in these sections, isolation values above 30 dB are difficult to achieve at frequencies above 1000 MHz. In actual practice, other sources often limit

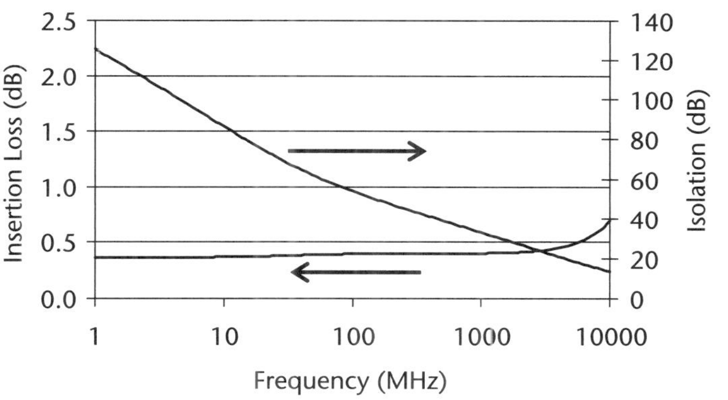

Figure 7.18 Two-MESFET SPDT TR switch frequency response using the Statz SPICE model and the parameters in Table 7.5.

the isolation and include inadequate shielding and unwanted radiation effects [10]. The addition of two shunt devices, one in each arm of the switch, can improve this high-frequency isolation by reflecting leakage energy away from the output switch node. This increased isolation is especially important on the receive arm for protection of the receiver front-end from a high-power transmit signal.

The general structure for the four-device SPDT TR switch is shown in Figure 7.2(b) where the control lines are connected to an opposing series-shunt pair; that is, if the series control device connecting Port 1 to COM is in the low insertion loss state, then the shunt control device at Port 2 is also biased into its low insertion loss state. The converse is true in the complementary switch state. This phenomenon can be seen in Figure 7.19(b) ('netlist7-14.net') where the previous pure lumped element illustrative example (5Ω for the low insertion loss state, 5Ω in series with a 0.5 pF capacitor for the high insertion loss state, Figure 7.19(a)). Comparing Figure 7.19(b) with Figure 7.4 shows little change in the insertion loss at low frequencies, including the frequency where the insertion loss begins to degrade due to increased capacitive coupling to the off port (Port 2). The large change for this circuit can be seen with the isolation performance. For the two-device TR switch, the 1000 MHz isolation is 18 dB, whereas for the four-device TR switch, the isolation is approximately 40 dB, a 20 dB improvement in isolation. Further improvement in isolation can be obtained by decreasing the resistance of the off-port shunt control element

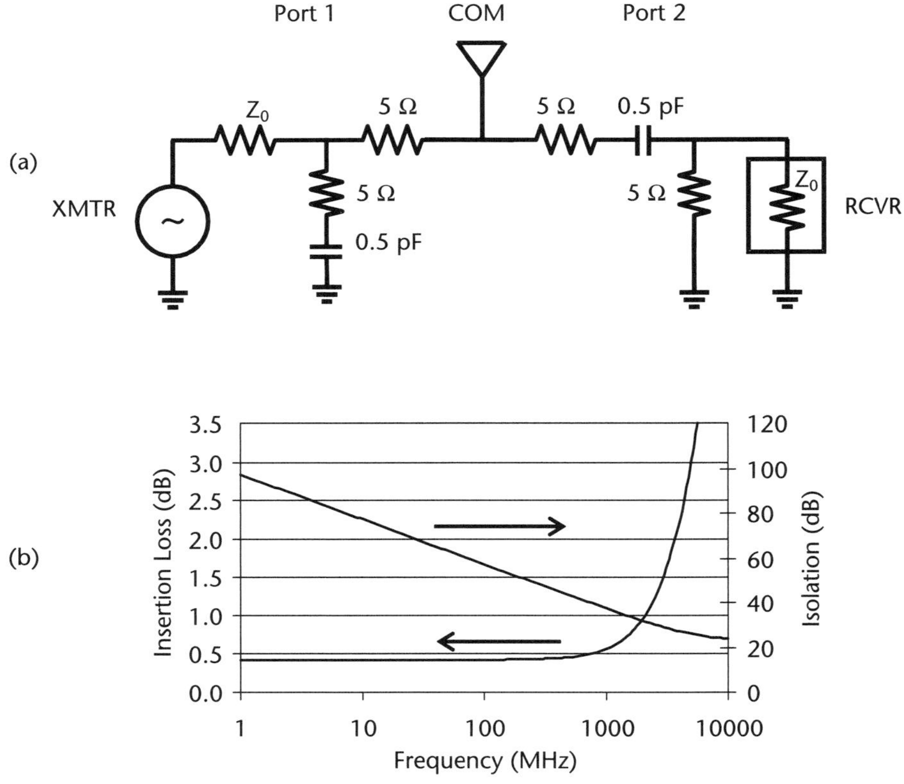

Figure 7.19 (a) Four-device illustrative example TR switch using lumped control elements. (b) Insertion loss and isolation for the example four-device TR switch using lumped control elements.

to increase the level of reflected energy. A similar decrease in the series resistance decreases the insertion loss in the on-arm as well. For PIN diode switches, increasing the dc forward current lowers the RF and microwave resistance but at the expense of increased power supply requirements and potentially an increase in switching speed. In FET-based switches, a physically larger device aspect ratio (W/L) provides this lower resistance, but at the expense of increased capacitance; this increased capacitance negatively impacts the overall frequency response.

The next two sections will separately cover this four-device TR switch using PIN diodes and FET-based control elements. The discussion will not be as detailed as in the two-device TR switch. Instead, various device parameters will be varied to observe the interplay between control device and circuit performance and to show the utility of the various models in control circuit simulation. The PIN diode discussion will use the same physical SPICE model parameters as shown in Table 7.4 and similarly for the FET-based four-device TR switch. Using the same models will aid in comparing the performance between the two TR switch topologies.

7.2.2.1 PIN Diodes as Switch Elements

The ideal control elements in the four-device SPDT TR switch discussed in the previous section are now substituted by PIN diodes using the physical PIN diode model as shown in Figure 7.20. Because of the added shunt-connected diodes and their locations in the RF path, several additional inductor and capacitors for dc decoupling are required. The choice of inductance and capacitance value, as previously discussed for the two-PIN diode TR switch, will influence the bandwidth and switching. For the simulation example, the same list of PIN diode parameters in Table 7.4 are used in the SPICE netlist file 'netlist7-15.net'. Figure 7.21(a) shows the results of the SPICE simulation for the four-PIN diode SPDT TR switch using the physical PIN diode model. The modeling results indicate that the insertion loss has not changed in terms of value or frequency from that observed in the two-PIN diode TR switch (Figure 7.8). Significant improvement in isolation, however, is noted with the four-PIN diode switch, with the 1000 MHz isolation approximately 20 dB better than that modeled using the two-device TR switch.

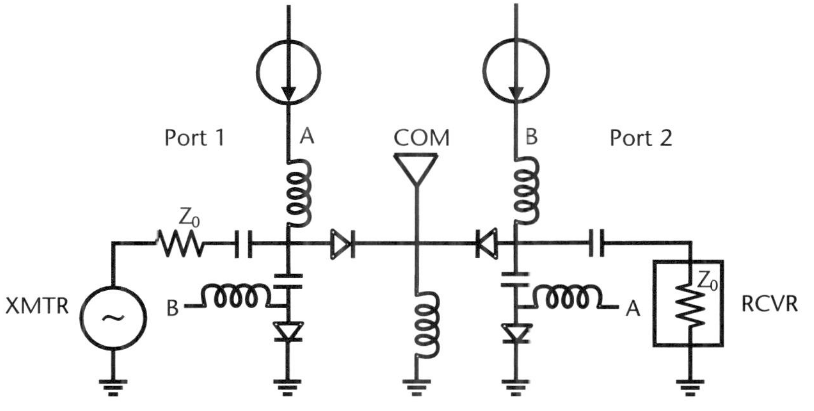

Figure 7.20 Four-PIN diode SPDT TR switch. The port labels A and B indicate identical dc biasing points.

The insertion loss is directly related to the RF impedance of primarily the series dc forward biased PIN diode and therefore can be changed by adjusting the dc forward current through the device. Figure 7.21(b) shows the influence of the dc forward current on the SPDT switch. Increasing the dc forward current reduces the insertion loss at all frequencies, with an insertion loss of 0.2 dB observed for 100 mA dc forward current. However, this considerably larger dc forward current may be a dc power budget issue in battery-operated or other dc power-limited systems. Lower dc forward current will cause a corresponding increase in the insertion loss.

In most commercial PIN diodes, I-region carrier lifetime τ tends to increase with I-region width *W*. Wider PIN diodes tend toward higher reverse breakdown voltages and hence higher power handling. For high-power TR switches, these wider I-region diodes may be necessary, and so it is of interest to see how changing the PIN diode characteristics can impact TR switch performance. For this study, the physical PIN diode model will be used with a 200 μm I-region width, 10 μS carrier lifetime PIN diode with 1.0 pF reverse bias capacitance and a 1000 volt reverse breakdown voltage. The same dc forward currents used in the simulation of Figure 7.21(b) will also be used for comparison purposes. These simulations, shown in Figure 7.22, indicate a higher level of insertion loss for each of the dc forward currents for the larger I-region width PIN diode compared with its smaller I-region counterpart. The

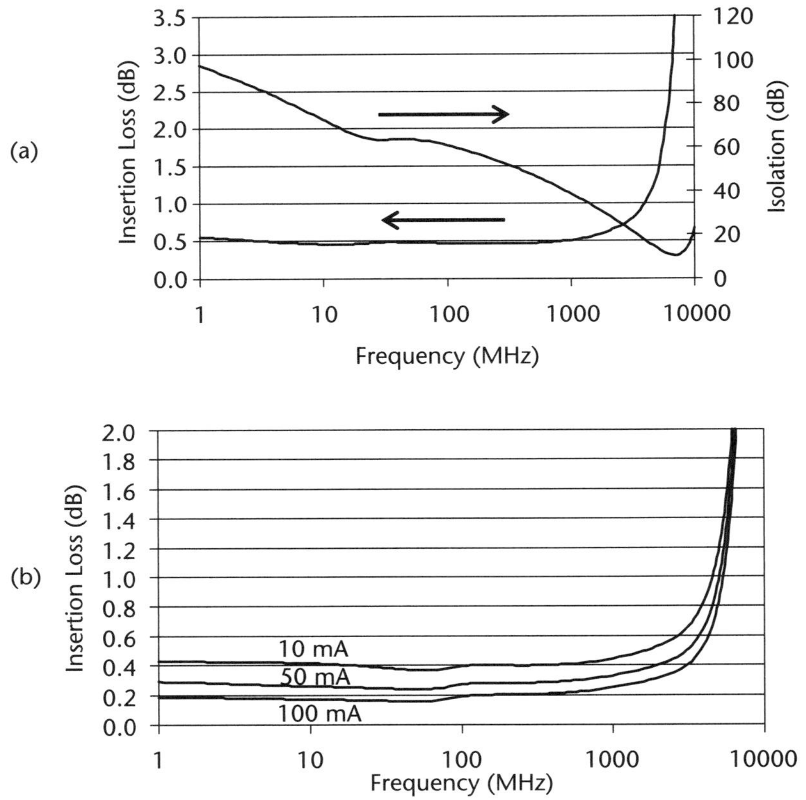

Figure 7.21 (a) Simulated insertion loss and isolation of the four-PIN diode SPDT TR switch using the physical PIN diode model. (b) Insertion loss of the four-PIN diode SPDT switch as a function of frequency with dc forward current as a parameter.

bandwidth of the switch is also reduced; a significant increase in insertion loss is observed at frequencies above approximately 500 MHz compared with 1000 MHz for the smaller PIN diode. This reduction in bandwidth is primarily caused by the increased capacitance in the larger PIN diode. For 10 mA dc forward current, the insertion loss is approximately 0.7 dB, which corresponds to a PIN diode on-state resistance of 8.7Ω. This high value of resistance implies that approximately 17% of the applied RF power is dissipated by the PIN diodes. In a high-power application, survivability of the PIN diode would need to be questioned. Increasing the dc forward current lowers the PIN diode resistance and insertion loss and lowers the power dissipation needs of the PIN diode. A dc forward current of 100 mA lowers this insertion loss to less than 0.2 dB.

7.2.2.2 FET-Based Switch Elements

The ideal control elements in the four-device SPDT TR switch discussed in the previous section can also be substituted with FET-based switch elements as in the two-device SPDT TR switch case (Figure 7.23(a)). For this discussion, only the MOSFET-based switch will be used because many of the same characteristics will be observed with the MESFET-based switch.

For the simulation example, the same SPICE BSIM4.6.1 example parameters used in the two-MOSFET SPDT TR switch will be used for comparison purposes (SPICE netlist file 'netlist7-16.net'). Figure 7.23(b) shows the results of the SPICE simulation for the four-MOSFET SPDT TR switch with floating substrate (R_B = 10 kΩ) using the BSIM4.6.1 example parameters in the model. The modeling results indicate that the insertion loss has not changed in terms of value or frequency from that observed in the two-MOSFET SPDT TR switch (Figure 7.13(b)). The 1000 MHz isolation is improved by approximately 20 dB in the four-MOSFET SPDT TR switch compared with its two-MOSFET counterpart.

Further improvement in isolation can be obtained by decreasing the MOSFET resistance of the off-port shunt control element to increase the level of reflected energy. A similar decrease in the MOSFET series resistance decreases the insertion loss in the on-arm as well. The gate width *W*, and the corresponding device aspect ratio *W*/*L*, for the MOSFET can be increased to create this lower RF resistance.

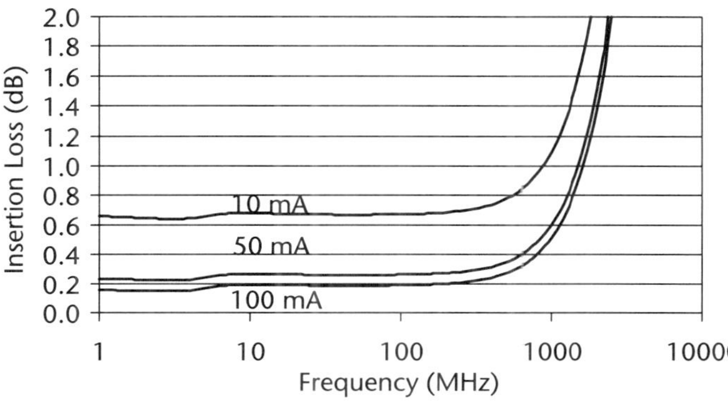

Figure 7.22 Simulated insertion loss of the 200 μm I-region width PIN diode SPDT TR switch as a function of frequency with dc forward current as a parameter.

Doing so, however, will increase both the gate area and source and drain area and perimeters, which will increase the overall device capacitance and negatively impact the overall frequency response. This effect can be seen if the MOSFET gate width W is doubled, which will reduce the RF resistance by a factor of two with a theoretical insertion loss of approximately 0.2 dB. Figure 7.23(c) shows the

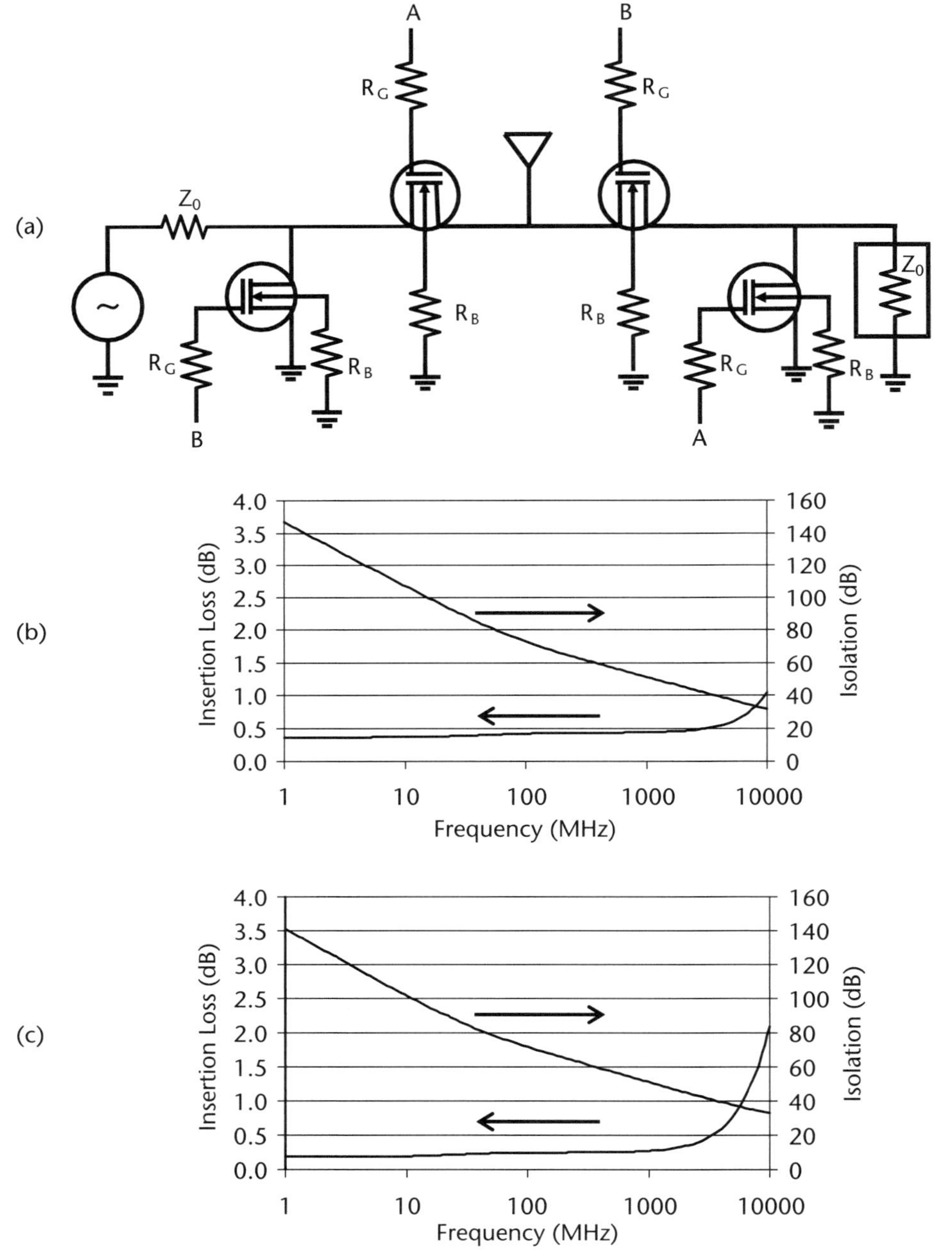

Figure 7.23 Four-MOSFET SPDT TR switch: (a) SPDT TR switch schematic, (b) simulated insertion loss and isolation over frequency, and (c) simulated insertion loss and isolation over frequency for the increased size MOSFETs. The port labels A and B indicate identical dc biasing points.

impact of this increase in MOSFET aspect ratio. In the simulation, the gate width W was increased by a factor of two and the source and drain geometries have been increased based on the layout and geometry effects discussed in Chapter 5 (SPICE netlist file 'netlist7-17.net'). The anticipated reduction in insertion loss to 0.2 dB is seen in the simulation results. The insertion loss begins to degrade at a lower frequency than the smaller MOSFET-based TR switch, although 0.5 dB insertion loss occurs at approximately the same frequency (3 GHz) for both TR switches. Above this frequency, the insertion loss degrades more rapidly because of the increased capacitance. The low-frequency isolation shows degradation as well (approximately 6 dB), but as the frequency increases, the impact of the on-state shunt MOSFET in the off-arm causes the isolation to become comparable.

7.2.3 Tuned $\lambda/4$ Transmission Line SPDT TR Switches

The tuned SPDT TR switch uses two shunt-connected control devices with two $\lambda/4$ transmission lines connecting the devices to the COM port (Figure 7.2(c)). The control devices are controlled by complementary signals so that the devices are not on at the same time. Ideally, when the shunt device is in its low impedance state (on-state), the low shunt impedance is transformed to a high shunt impedance by the $\lambda/4$ transmission line, which effectively removes the shunting element from the opposite switch arm. The shunt element, however, still protects the isolated port since the low impedance creates a high reflection coefficient at the port. This impedance transformation only occurs over a relatively narrow bandwidth (approximately 10%) because of the nature of the $\lambda/4$ transmission lines; hence, this is a TR switch "tuned" to a narrow operating frequency range.

7.2.3.1 Lumped Element Example

Some fundamental features of simulations of the SPDT TR switch can be discussed by replacing the generalized control elements with ideal switch elements shown in Figure 7.24. For illustrative purposes consistent with the other TR switches described in the section, an on-state impedance of 5Ω (Port 1) and off-state impedance of 5Ω in series with a 0.5 pF capacitor (Port 2) will be used. The $\lambda/4$ transmission lines have a 50Ω characteristic impedance and are based on a design frequency of 100 MHz. Figure 7.25 shows insertion loss (Figure 7.25(a)) and isolation (Figure 7.25(b))

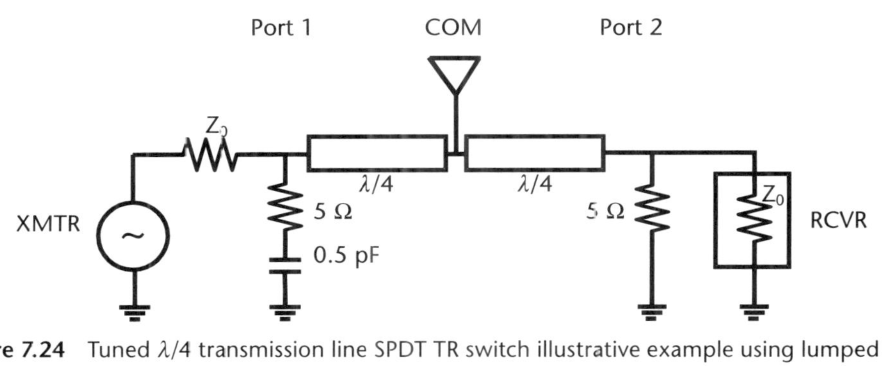

Figure 7.24 Tuned $\lambda/4$ transmission line SPDT TR switch illustrative example using lumped control elements.

results of a SPICE frequency sweep analysis using the SPICE netlist file 'netlist7-18.net'. This frequency sweep shows an insertion loss of less than 0.5 dB only in the vicinity of the design frequency of 100 MHz; other low insertion loss frequencies occur at odd multiples of the design frequency because of the repeating impedance transformation behavior of the 100 MHz $\lambda/4$ transmission lines. The isolation is relatively low at all frequencies, with only 21 dB of isolation occurring at the design frequency. Improved isolation can be obtained by decreasing the shunt resistance on the isolated arm; a 1Ω resistance provides better than 30 dB isolation at 100 MHz.

The $\lambda/4$ transmission line may also be replaced by a lumped element equivalent as defined in (1.40) for a more compact design if low-frequency applications are anticipated. Other variations of this TR switch use, for example, a series device without the associated $\lambda/4$ transmission line in the on-arm and a shunt-connected device with the $\lambda/4$ transmission line in the off arm [10].

7.2.3.2 Active Control Elements—PIN Diode and MOSFET

The ideal lumped control elements in the tuned $\lambda/4$ transmission line SPDT TR switch discussed in the previous section can be replaced by active control elements,

(a)

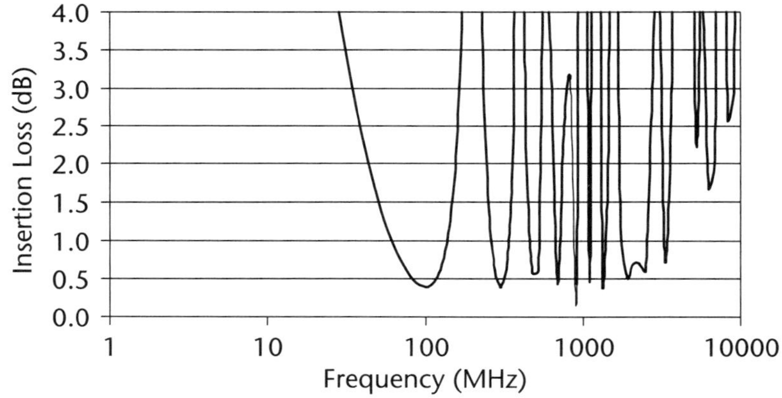

(b)

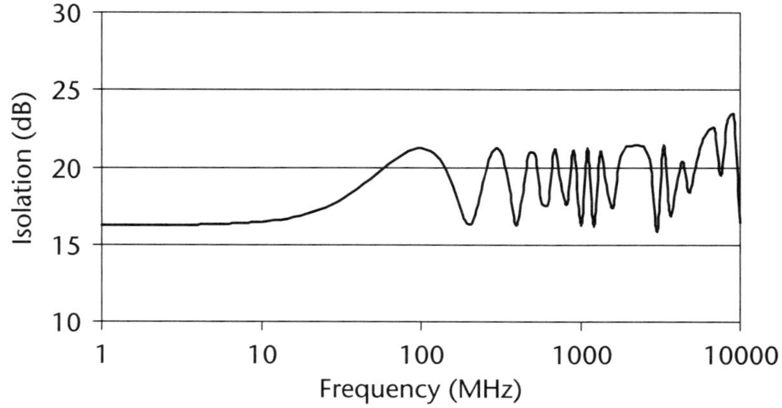

Figure 7.25 Simulation results for the example tuned $\lambda/4$ transmission line SPDT switch using lumped control elements: (a) insertion loss and (b) isolation.

such as PIN diodes and FETs to actively control the switch state (Figure 7.26). This section will look at simulations of the response of this switch topology by replacing the lumped elements with PIN diodes using the physical PIN diode model parameters in Table 7.4 and MOSFETs using the previously described SPICE BSIM4.6.1 example parameters (SPICE netlist files 'netlist7-19.net' and 'netlist7-20.net').

Figure 7.27 shows insertion loss and isolation results of the SPICE simulation for the $\lambda/4$ transmission line SPDT TR switch using PIN diodes with the physical PIN diode model. Three different dc forward currents, 10, 50, and 100 mA, were simulated to show the impact of decreasing PIN diode RF resistance on switch performance. Figure 7.27(a) shows insertion loss centered at the 100 MHz design point. Increasing the dc forward current from 10 to 50 mA reduces the insertion loss at the design frequency from approximately 0.4 dB to 0.2 dB. Further increasing the dc forward current (100 mA) shows only marginal improvement in insertion loss. In a real application, this marginal improvement in insertion loss may not warrant the doubling of the current required from the dc power supply. Significant improvement in isolation, however, is noted by increasing the dc forward current in the off-arm shunt devices as shown in Figure 7.27(b). A 10 dB improvement in the 100 MHz isolation is noted when the dc forward current is increased from 10 to 100 mA, indicating the engineering tradeoffs that may need to be made between insertion loss, isolation, and the required dc forward current and associated power consumption of the power supply.

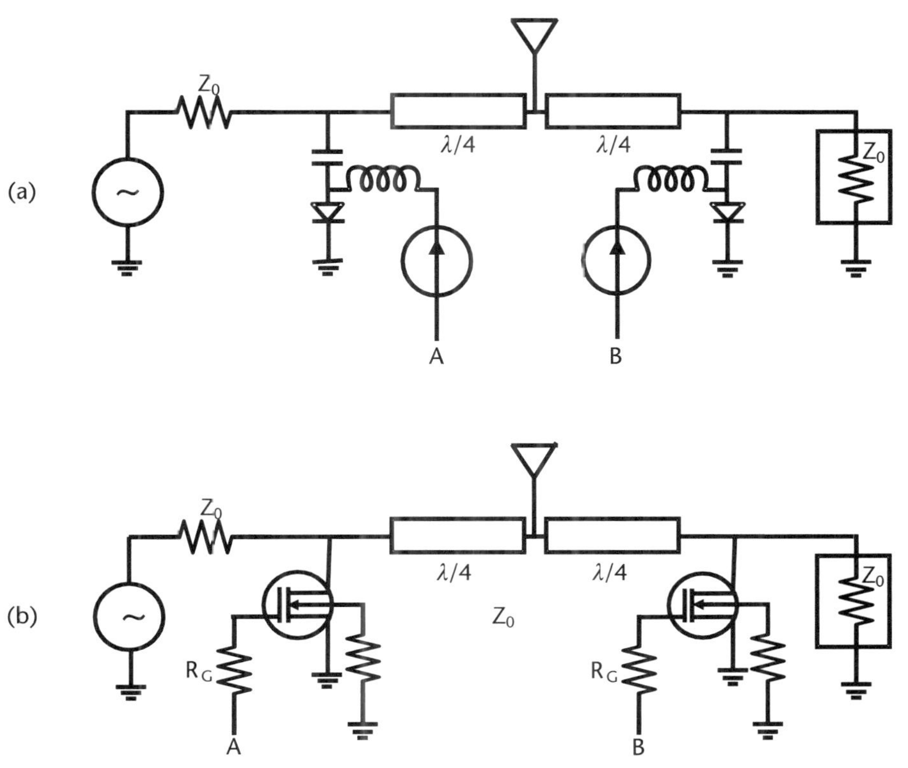

Figure 7.26 Tuned $\lambda/4$ transmission line SPDT TR switch: (a) PIN diode control elements, and (b) MOSFET control elements. The port labels A and B indicate complementary dc biasing points.

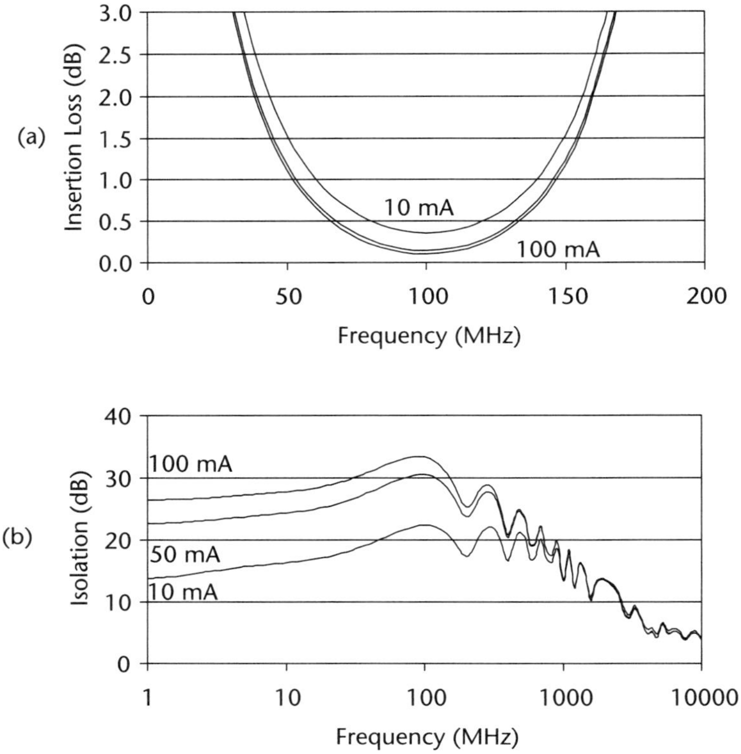

Figure 7.27 Simulated frequency response for the example tuned $\lambda/4$ transmission line SPDT switch using PIN diode control elements with dc forward current as a parameter: (a) insertion loss centered at the 100 MHz design frequency and (b) isolation over a 1 MHz to 10 GHz frequency span.

MOSFETs can also be used in the $\lambda/4$ transmission line SPDT switch as shown in Figure 7.26(b). For the simulation example, the same SPICE BSIM4.6.1 example parameters and a floating substrate (R_B = 10 kΩ) used in the two-MOSFET SPDT TR switch will be used for comparison purposes (SPICE netlist file 'netlist7-21.net'). The two MOSFET aspect ratios used in the previous section to show the impact on changing device geometry in the four-device SPDT TR switch will also be used in this analysis. Figure 7.28 shows the results of the SPICE simulation for the two-MOSFET $\lambda/4$ transmission line SPDT switch with floating substrate using the BSIM4.6.1 example parameters in the model. The modeling results indicate that the insertion loss has been reduced by approximately 0.15 dB for the larger MOSFET; the isolation has improved by approximately 6 dB for the larger device.

7.2.4 Linear Balanced Duplexer-Based Switch for Magnetic Resonance Imaging (MRI)

In the medical field, there is a need for TR switches in instruments such as magnetic resonance (MR) imagers [12–15]. In MR transmit coils, the high-power RF signal (in the range of kilowatts of pulsed power) must be switched on/off rapidly so that the receive circuitry will be able to detect the weak signal from the relaxing flipped

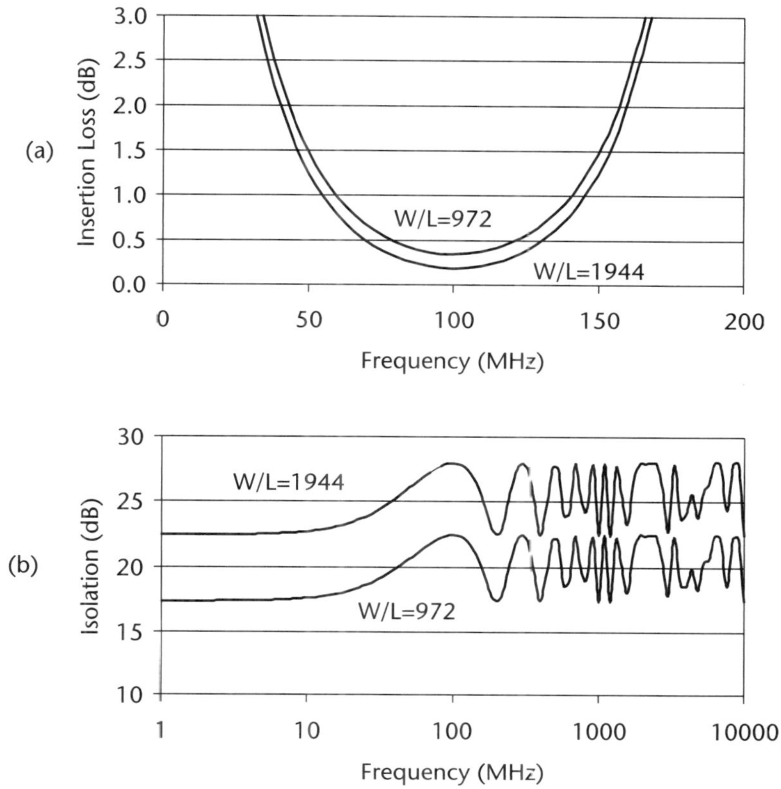

Figure 7.28 Simulated frequency response for the example tuned $\lambda/4$ transmission line SPDT switch using two different MOSFET control element geometries: (a) insertion loss centered at the 100 MHz design frequency and (b) isolation over a 1 MHz to 10 GHz frequency span.

atomic spins, or to detune the transmit coil so that it has minimal coupling to the receive coils(s). On the receive side, high-speed switching is required for detuning and signal blocking of the MR receive coils for both protection of the patient as well as the sensitive low noise, high gain receive pre-amplifiers. MR coils (both transmit and receive) are often of high Q, so the switches must be very low loss in both states so as not to degrade the Q of the MR coils.

In the MR system, a high-power transmit pulse excites the atomic spins; when the pulse is removed, the very weak signal originating from the relaxing spins is then received. Transmit and receive signals are within the same narrow frequency band, and so the MR system resembles a classical narrowband communication system with a standard TR switch based on the linear balanced duplexer (LBD) [6]. The LBD is often used in an MR T/R switch application where a single coil is used for both transmit and receive [13]. Figure 7.29 shows the basic arrangement of two identical and ideal matched ($S_{ii} = 0$) 90° quad hybrids connected in series, with Port 1 the input port, Port 2 the through port, Port 3 the coupled port, and Port 4 the isolated port (the quad hybrid was discussed in Section 1.3.3.3.1). Z_{CTL} are the control elements to ground. As shown in Figure 7.29, in transmit mode (source path to RF coil), both Z_{CTL} are in the low impedance state. In this case, all transmit power is circulated in the left half and reflected to the RF coil. In receive mode,

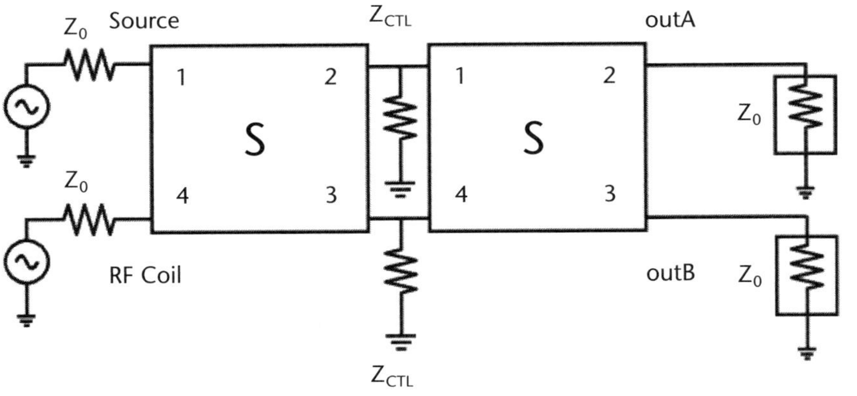

Figure 7.29 Block diagram of the basic MR LBD-based TR switch.

both Z_{CTL} are in the high impedance state, and the RF coil signal originating from the relaxing flipped spins of the specimen is split into 90° components, passing by Z_{CTL}, and then recombining at output Port outA. Port outB (Idle) absorbs any RF leakage or unbalanced power and is terminated in the system impedance (usually $Z_0 = 50\Omega$). Often, an additional control element is placed on Port outA to provide added isolation during transmit.

The voltage transmission ratios between the important switching ports can be computed assuming control elements with impedance Z_{CTL}:

$$\frac{V_{outA}}{V_{\text{RF-Coil}}} = (1+\Gamma)\left(S_{21}\frac{S_{24} + \Gamma S_{23}S_{24}}{1-\Gamma^2 S_{23}S_{32}} + S_{24}\frac{S_{34} + \Gamma S_{32}S_{24}}{1-\Gamma^2 S_{23}S_{32}} \right) \tag{7.2a}$$

$$\frac{V_{outB}}{V_{\text{RF-Coil}}} = (1+\Gamma)\left(S_{31}\frac{S_{24} + \Gamma S_{23}S_{24}}{1-\Gamma^2 S_{23}S_{32}} + S_{34}\frac{S_{34} + \Gamma S_{32}S_{24}}{1-\Gamma^2 S_{23}S_{32}} \right) \tag{7.2b}$$

$$\frac{V_{\text{RF-Coil}}}{V_{\text{Source}}} = S_{41} + \Gamma\left(S_{42}\frac{S_{21} + \Gamma S_{23}S_{31}}{1-\Gamma^2 S_{23}S_{32}} + S_{43}\frac{S_{31} + \Gamma S_{32}S_{21}}{1-\Gamma^2 S_{23}S_{32}} \right) \tag{7.2c}$$

where the reflection coefficient term Γ at the control elements is defined as

$$\Gamma = \frac{-Z_0}{2Z_{CTL} + Z_0} \tag{7.3}$$

A common implementation of the general LDB structure for TR switches in MR imagers uses two quasi-lumped element quad 90° hybrids fabricated using two $\lambda/8$ lines with capacitive coupling (see Section 1.3.3.3). Using this ideal four-port network representation for the quad hybrid allows estimation of the isolation and insertion loss afforded by the TR switch. With this configuration, outA becomes the receive Port and outB becomes the Idle Port.

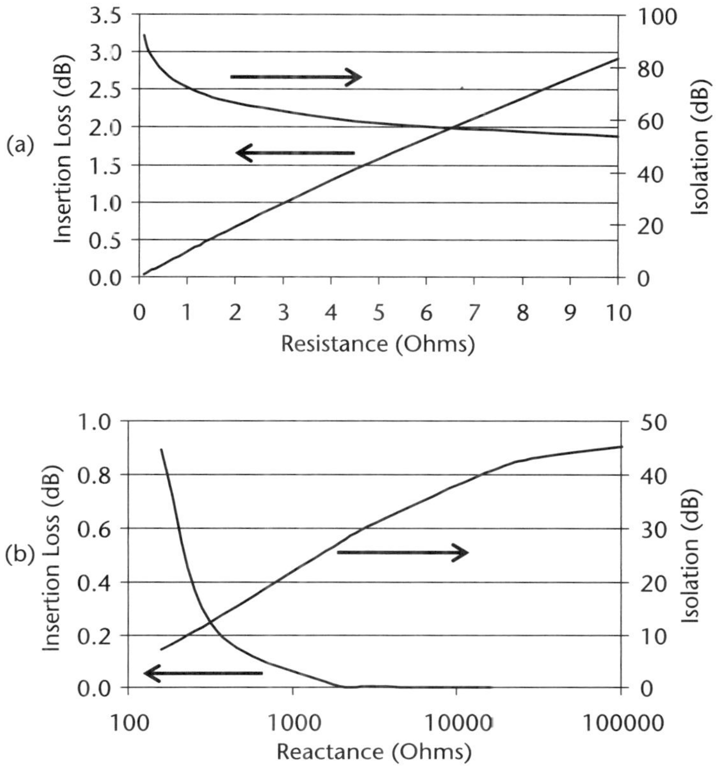

Figure 7.30 LBD-based TR switch using lumped elements for Z_{CTL}: (a) transmit state RF coil insertion loss and Preamp isolation; (b) receive state Preamp insertion loss and RF coil port isolation.

Figure 7.30(a) shows the degree of insertion loss and isolation of the various ports during the transmit and receive states using the quad hybrid and ideal lumped elements for Z_{CTL}. Isolation values at the Preamp port of approximately 70 dB with insertion losses on the coil port below 0.5 dB can be obtained if Z_{CTL} is purely resistive with a resistance of 1.0Ω. Lower Z_{CTL} resistance improves both the insertion loss and isolation results. Figure 7.30(b) shows Preamp port insertion loss and RF coil port isolation as a function of Z_{CTL} in its off, capacitive reactance state (switch in receive mode). For capacitive reactances higher than 1,000Ω, Preamp insertion loss is better than 0.1 dB with greater than 20 dB RF coil port isolation. The 1,000Ω reactance at 298 MHz (7 Tesla Larmor frequency for hydrogen, the most commonly imaged element [14, 15]) corresponds to approximately 0.5 pF.

By using a pair of quasi-lumped element quad hybrids and PIN diodes for Z_{CTL} (Figure 7.31), the general LBD circuit structure for a 7 Tesla MR system can be developed (SPICE netlist 'netlist7-22.net'). Each of the transmission line sections is $\lambda/8$ (50Ω) at 298 MHz. Capacitors C1 and C3 exhibit a 50Ω reactance (10.68 pF) and C2 a 25Ω reactance (21.27 pF) at 298 MHz. The PIN diodes in this example use the physical PIN diode model parameters listed in Table 7.4. In this circuit, when diodes D1/D2 are turned on, all the transmit power is directed to the RF Coil; in the opposite switch state (D1/D2 off), the RF signal coming from the coil is split between the two hybrids and recombines in-phase at the receive port while the

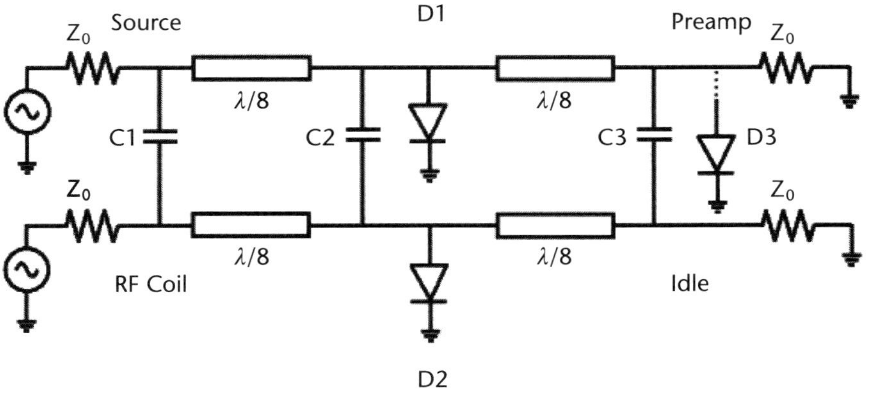

Figure 7.31 Implementation of the LBD-based MR TR switch using PIN diodes (dc biasing structures not shown).

idle port absorbs leakage or unbalanced power [14, 15]. PIN diode D3 is not used in this simulation, although it could be added for increased Preamp isolation. Dc forward currents of 10, 50, 100, and 500 mA were used to observe the impact of reduced resistance on insertion loss and isolation in the TR switch for comparison with Figure 7.32. Figure 7.32(a) and 7.32(b) shows the simulation results of applying the physical PIN diode at the four dc forward currents on the insertion loss between the source and RF coil ports and the Source and Preamp ports. The tuned nature of the LBD TR switch is especially noted in the isolation, Figure 7.32(b) and 7.32(c), where the Source-Preamp isolation on transmit shows a sharp peak and the receive RF Coil-Preamp insertion loss shows a minimum at the design frequency of 298 MHz. These results show good agreement with the theoretical results in Figure 7.30(a), indicating the robustness of the physical PIN diode model in simulating more complex TR switch structures such as the LBD.

This design can be scaled in frequency by simply scaling the length of the LBD transmission lines and recalculating C1, C2, and C3 for the new design frequency.

7.3 Switched Passive Element for Tuning and Matching

The semiconductor control elements covered in this book can be used in a wide variety of electronically reconfigurable circuits such as automatic antenna tuners and impedance matching networks in both low- and high-power applications in areas as diverse as satellite communications and cognitive radio [16–21]. These applications require switching in banks of inductors and/or capacitors, often under the guidance of an optimization algorithm that quickly accomplishes the tuning or matching procedure. Reconfigurable impedance tuners significantly increase the bandwidth of a matching network by switching in appropriate combinations of capacitors (and to a lesser extent, inductors) [22–24] or transmission line tuning stubs [25]. For power amplifiers, changing power levels change the output impedance of the amplifier, often requiring retuning of the output matching network. Care must be taken in these applications to ensure that impedances presented to the amplifier do not cause instability. In MRI applications, the transmit coils in the MR scanner

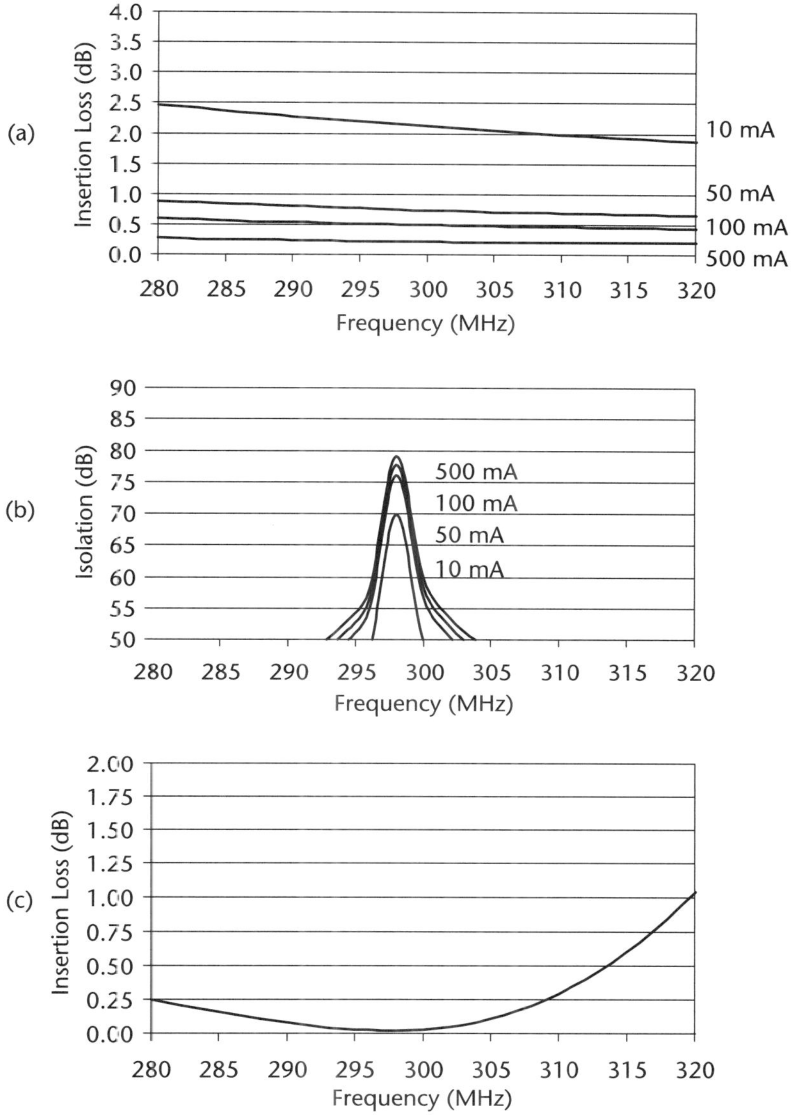

Figure 7.32 LBD-based TR switch using the physical PIN diode model: (a) transmit state RF coil insertion loss, (b) transmit state Preamp isolation, and (c) receive state Preamp insertion loss.

need to be retuned with different patients because the different body masses and sizes cause the coil to slightly detune [16, 17]. On-chip solutions use MOSFET and MESFET-based switches, whereas discrete and high-power applications often use discrete PIN diodes and PIN diode arrays.

7.3.1 Capacitor and Inductor Bank Switching

PIN diodes have been used in a variety of matching and tuning applications. These devices are especially important in high-power applications and are often the only semiconductor devices that have the combination of power handling and switching

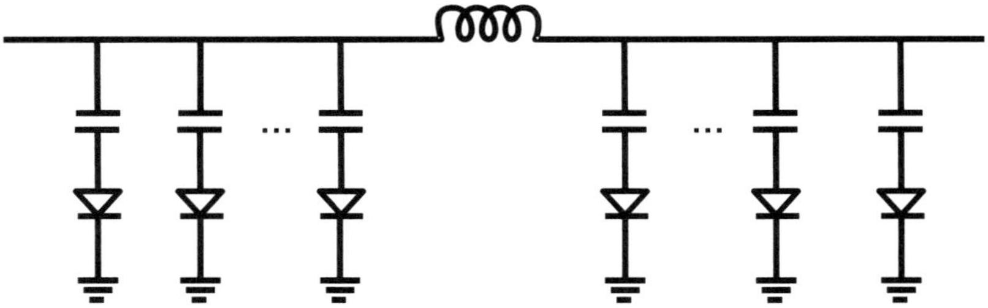

Figure 7.33 PIN diodes can be used to switch in capacitor banks for matching (control and dc biasing structures not shown) (after [21]).

speed for many applications. Figure 7.33 shows an automatic impedance matching network with capacitor banks being reconfigured by PIN diodes. Each PIN diode can be individually addressed by a control circuit to switch the appropriate capacitor(s) for best impedance match. In Figure 7.33, high reactance inductors will be needed at the anode of each PIN diode to the control system to vary the device's state; the dc bias is isolated from the remainder of the circuit by the series capacitors in the array.

MOSFET switches have been used with stacked inductor technologies for on-chip applications [26] to implement a voltage-variable inductor that can be changed in discrete inductance steps. In Figure 7.34(a), the total inductance ranges between L and 3L, depending on the state of the control voltages V_A and V_B. MOSFETs can

(a)

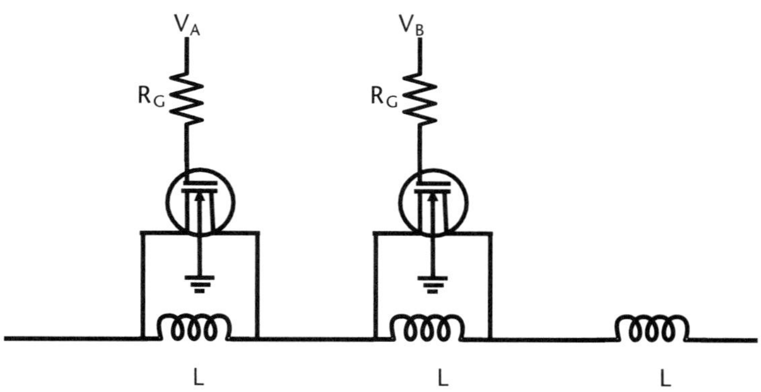

(b)

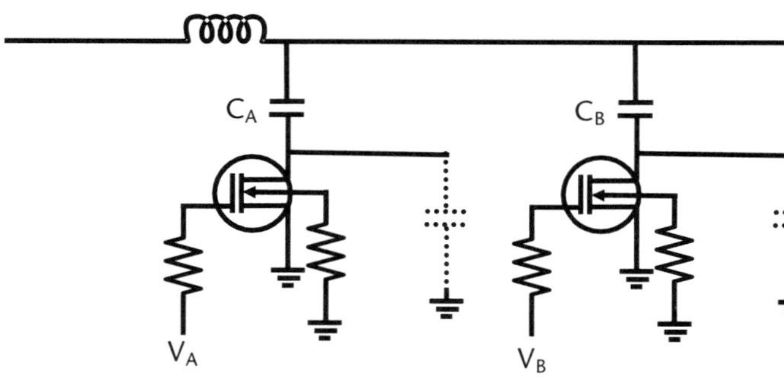

Figure 7.34 (a) MOSFET switching of lumped element components (after [26]). (b) MOSFET switching of lumped capacitor banks for filter or tank circuit tuning.

also be used to switch in banks of capacitors, effectively placing all the elements in parallel (Figure 7.34(b)). For low-loss ground paths, the MOSFETs must exhibit a low impedance in their on-state, requiring large aspect ratio devices with a corresponding high parasitic capacitance (dotted capacitor in Figure 7.34(b)), a major drawback for this technique.

References

[1] LT-SPICE IV, Linear Technology Corporation, 2015 [http://www.linear.com/designtools/software/#Ltspice].

[2] Thompson, R.E., "The Uni-Control Receiver," *IRE Proc.*, Vol. 7, No. 5, October 1919, pp. 499–514.

[3] Jones, E. T., *Radio Controlling Device*, U.S. Patent 1,415,179-, filed July 21, 1920, issued May 9, 1922.

[4] Brady, J. B., "Digests of United States Patents Relating to Radio Telegraphy and Telephony," *IRE Proc.*, Vol. 10, No. 4, 1922, pp. 290–311.

[5] Samuel, A. L., J. W. Clark, and W. W. Mumford, "The Gas-Discharge Transmit-Receive Switch," *The Bell System Technical Journal*, Vol. 25, No.1, January 1946, pp. 48–101.

[6] Smullin, L. D., and C. G. Montgomery, "Microwave Duplexers," M.I.T. Rad. Lab. Ser., New York: McGraw-Hill Book Co., Inc., Vol. 14, Ch. 8, 1948 (see also Jones, C.W., "Broad-Band Balanced Duplexers," *IRE Trans. on Microwave Theory and Techniques*, Vol. 5, No.1, January 1957, pp. 4–12).

[7] Walston, J., *SPICE Circuit Yields Recipe for PIN Diode, Microwaves & RF*, November 1992, pp. 78–89.

[8] Ardizzoni, J., "Driving PIN Diodes: The Op-Amp Alternative," *Analog Dialogue* 44-02, February 2010. [http://www.analog.com/library/analogDialogue/archives/44-02/pin_diode.pdf]

[9] Doherty, W. E., and R. Joos, *PIN Diode Circuit Designers Handbook*, Watertown, MA: Microsemi Corp, 1998.

[10] Hiller, G., *Design with PIN Diodes*. M/A-COM Application Note AG312, 2010 (see also *Design with PIN Diodes*, Skyworks Application Note 200312 Rev. C, April 2009).

[11] "Understanding RF/Microwave Solid State Switches and their Applications," Agilent Application Note 5989-7618EN, 2010 [http://cp.literature.agilent.com/litweb/pdf/5989-7618EN.pdf]

[12] Caverly, R., W. E. Doherty Jr., and R. Watkins, "298MHz Micro-Miniature 2KW Transmit Receive Switch for 7.0 Tesla TR Arrays," *2012 Intl. Symp. Magnetic Resonance in Medicine (ISMRM-12)*, May 2012.

[13] Caverly, R., W. E. Doherty Jr., and R. Watkins, "Modeling High Speed MRI Coil Switching using PIN Diodes," *Proc. 2011 IEEE Intl. Microwave Symp.*, May 2011, pp. 1–4.

[14] Omar, A., et al., "A Microwave Engineer's View of MRI," *IEEE Microwave Magazine*, Vol. 12, No. 3, May 2011, pp. 78–86.

[15] Caverly, R., "MRI Fundamentals: RF Aspects of Magnetic Resonance Imaging (MRI)," *IEEE Microwave Magazine*, Vol. 16, No. 6, July 2015, pp. 20–33.

[16] Sohn, S-M., et al., "Design of an Electrically Automated RF Transceiver Head Coil in MRI," *IEEE Trans. on Biomedical Circuits and Systems*, 2015.

[17] Sohn, S-M., et al., "8-Channel RF Head Coil of MRI with Automatic Tuning and Matching," *Microwave Symposium Digest (IMS), 2013 IEEE MTT-S International*, June 2013, pp. 1–3, 2–7.

[18] Ranta, T., et al., "Next-Generation CMOS-on-Insulator Multi-Element Network for Broadband Antenna Tuning," *Microwave Conference (EuMC), 2013 European*, Vol. 1, No. 1, October 2013, pp. 1567–1570.

[19] Franco, M. J., and D. Dening, "Broadband Reconfigurable Matching Network of Reduced Dimensions for the UHF Military Satellite Communication Band," *Microwave Symposium Digest (MTT), 2011 IEEE MTT-S International*, June 2011, pp. 1–4, 5–10.

[20] Sun, Y., J. Moritz, and Xi Zhu, "Adaptive Impedance Matching and Antenna Tuning for Green Software-Defined and Cognitive Radio," *2011 IEEE 54th International Midwest Symposium on Circuits and Systems (MWSCAS)*, Vol. 1, No. 1, August 2011, pp. 1, 4, 7–10.

[21] Song, H., et al., "Automatic Antenna Tuning Unit for Software-Defined and Cognitive Radio," *2007 IEEE Antennas and Propagation Society International Symposium*, June 2007, pp. 85–88, 9–15.

[22] Vaha-Heikkila, T., and G. M. Rebeiz, "A 20–50 GHz Reconfigurable Matching Network for Power Amplifier Applications," *Proc. 2004 IEEE Intl. Microwave Symp.*, 2004, p. 717.

[23] Lu, Y., et al., "A MEMS Reconfigurable Matching Network for a Class AB Amplifier," *IEEE Microwave Wireless Components Let.*, Vol. 13, No. 10, October 2003.

[24] Brown, E. R., "RF-MEMS Switches for Reconfigurable Integrated Circuits," *IEEE Trans. on Microwave Theory and Techniques*, Vol. 46, No. 11, November 1998, p. 1868.

[25] Papapolymerou, J., et al., "Reconfigurable Double-Stub Tuners Using MEMS Switches for Intelligent RF Front-Ends," *IEEE Trans. on Microwave Theory and Techniques*, Vol. 51, No. 1, January 2003.

[26] Park, P., et al., "Variable Inductance Multilayer Inductor with MOSFET Switch Control," *IEEE Electron Dev. Lett.*, Vol. 25, No. 3, March 2004.

[27] Sheu, B., et al., "BSIM: Berkeley Short-Channel IGFET Model for MOS Transistors," *IEEE J. of Solid State Circuits*, Vol. SC-22, 1987, pp. 558–566.

[28] Agarwal, H., et al., *BSIM6.1.0 MOSFET Compact Model Technical Manual*, Regents of the Univ. Calif., 2014.

[29] BSIM4 Official Release, BSIM Research Group at Univ. Calif. Berkeley, 2007. http://www-device.eecs.berkeley.edu/bsim/?page=BSIM4_Arc

CHAPTER 8

Control and Attenuator Applications

8.1 Introduction

The last chapter presented modeling examples of semiconductor control in directing, routing, or switching RF and microwave energy. These same solid-state devices find uses in many other microwave and RF applications, such as attenuators, phase shifters, and limiter/circuit protectors. Each of these applications will be discussed in turn in this chapter. Following the theme of Chapter 7, SPICE will be used with the same semiconductor control elements described in detail in Chapters 3 through 6 in the various examples. The reader may need to iteratively move back and forth between this chapter and the relevant preceding chapters during the development of the circuit model that fits their needs. The SPICE netlists used in the examples in the chapter can be found at the repository website listed in Chapter 1. The files are fully editable, allowing the reader to explore the impact of various changes to circuit elements in the files or to add other elements for study of new circuit topologies. The netlist files in this chapter are run using the LT-SPICE version of the generic SPICE simulator [1].

8.2 Attenuators

Attenuators, or pads, refer to circuitry used to reduce signal strength, usually in a controlled manner. Attenuators can be fixed or variable using lossy material or circuit elements to provide the controlled amount of signal reduction as well as to fit inside a required footprint. Microwave and RF measurement laboratories usually have a wide variety of passive attenuators of fixed value ranging from 3 dB up through 30 dB and with various power handling capacities and connector styles. Attenuators in these applications might be at the input of a spectrum analyzer to reduce signal levels to protect the instrument, or at the output of a signal generator to provide signal level control. Passive attenuators are also useful in impedance matching circuits since any reflection coefficient seen by a source will be reduced by introduction of the attenuator. Variable attenuators can be continuously varied over a specific attenuation range or "stepped" by switching in fixed attenuator elements of specific values. Attenuators made of lossy material or circuit elements are frequently modeled as resistive networks as outlined in Section 1.3 or in [2], with straightforward mathematical relationships relating the circuit topology to the circuit resistor values. Changing the attenuation value of passive attenuators is done mechanically and is therefore a slow process and unsuitable for high-speed RF control applications. From the material presented in the previous chapters, PIN diodes, MOSFETs, and MESFETs all have controllable resistive characteristics that allow their use as signal attenuators. Furthermore, the resistive nature of the devices can be varied rapidly by their control signals. The electronic controllability of these semiconductor control elements allows their use not just in standard attenuator applications but also as signal modulators, signal levelers, and automatic gain control (AGC) circuits [3–14].

There are two basic types of attenuators, reflective and matched. As their names imply, reflective attenuators operate by both reflecting RF energy back toward the source and dissipating some of this incident energy, thereby reducing the amount of energy arriving at the load. Matched attenuators operate by directly dissipating the RF energy while presenting an impedance match (usually to both the source and the load). In both types of attenuators, semiconductor control elements can be placed in series and/or shunt configurations depending on attenuator type. The simplest reflective attenuators usually contain one control element connected either in series or shunt (Figure 1.4), whereas matched attenuators contain several control elements in a combination of series and shunt control elements (Figure 1.12). Modeling examples in this section will focus on a series reflective attenuator and Π-connected matched attenuator; the techniques presented can be easily applied to other circuit topologies. A special class of variable attenuator, the so-called digital stepped attenuator, is based on a collection of matched attenuator pads and will be covered separately. These stepped attenuators typically use FET-based control elements to control the attenuation, although PIN diodes can also be used [15–20].

8.2.1 Reflective Attenuator

The basic microwave and RF signal attenuator is the reflective attenuator because it can be created using a single control element. For the series reflective attenuator, the attenuation level in dB (8.1) is determined by the on-state resistive characteristics of the control element, which in turn is a function of the dc bias voltage or current applied to the device. Figure 8.1 shows the attenuation in decibels as a function of on-state resistance R_{ON} for the series reflective attenuator from (8.2), showing approximately 20 dB attenuation with an R_{ON} of 1,000Ω. For practical purposes, 20 dB attenuation is approximately the upper limit that can be achieved using a single semiconductor control element because the element's on-state RF resistance becomes very sensitive to small variations in the applied dc control bias.

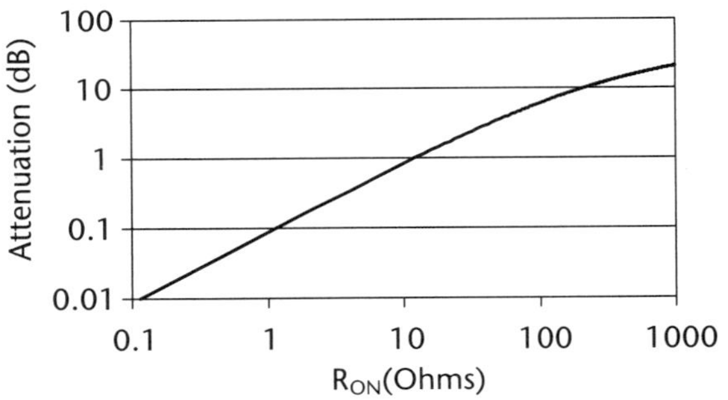

Figure 8.1 Attenuation ATT in a series reflective attenuator as a function of resistance R_{ON}.

$$ATT = 10\log_{10}\left(1 + \frac{R_{ON}}{2Z_0}\right)^2 \text{ dB} \tag{8.1}$$

The control resistance R_{ON} can originate with a passive resistor or a PIN diode or FET-based control element. The next two sections look at the series reflective attenuator's design and operation using these two semiconductor control elements.

8.2.1.1 PIN Diode Reflective Attenuator

The attenuator resistance R_{ON} in the ideal series reflective attenuator can be substituted by a PIN diode to provide electronic control over the level of attenuation (Figure 8.2(a)). Figure 8.2(a) shows two additional inductances and capacitances; these reactive elements are needed to decouple the dc current source from the RF path. The design guidelines for these decoupling elements were previously covered in Chapter 7.2. For the simulation, the physical PIN diode model will be used, with a full list of parameters shown in Table 7.4. The full SPICE subcircuit implementation of this model has been discussed previously and can be found in Table 4.3 and SPICE netlist file 'netlist8-1.net'. In this simulation, the dc current was varied from 10 to 300 μA to see the impact on the changing PIN diode resistance on attenuation over the frequency range 10 to 1000 MHz, with the results shown in Figure 8.2(b). Focusing on the attenuation level at an operating frequency of 100 MHz, attenuation levels greater than 5 dB can be obtained for dc forward currents less than approximately 300 μA, with 26 dB attenuation at 10 μA of dc forward current. For high levels of attenuation, small dc forward currents are required, and keeping this attenuation level constant requires tight control of the dc forward current biasing the diode. For example, doubling the current from 10 to 20 μA causes more than a 4 dB decrease in attenuation. Larger dc forward currents may be easier to control, and since the PIN diode RF resistance is inversely related to the dc forward current,

$$R_{ON} = \frac{W^2}{2\mu I_{dc}\tau} \tag{8.2}$$

larger I-region devices may be used. Larger I-region PIN diodes also tend to have longer I-region carrier lifetimes τ, and so larger dc forward currents will also increase the I-region stored charge $I_{dc}\tau$. This increased stored charge can reduce the distortion introduced by the PIN diode into the circuit. The impact of increased I-region stored charge will be discussed in the next paragraph. Figure 8.2(c) shows the simulation results of a larger PIN diode with I-region thickness of 200 μm and carrier lifetime of 2 μS (all other parameters remain the same in 'netlist8-2.net'). For the same dc forward current of 10 μA, the thicker PIN diode yields approximately 8 dB more attenuation than its thinner counterpart. The lowest attenuation level of 5 dB, however, requires 2 mA of dc forward current, almost an order of magnitude more than the thinner device. Higher levels of attenuation can therefore be obtained using thicker PIN diodes for a given dc forward current.

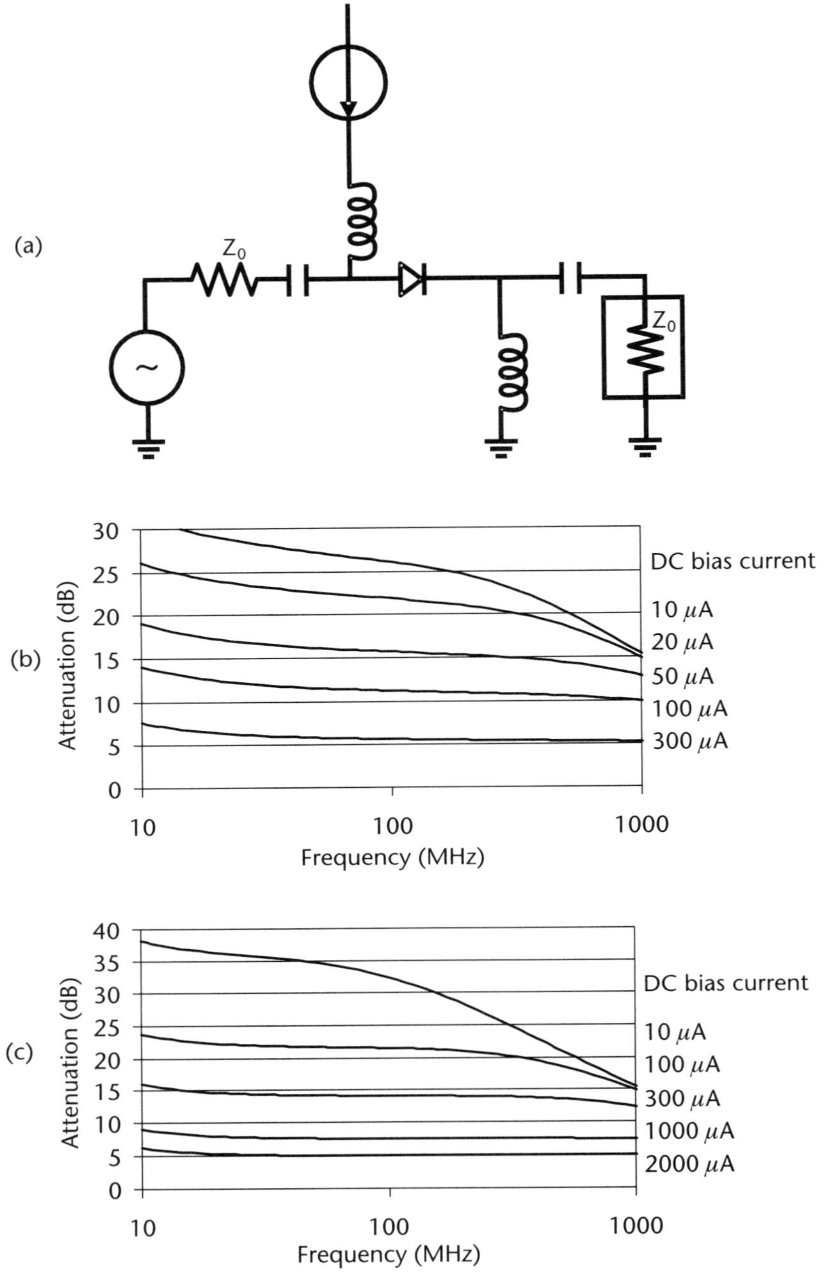

Figure 8.2 (a) PIN diode series reflective attenuator circuit, (b) attenuation as a function of frequency with dc bias current as a parameter ($W = 25\ \mu$m, $\tau = 150$ nS), and (c) attenuation as a function of frequency with dc bias current as a parameter ($W = 200\ \mu$m, $\tau = 2\ \mu$S).

An advantage of using thicker PIN diodes in attenuator circuits can be seen by studying the distortion intercept points IP2 and IP3 defined earlier in Chapter 4 in (4.21). The key factor in the intercept point relationships is the ratio QF/R_{ON} where Q is the stored charge ($I_{dc}\tau$) and F is the frequency of operation. A large QF/R_{ON} ratio gives rise to high distortion intercept points and hence lower generated

PIN diode distortion. Heavily biased PIN diodes (with large stored charge) exhibit low resistances (R_{ON}) and hence provide low distortion operation. In attenuator applications, however, the dc forward current is kept deliberately small to achieve high levels of attenuation, and so the stored charge Q is lower than it would be in a switch situation. In addition, the associated large values of R_{ON} required for a given level of attenuation can make the factor QF/R_{ON} small, implying lower intercept points and therefore higher levels of distortion. A modified version of the IP2 and IP3 relationships of (4.21) applicable to series reflective attenuator circuits has been advanced by looking at the relationship between R_{ON} and the attenuation level [21]. For two closely spaced frequencies ($F_1 = F_2 = F$) and R_{ON} as defined in (8.2), IP2 and IP3 may be written as [21]

$$\text{IP2} = \frac{1}{32Z_0}\left(\frac{\pi F W^2}{\mu}\right)^2 \frac{A}{\left(1-\sqrt{A}\right)^4} \tag{8.3}$$

$$\text{IP3} = \sqrt{\frac{kT}{q}}\,\frac{1}{6Z_0}\left(\frac{\pi F W^2}{\mu}\right)^{1.5} \frac{A}{\left(1-\sqrt{A}\right)^3}$$

where A is the absolute attenuation level ($A = [2Z_0/(R_{ON} + 2Z_0)]^2$). Equation 8.3 shows that for a given frequency of operation F and attenuation level A, the distortion intercept point is governed strictly by the I-region thickness, W, and with no dependence on carrier lifetime τ. The carrier lifetime τ does, however, govern the RF resistance-current relationship as seen in (8.2). For example, devices of the same W that differ in lifetime τ by a factor of 2 require one half the amount of forward bias current for the same resistance. Since the resultant stored charge Q is the same in both cases, the distortion generated by both devices will be the same. For the previous PIN diode attenuator examples (Diode1: $W = 25\ \mu$m, $\tau = 150$ nS; Diode2: $W = 200\ \mu$m, $\tau = 2\ \mu$S), second and third order distortion should differ by 36 and 27 dB, respectively, at the same frequency and attenuation level. This phenomenon is observed in Figure 8.3, where the distortion spectrum is shown at 100 MHz for approximately 10 dB attenuation, which occurs at 100 μA for Diode1 and 600 uA for Diode2 ('netlist8-3.net'). Close analysis of the result shows differences in second and third order distortion levels of 33 and 19 dB, respectively.

The results presented in (4.21) and (8.3) show the different distortion figures of merit for PIN diodes used in switch and attenuator applications. In switch circuits, the ratio τ/W (high I-region lifetime and/or thin I-region thickness) should be maximized for optimum distortion performance. For attenuator circuits, on the other hand, thick I-region diodes result in improved distortion performance.

8.2.1.2 FET-Based Reflective Attenuator

The variable resistance required for attenuator circuits can also be generated by both MOSFET- and MESFET-based devices. In these cases, the control voltage on the gate varies the conducting characteristics of the drain-source channel, providing the necessary resistance for the attenuator. For the MOSFET, the gate bias variation changes the inversion channel carrier density, thereby varying the channel

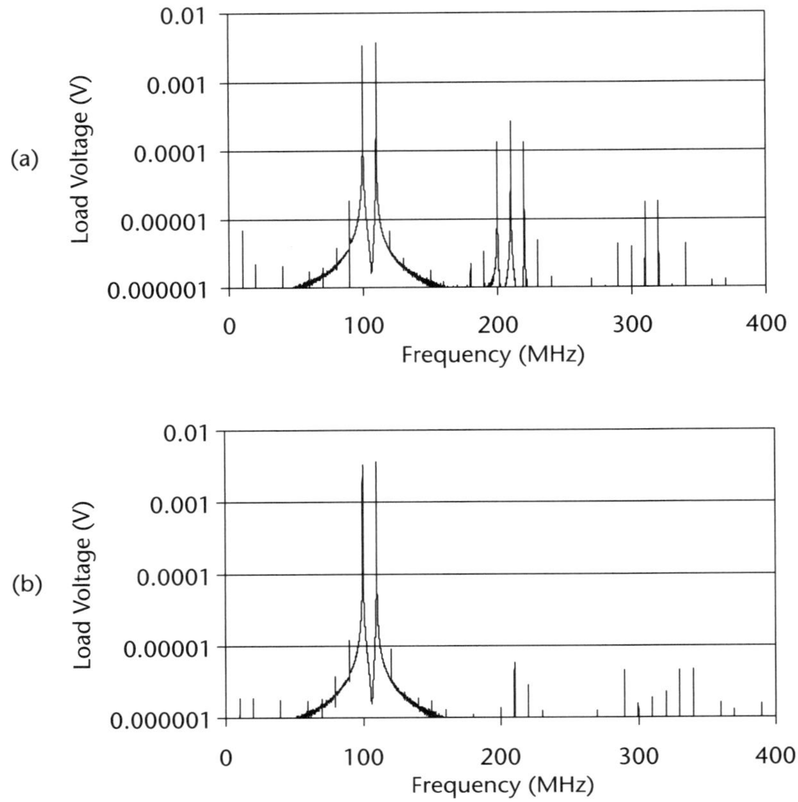

Figure 8.3 Voltage spectra at the load of two different PIN diodes used in a 10 dB series reflective attenuator: (a) thin PIN diode (Diode1) and (b) thick PIN diode (Diode2).

resistance. In the MESFET, the gate Schottky depletion layer can be varied by the applied gate bias, which modulates the channel depth and hence varies the channel RF resistance. One of the advantages of FET-based attenuator elements over PIN diode-based elements is the same as in the switch case: the dc decoupling network is dramatically reduced to a single resistor between the FET gate and the dc control source. Other capacitors may be needed to dc decouple the source and load or to float the MESFET-based control element in the case of unipolar supply operation (see Chapter 6). One of the disadvantages of using FET-based attenuators is that for high resistance values, both MOSFETs and MESFETs operate with higher V_{DS} values compared with the switch case, moving their operating point out of the triode region and closer to the "knee" of their current-voltage characteristic, an area of increasingly nonlinear operation.

This following set of simulations of the series reflective attenuator will show both MOSFET and MESFET operation using the same model parameters as the illustrative examples of these control components in Chapters 5 and 6, respectively. Both SPICE Level 3 and BSIM parameters for a 180 nm CMOS process will be used for the MOSFET discussion [22]; the Statz model will be used in the MESFET discussion. For the MOSFET, the bulk node will be fully RF grounded.

8.2.1.3 MOSFET Model Example

The first part of this section is a study of series reflective attenuator performance using both Level 3 and BSIM models for a series connected 180 nm MOSFET (Figure 8.4) with the bulk node connected to ground through bulk resistance R_B ('netlist8-4.net'). The Level 3 parameters are listed in Table 5.3, and the BSIM parameters are BSIM4.6.1 example parameters available from the UC Berkeley BSIM Research [22]. A 10 kΩ fixed resistance for gate bias resistance R_G will be assumed with the gate control voltage varying between zero and the threshold voltage.

Figure 8.5 shows the results of the SPICE simulation for the MOSFET series reflective attenuator using the SPICE Level 3 (Figure 8.5(a)) and BSIM4.6.1 (Figure 8.5(b)) parameters for a fully grounded bulk node (R_B = 0). There are significant differences between the two modeled attenuation results, as can be seen by comparing the results in the two figures. The most significant difference is that there are few simulated attenuation results for the SPICE Level 3 model, whereas the BSIM model results show the full range of gate voltages and associated attenuation levels. Since the dc gate voltage is varied in 0.1V steps from zero to 0.8V, the MOSFET is operating in the subthreshold region for some applied gate voltages (threshold voltage is approximately 0.35V). The SPICE Level 3 model performs poorly in subthreshold because the model assumes the conducting channel is only present for gate voltages above the threshold voltage [23]. Only signal leakage through the gate-drain and gate-source capacitances is then present, as seen by the strong frequency dependence of the frequency response in Figure 8.5(a). The BSIM model, however, does not make that same channel assumption but rather models channel and device behavior through subthreshold. For attenuator applications, the BSIM model is the preferred SPICE model. Even the BSIM model shows a strong frequency dependence at higher levels of attenuation since the subthreshold region of operation has reduced capacity to respond to signal swings [24].The effects of a floating bulk are negligible for the MOSFET attenuator as shown in Figure 8.5(c) where very little difference is seen in the attenuation results compared with the RF-shorted bulk in Figure 8.5(b) (R_B = 10 kΩ, 'netlist8-5.net).

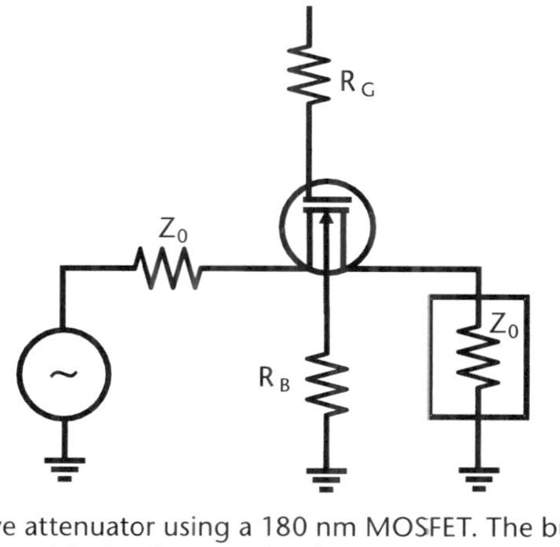

Figure 8.4 Series reflective attenuator using a 180 nm MOSFET. The bulk resistance R_B is used for floating bulk studies. Point A is the dc control point.

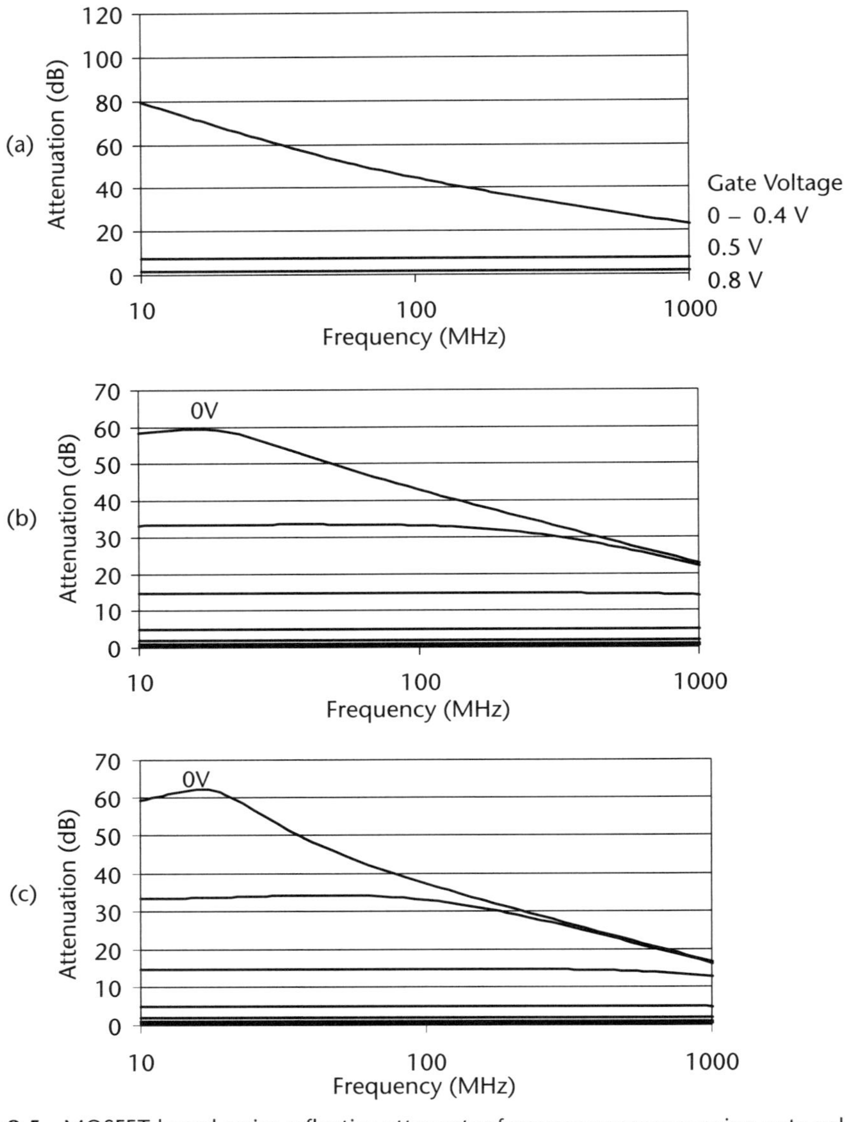

Figure 8.5 MOSFET-based series reflective attenuator frequency response using gate voltage as a parameter: (a) SPICE Level 3 ($R_B = 0$), (b) BSIM4.6.1 parameters ($R_B = 0$), and (c) BSIM4.6.1 parameters ($R_B = 10\ \text{k}\Omega$).

Operation at high levels of attenuation implies operation of the MOSFET near the knee of the I-V curve, which is a highly nonlinear region that causes the generation of distortion components that will eventually appear at the load (Chapter 5.5). From (5.11), increasing R_{ON} as well as the I-V expansion coefficient α_i (which happens as operation near the I-V "knee" and the onset of saturation) causes a reduction in the distortion intercept points, implying an increase in signal distortion at the load. SPICE (BSIM) can model these effects as shown in an example simulation of the MOSFET series reflective attenuator operating at 100 MHz with 32 and 15 dB attenuation using applied dc gate voltages of 0.1 and 0.2V, respectively ('netlist8-6.net'). The load voltage spectral response for these two attenuation levels is shown in Figure 8.6. Figure 8.6 shows that for the 15 dB attenuation case, the second and third order distortion components are 16.5 and 33 dB, respectively, below the 100 MHz fundamental (dBc). For the 32 dB attenuation case, the second and third order components are 12 and 22 dBc, respectively, indicating increased distortion levels at the higher attenuation.

8.2.1.4 MESFET Model Example

The second part of this section is a study of MESFET-based series reflective attenuator performance (Figure 8.7(a)) using the Statz model, with model parameters listed in Table 6.5 ('netlist8-7.net'). A 10 kΩ gate bias resistance R_G will be assumed with the gate control voltage varying between −2.0 and −2.5V in 0.1V steps. Shunt-connected resistors R_{DC} provide the proper drain and source dc returns and are of sufficiently large value (10 kΩ) to have minimal loading on the RF path. The grounded end of R_{DC} (V_{bias} = 0 in Figure 8.7(a)) may also be placed at a dc potential to change the voltage range of the gate bias potential at point A. Figure 8.7(b) shows the results of the SPICE simulation for the MESFET series reflective attenuator. For voltages in the range −2.0 to −2.4V (open channel conditions), the purely resistive behavior of the MESFET channel is seen by the constant attenuation at

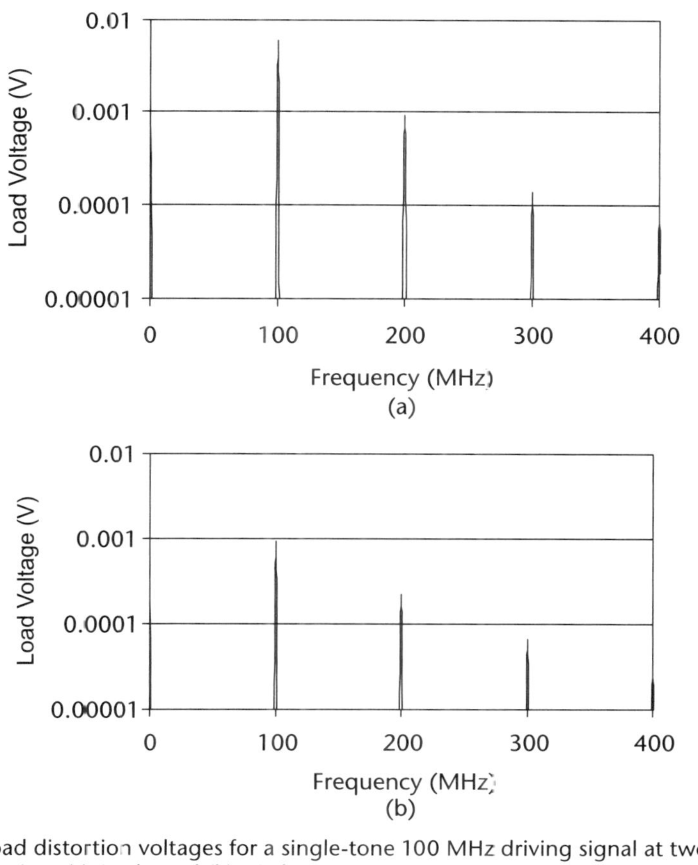

Figure 8.6 Load distortion voltages for a single-tone 100 MHz driving signal at two different levels of attenuation: (a) 15 dB and (b) 32 dB.

all frequencies. For the −2.5V gate control voltage, the channel is just pinched off and the frequency response is governed by signal leakage through the gate-drain and gate-source capacitances, as evident by the strong frequency dependence of the frequency response in Figure 8.7(b). Positive control voltages may be used if the

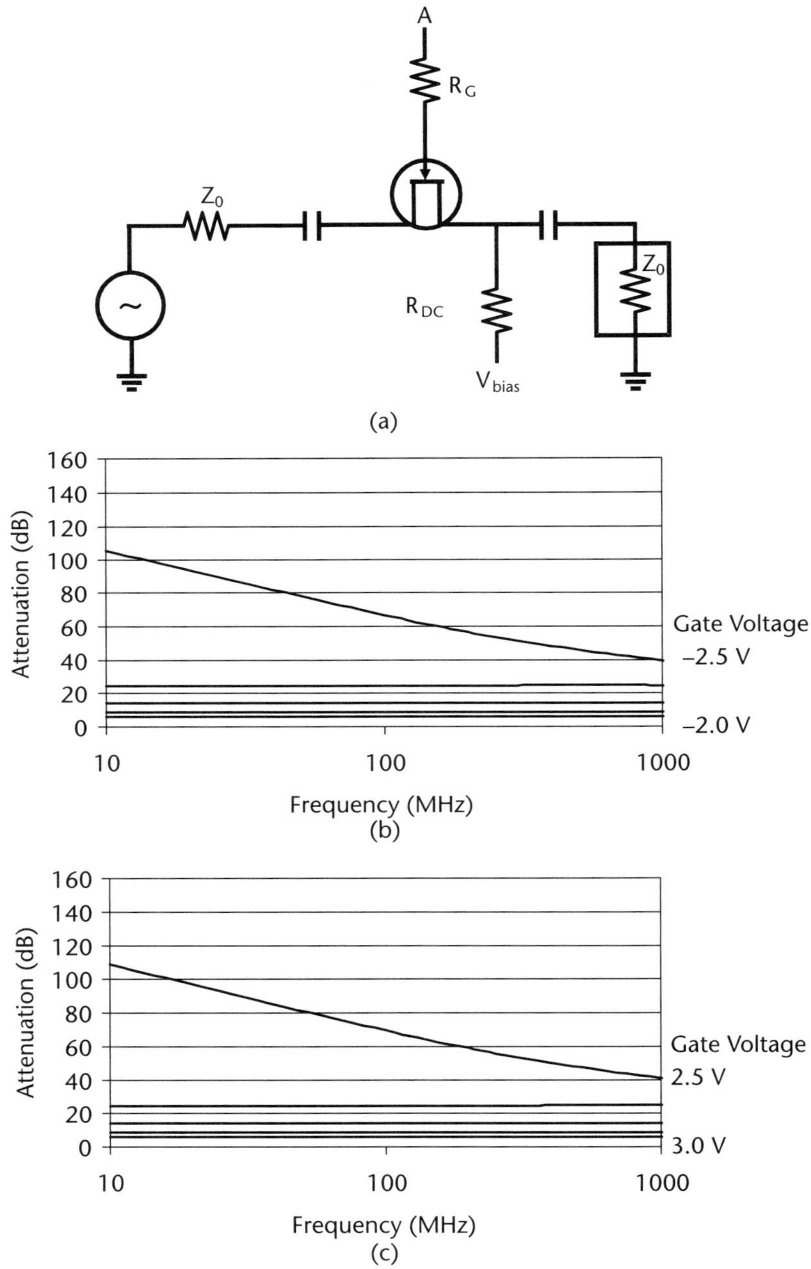

Figure 8.7 (a) Series reflective attenuator using a MESFET. Point A is the dc control point. (b) Attenuation versus frequency using applied gate voltage as a parameter with grounded R_{DC}. (c) Attenuation versus frequency using applied gate voltage as a parameter with R_{DC} connected to a +5.0V pull-up dc source.

R_{DC} pull-up resistor (B in Figure 8.7(a)) is biased by a dc voltage source, V_{bias}. For example, if R_{DC} is connected to a + 5.0V source at V_{bias} ('netlist8-8.net'), control voltages at A can vary between +2.5 and +3.0V and will yield the same attenuation curve (Figure 8.7(c)) as shown in Figure 8.7(b).

8.2.2 Π-Connected Matched Attenuator

A major problem of the reflective attenuator is that the signal attenuation at the load is achieved by reflection of RF power back toward the generator. This reflective power would then have to be dissipated by the generator and could cause undesired circuit operation. This reflected energy is minimized if the generator looks into a matched or other constant impedance, requiring the development of attenuators that provide both the necessary match and the desired signal attenuation.

Matched attenuators achieve this dual goal with specific resistance values dependent on the attenuator circuit topology. Chapter 1.3 outlined three matched attenuator types, the Π-connected attenuator, the T-connected attenuator, and the bridged-T attenuator, and included design equations for the resistor values as a function of attenuation (Table 1.1). The fixed resistances in all the matched attenuator cases can be replaced with PIN diodes or FETs, each operating with a specific applied dc bias that places the semiconductor control element in the proper operating regime to provide the equivalent resistance.

This section will look at only one of these matched attenuator cases, the Π-connected attenuator (Figure 8.8), since the other attenuator types use very similar modeling techniques. The discussion will focus on three levels of attenuation: 6, 10, and 20 dB. Limiting the discussion to these three attenuation levels also limits the number of required resistance values for the matched attenuator and, in turn, also limits the number of required dc bias conditions. For the Π-connected attenuator, Table 8.1 shows the necessary resistance values for 6, 10, and 20 dB attenuation assuming a $Z_0 = 50\Omega$ system.

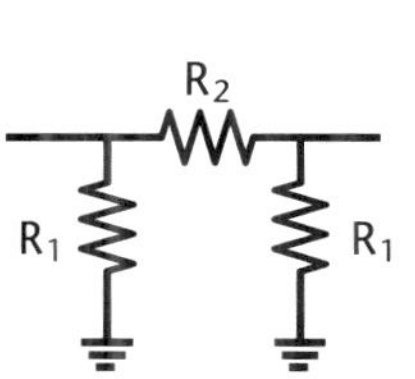

Figure 8.8 The Π-connected matched attenuator with values for specific attenuation values in a 50Ω system shown in Table 8.1.

Table 8.1 Π-Connected Matched Attenuator Resistance Values

Attenuation (dB)	*R_1 (Ω)*	*R_2 (Ω)*
6	150.4	37.35
10	96.24	71.15
20	61.11	247.50

Table 8.2 Π-Connected Matched Attenuator PIN Diode Forward Current Values

Attenuation (dB)	*PIN_1 (mA)*	*PIN_2 (mA)*
6	0.2	0.9
10	0.3	0.4
20	0.5	0.1

8.2.2.1 PIN Diode Π Matched Attenuator

The resistors R_1 and R_2 in the Π-matched attenuator can be replaced by PIN diodes as shown in Figure 8.9(a) (with dc decoupling elements shown). For the design and subsequent simulation, the physical PIN diode model will be used, with the full list of parameters shown in Table 7.4. The full SPICE subcircuit implementation of this model has been discussed previously and can be found in Table 4.3; the file 'netlist8-9.net' will be used for SPICE simulation.

The first step in designing the attenuator is to determine the dc forward currents needed to place the series and shunt PIN diodes into the proper RF resistance state so that they act as exact replacements for resistors R_1 and R_2. Figure 8.9(b) shows simulated RF resistance as a function of dc forward current over the range of 10 μA to 10 mA for the PIN diode with parameters in Table 7.4. SPICE and the physical PIN diode model were used to generate this data. The necessary dc forward currents for the PIN diodes to exhibit the resistance values required in Table 8.1 are shown in Table 8.2 based on the simulated resistance values shown in Figure 8.9(b).

Figure 8.9(c) shows the results of the SPICE simulation for the Π-connected matched attenuator with PIN diodes using the physical PIN diode model, Table 7.4 model parameters, and the dc forward currents shown in Table 8.2. The results show excellent agreement with the design attenuations of 6, 10, and 20 dB, with the attenuation levels constant over a two-decade frequency range, although some degradation for the 20 dB attenuation level is starting to appear at 1000 MHz.

One of the complexities in matched attenuators such as the Π, *T*, and others is that specific dc forward currents are required for the circuit to exhibit both a matched impedance and the desired attenuation level. The dc current sources have to be accurately controlled so that the specific PIN diode resistance is obtained in a particular branch of the attenuator. This resistance value also must be held constant under wide temperature variations, both ambient and due to power dissipation (and subsequent heating) in the PIN diodes [25, 26]. For the Π-connected matched attenuator, the series and shunt PIN diodes (PIN_1 and PIN_2) are often biased with separate currents sources; these two sources must track to maintain the desired resistance and therefore the match and attenuation [13, 14]. The same is true in the T-connected attenuator for the series and shunt branch PIN diodes. In the Π-connected matched attenuator, two PIN diodes are often used in the series branch to provide a more symmetric circuit, with each port seeing one series and one shunt PIN diode [13, 14, 25, 26]. If two PIN diodes are used in place of PIN_2, each device needs twice the dc forward current shown in Table 8.2. Each PIN diode resistance is then halved, with the sum of the two series PIN diode resistances then providing the necessary R_2 value.

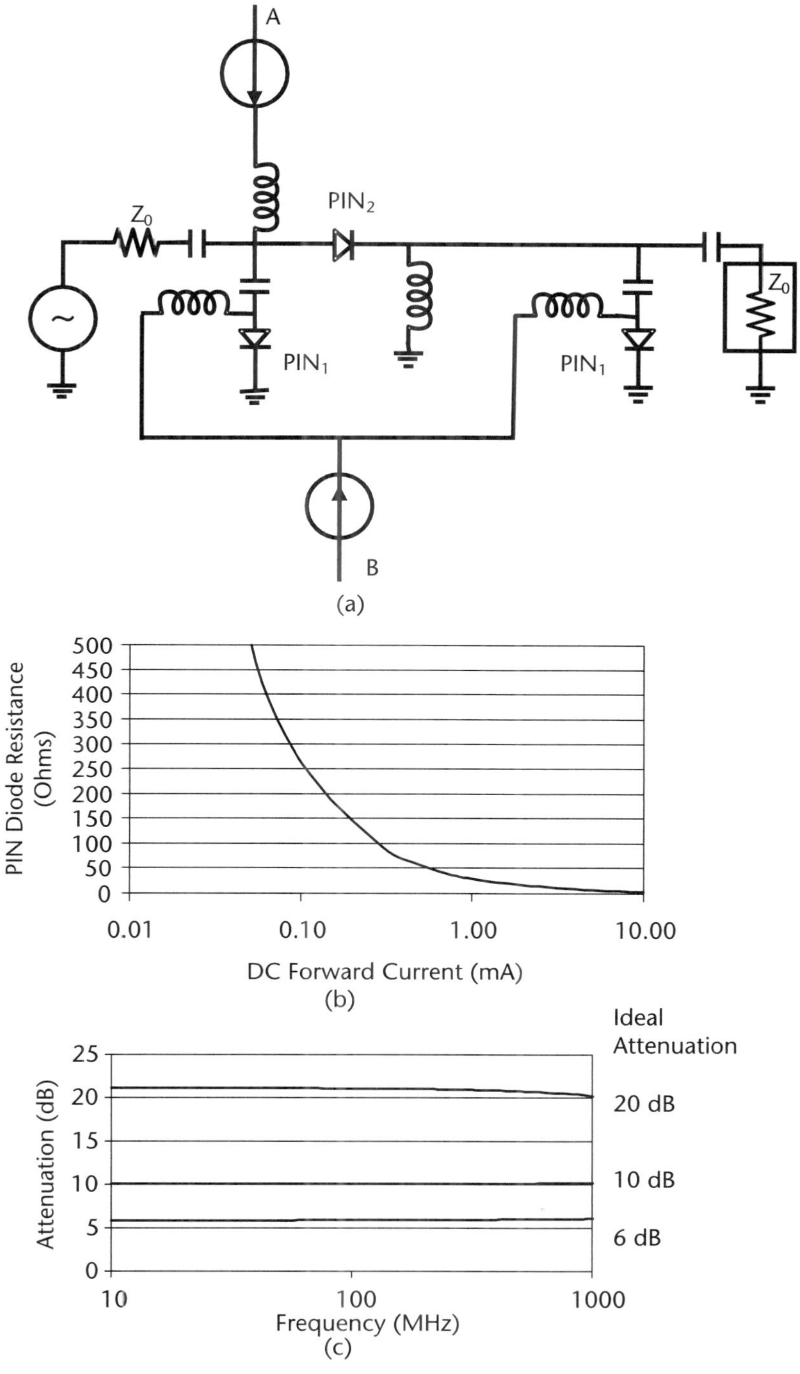

Figure 8.9 (a) Π-connected matched attenuator with PIN diodes. (b) Individual PIN diode resistance as a function of dc forward current for the PIN diode parameters in Table 7.4. (c) Simulated results of the Π-connected matched attenuator with PIN diodes with dc forward currents in Table 8.2.

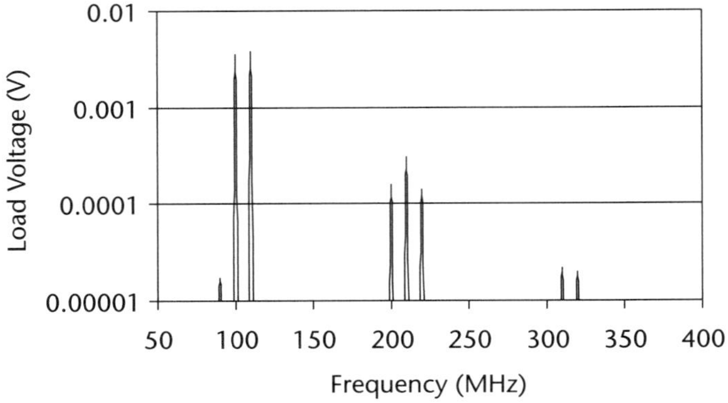

Figure 8.10 Frequency spectrum from the Π-connected matched attenuator with 10 dB of attenuation. Excitation signals at 100 and 110 MHz have been used.

Distortion in attenuators such as the Π-connected matched attenuator follows the same thinking as in the series reflective attenuator case in that small dc forward currents are required for high resistances, with associated low I-region stored charge and therefore higher levels of generated distortion. Distortion effects in both Π- and T-connected matched attenuators have been previously studied [21] and the physical PIN diode model can be used to compare the theoretical analysis in this reference with simulated results. For this comparison, the Π-connected matched attenuator with 10 dB of attenuation is chosen ('netlist8-10.net') with two closely spaced frequencies near 100 MHz investigated. The resulting frequency spectrum of the load voltage is shown in Figure 8.10 out to 400 MHz. From data provided in frequency spectra such as shown in this example, distortion intercept points for specific attenuation levels and circuit topologies can be determined. From the spectrum shown in Figure 8.10, the second and third order intermodulation components are 22 and 45 dBc, consistent with power levels shown in [21] for the Π-connected matched attenuator.

8.2.2.2 FET-Based Π Matched Attenuator

The variable resistance required for the Π matched attenuator can also be generated by both MOSFET and MESFETs. The design and simulation of the attenuator is similar for both FET types, and so this discussion will be limited to the use of the MOSFET with a fully grounded RF bulk ($R_B = 0$). The MOSFET-based Π matched attenuator (Figure 8.11(a)) will be designed for the same attenuation levels as the PIN diode design: 6, 10, and 20 dB. Only the BSIM parameters for the 180 nm CMOS process will be used for the MOSFET discussion, as it was shown in the discussion on the MOSFET-based series reflective attenuator that the SPICE Level 3 model did not accurately describe MOSFET operation in the important sub-threshold region.

In a similar manner as the PIN diode, the same first step in designing the attenuator is to determine the gate voltage needed for the MOSFET resistance needed to replace resistors R_1 and R_2. Figure 8.11(b) shows this RF resistance as a function of gate voltage over the range 0.2 to 0.7 V for the MOSFET with parameters in Table 5.3 and the example BSIM parameters ('netlist8-11.net'). The necessary

Table 8.3 Π-Connected Matched Attenuator MOSFET Gate Voltage Values

Attenuation (dB)	*M1 (V)*	*M2 (V)*
6	0.25	0.34
10	0.29	0.30
20	0.31	0.22

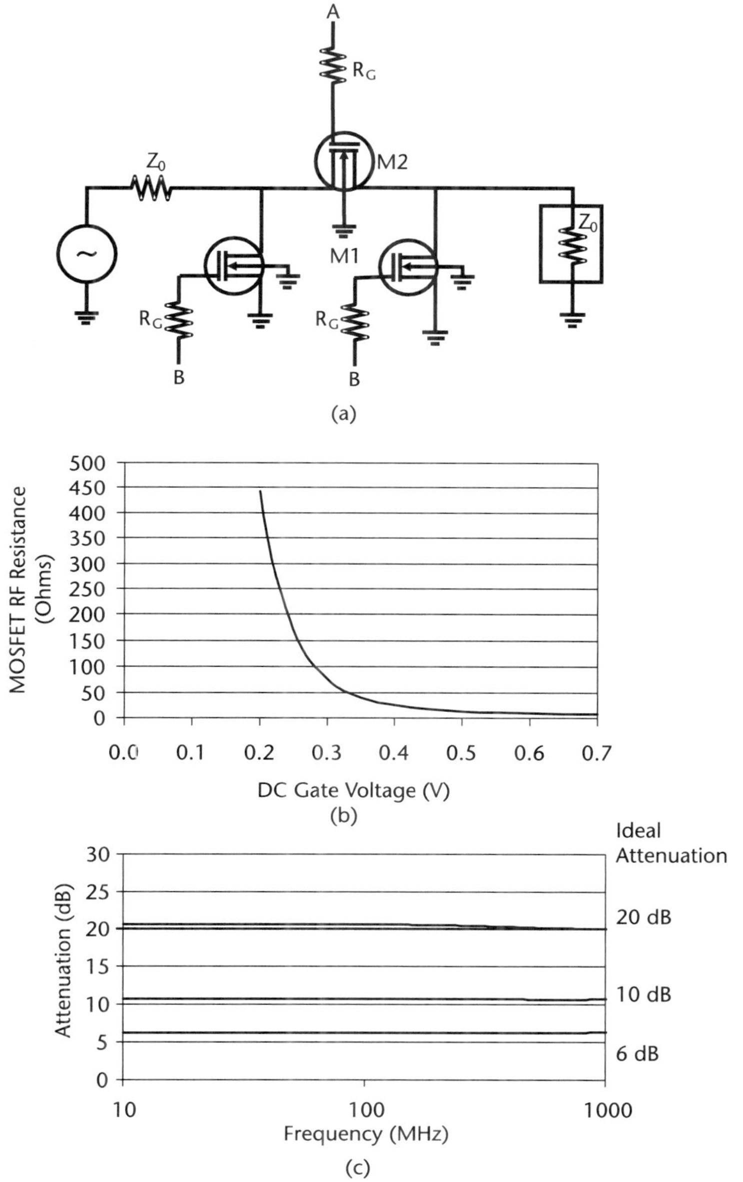

Figure 8.11 (a) Π-connected matched attenuator with MOSFETs. (b) Individual MOSFET resistance as a function of gate voltage for the MOSFET parameters in Table 5.3. (c) Simulated results of the Π-connected matched attenuator with MOSFETs using gate voltages in Table 8.3. Points A and B are the dc control points.

MOSFET gate voltages needed to create the RF resistance values required in Table 8.1 are shown in Table 8.3.

Figure 8.11c shows the results of the SPICE simulation for the Π-connected matched attenuator with MOSFETs using the BSIM model parameters and the gate voltages shown in Table 8.3. The results show excellent agreement with the design attenuations of 6, 10, and 20 dB and are constant over a two-decade frequency range, although some degradation for the 20 dB attenuation level is starting to appear at 1000 MHz.

For the 6 to 20 dB attenuation range, RF resistances for the Π-connected matched attenuator range from 37.6 to 247.8Ω. The MOSFET RF resistance curve in Figure 8.11(b) (or Table 8.3) shows that the required gate voltages span only a range of 120 mV, requiring very stable and accurate voltage sources to ensure that the RF resistance remains constant under all operating conditions.

MOSFET (as well as MESFET) attenuators can have large RF voltages on the various nodes, and under large signal conditions the RF swing may drive the FET away from its desired RF resistance value. This can be studied for the MOSFET-based attenuator in a simulator by looking at the attenuator response as a function of signal level, focusing on the attenuator set at 6 dB at 100 MHz. SPICE simulations as a function of RF drive ('netlist8-12.net') show significant changes in attenuation once the input signal swing reaches approximately 0.1 V (−10 dBm in a 50Ω system). The increase in attenuation at the higher power levels results from the MOSFET switching off during part of the RF cycle, lowering the level of 100 MHz RF energy reaching the load due to increased distortion of the RF drive signal. This effect is observed in Figure 8.12(a) for a +30 dBm RF input signal, with the RF load voltage showing a significantly distorted waveform; the associated spectral response is in Figure 8.12(b). Second and third order distortion are only 15 and 12 dBc, respectively, indicating a significant amount of RF energy is at higher frequencies rather than the fundamental, with the result being increased attenuation at the 100 MHz fundamental. Higher power attenuation can be achieved by using technologies with higher MOSFET threshold voltages (with the required higher dc gate voltages) or by stacking several MOSFETs in both the series and shunt branches to reduce the RF voltage drop across the individual devices. This stacking, however, increases the device capacitance and can negatively impact the frequency response of the attenuator.

8.2.2.3 Stepped Attenuators

The previous attenuators all have the capability of providing continuously variable attenuation over a wide range, controlled by either a dc forward current (for PIN diode-based attenuators) or a dc control voltage (for MOSFET and MESFET-based attenuators). The voltages and currents change the RF resistance of the various devices to provide the signal loss for the attenuator, requiring accurate dc control sources to provide the correct control signal to place, and keep, the semiconductor control element at the proper resistance for the desired attenuation. Small variations in control voltage or current can cause large resistance swings and subsequent attenuation variations as well as unintended impedance mismatches (particularly a problem in matched attenuators). The high resistance often needed for the attenuators

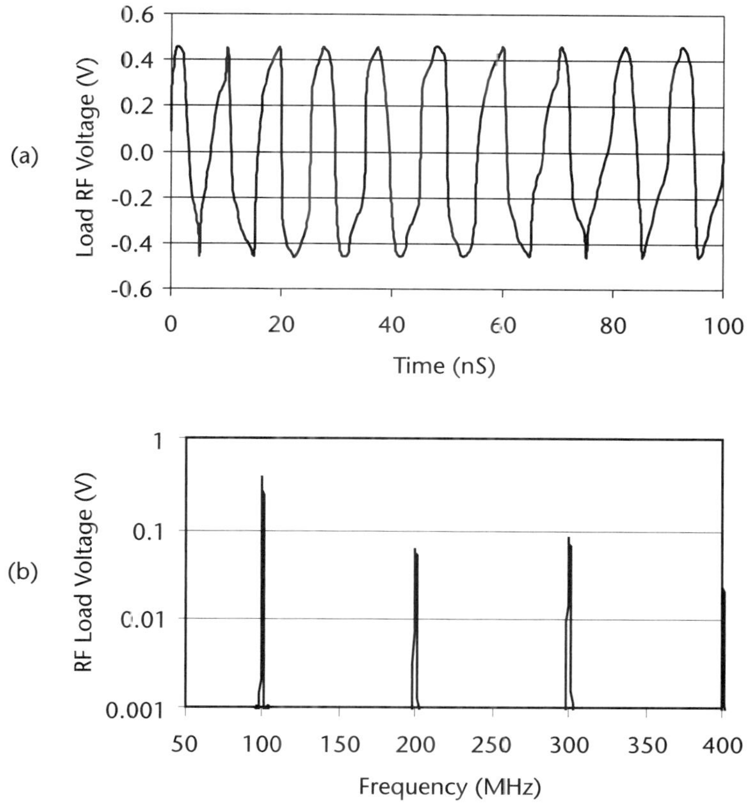

Figure 8.12 (a) Time and (b) frequency domain response of the MOSFET Π-connected matched attenuator set a 6 dB with 30 dBm input power level.

places these control devices into highly nonlinear regions of operation that can introduce distortion into the RF path.

An alternative attenuator type that provides good linearity with more stable attenuation is the stepped attenuator. The stepped attenuator uses the control device (PIN diode or FET) as a true low resistance switch rather than a variable resistor by reconfiguring an array of fixed attenuator pads to provide the desired attenuation. The switches can then be activated by control signals that toggle the device between the on- and off-state; these control signals are often digital in nature, with the attenuator then known as a digital stepped attenuator (DSA). The DSA is specified by the number of controlling bits N and the attenuation range R_{DSA}, which provides the number of unique attenuation states $M = 2^N$. The DSA "steps" through the various attenuation levels according to the digital control pattern, with a step size defined as $\Delta A = R_{DSA}/(M-1)$. For example, a 6-bit, 31.5 dB range DSA would require a step size $\Delta A = 0.5$ dB. The ΔA are generated by N fixed attenuator pads, each pad in the RF path increasing by a factor of two and toggled in and out of the RF path by switch elements. The DSA concept can be explained by assuming a 3-bit DSA with a 7 dB attenuation range and associated 1 dB attenuation steps (Figure 8.13). In Figure 8.13, the a_i are the digital control signals (a_0 is the least significant bit, LSB), and an a_i of logic 1 indicates the appropriate switch connects the RF

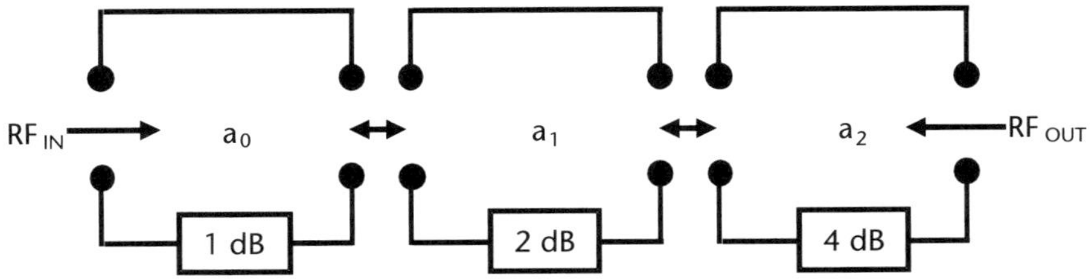

Figure 8.13 3-bit digital stepped attenuator with a 7 dB attenuation range.

path to the pad; a logic 0 connects to the through line. An attenuation table with the associated logic state is shown in Table 8.4. For example, the digital logic word 011 would initiate the switch elements to configure the pad array so that the 1 and 2 dB pads were both switched into the RF path while the 4 dB pad was bypassed, for a total of 3 dB attenuation.

In actual practice, the through line will exhibit some insertion loss because of the control elements in the through RF path. From the 3-bit example shown in Figure 8.13, the through path actually consists of six switch elements in series, each switch contributing to the insertion loss (the number of through switch elements in general is $2N$). Low insertion loss switches are needed to keep the "0 dB" attenuation state as low as possible. The total attenuation afforded by the DSA from module input to output is then this total through line insertion loss plus the switched attenuation. Using the 3-bit DSA example, if the through lines exhibited a 0.5 dB insertion loss, then the total attenuation of the DSA would range from 0.5 to 7.5 dB. In addition, the switches connecting to the pads introduce insertion loss that can be compensated by slightly reducing the pad attenuation from its integer value, with the remainder of the loss made up by the switches. The pads themselves are matched to the system impedance (usually 50Ω with 75Ω used for cable television applications) and can be modeled as T- or Π-connected resistive elements for simulation purposes. The frequency response of the DSA is governed by the frequency characteristics of the switch elements and the pads. The frequency characteristics of PIN diode and FET-based switching elements have been discussed in Chapters 3 through 6.

Table 8.4 Table of attenuation as a function of a_i logic states

a_2	a_1	a_0	*Attenuation (dB)*	*Action*
0	0	0	0 dB	RF path through
0	0	1	1 dB	1 dB pad switched in
0	1	0	2 dB	2 dB pad switched in
0	1	1	3 dB	1 and 2 dB pads switched in
1	0	0	4 dB	4 dB pad switched in
1	0	1	5 dB	1 and 4 dB pads switched in
1	1	0	6 dB	2 and 4 dB pads switched in
1	1	1	7 dB	1, 2 and 4 dB pads switched in

The large number of switches that need to be configured during any change of state has a direct impact on the overall switching speed of the DSA. Figure 8.14(b) shows a SPICE transient analysis of the 3-bit DSA using MOSFET switches (Figure 8.14(a)) and the example BSIM parameters with +10 dBm applied power at 100 MHz ('netlist8-13.net'). The three attenuator pads are modeled using a Π-connected resistive network with resistance values based on the expressions presented in Table 1.1. The simulation shows the switching transition from the 000 state (0 dB attenuation) to the 111 state (7 dB attenuation) occurring in approximately 200 nS. The switching speed is a function of the gate bias circuit RC time constant, as described in Chapter 5.

The DSA has several advantages over the continuously variable attenuators. The first advantage is that the linearity is significantly better for two reasons: one, the attenuation is governed by linear resistive elements (the pads); and second, the switches are in their fully on-state and therefore in their best state for linear operation. The second advantage is that the power level is governed by the pads and the

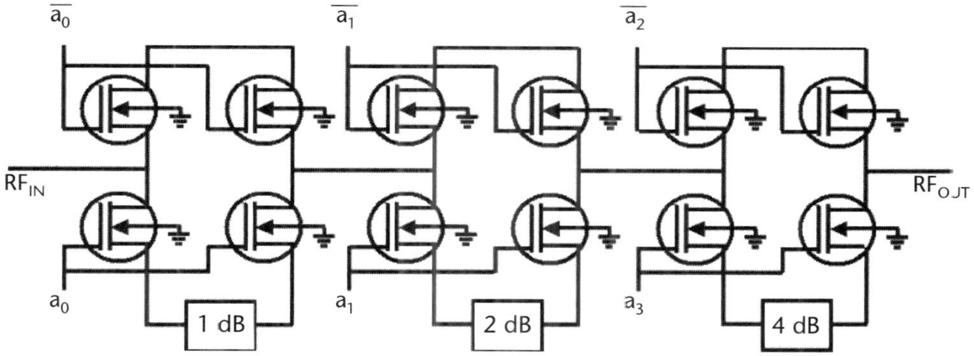

(a)

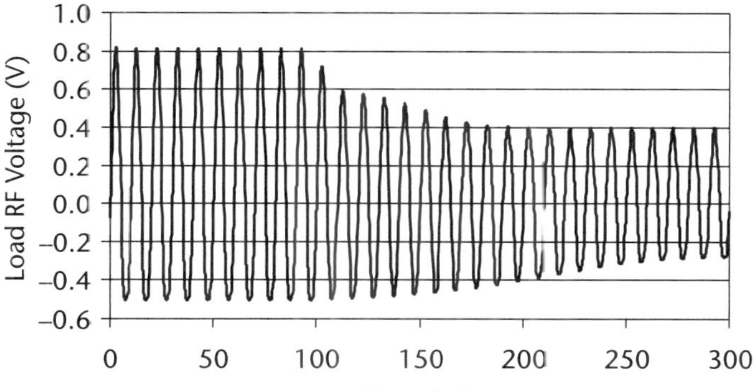

(b)

Figure 8.14 (a) 3-bit digital stepped attenuator using MOSFET-based switches. (b) Simulated digital stepped attenuator switching response from low to high attenuation with a 100-MHz excitation using MOSFET switches.

switches and so DSAs can operate at higher power levels. A 1 dB compression point, P_{1dB}, is often used to specify the power range of DSAs (see Chapter 1 for this definition), since this point describes the upper limit of linear operation.

8.3 Microwave and RF Limiters

Communication and radar receiver systems are designed to capture extremely weak signals and amplify them so that the desired information can be extracted. Strong signals, especially those on or near the desired receive frequency or frequency band (such as radar), with their associated large RF voltages can overwhelm the receiver front end. In the presence of strong signals, the first stages of the receiver can saturate and desensitize (so-called "receiver desense") the receiver, which takes time to return to normal operation after the large signal is removed and hence desired weak signals may be missed. Or, the energy in the large RF signal can be so large that the first receiver stages, designed for small signals, can exceed their safe operating limits and risk failure. A limiter is often added prior to the first stage of the receiver to prevent these large signals from negatively impacting receiver performance. Because the limiter appears as the receiver's first stage, the limiter should exhibit minimal loading so that the bandwidth is not adversely affected and low loss so that the weak signal amplitude is not affected and excessive noise not added to the receiver chain.

The ideal limiter acts as a self-actuated, amplitude-dependent switch or attenuator, reflecting or absorbing the incoming RF energy before it reaches the receiver circuitry. Actual limiters, however, do not respond instantaneously, and so a small amount of RF energy can reach the receiver prior to the limiter being fully activated; this energy is referred to as spike leakage (Figure 8.15). Even after the limiter has been fully activated, not all the RF energy is reflected or absorbed, and the amount that reaches the receiver is referred to as flat leakage (Figure 8.15). Once the RF pulse has been removed, there is a short period of time that the residual RF energy must be removed from the circuit, and this is referred to as the recovery time.

Early microwave and RF limiters used gas discharge tubes to provide this protection but were bulky and had a relatively limited lifespan compared with

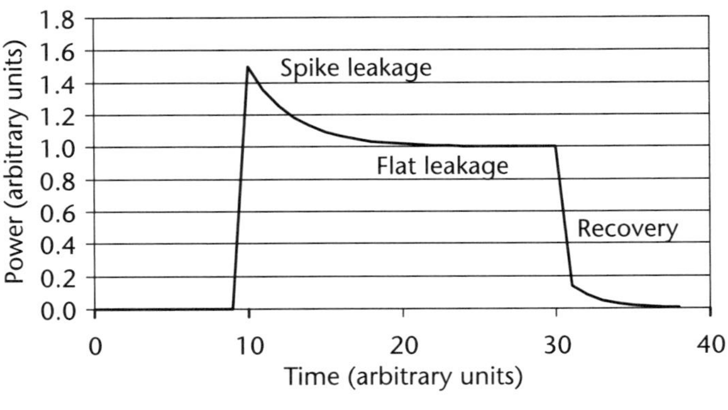

Figure 8.15 Spike and flat leakage power in a limiter circuit. The large RF signal is applied at time-unit 10 and turns off at time-unit 30.

semiconductor devices [27]. Since that time, silicon and GaAs PIN diodes have found widespread use in microwave and RF limiter circuits for receiver protection (Figure 8.16(a)) [28–33]. For fully integrated circuit applications, FET-based limiter and over-voltage protection circuits can be employed [34]. The simplest limiter circuits using FETs configures the devices as a diode-connected pair where the gate and drain are tied to the same potential (Figure 8.16(a)). RFID is one recent application of this technique [34].

The limiting level (LL) or flat leakage is dependent on the impedance presented by the limiting elements, with the limiting expression resembling the shunt-reflective attenuator from (1.27b) [28]:

$$LL = 10\log_{10}\left|1 + \frac{Z_0}{2Z_{\text{LIM}}}\right|^2 \text{ dB} \tag{8.4}$$

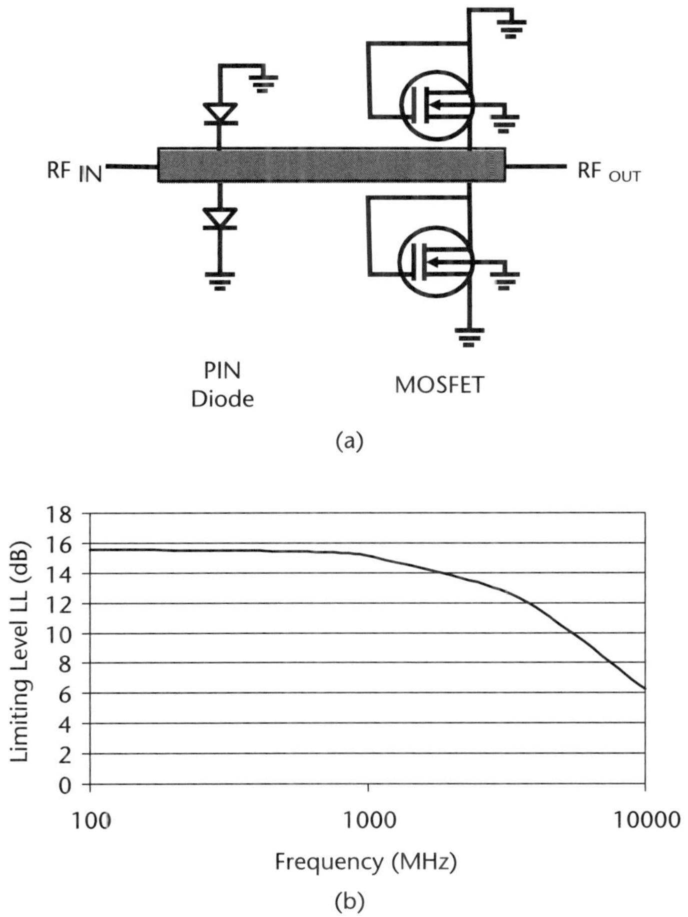

Figure 8.16 (a) Simple microwave and RF limiter structures showing anti-parallel pair of PIN diodes and antiparallel pair of diode-connected MOSFETs. (b) Limiting level LL (dB) for shunt limiter impedance Z_{LIM} consisting of a 5Ω resistive element and a 0.25 nH inductor.

where Z_{LIM} is the impedance of the shunt-connected limiting element. In the high-power limiting case, Z_{LIM} is often modeled as an on-state resistive part R_{LIM} and a series inductance. The series inductance usually originates from the connection from the RF path to the semiconductor control element. The influence of the inductance can be seen by assuming the limiter in its high-power limiting state exhibits a 5Ω resistance and 0.25 nH inductance. Figure 8.16(b) shows LL as a function of frequency for this element in a 50Ω and shows that the LL decreases rapidly above 100 MHz, a frequency where the inductive reactance is approximately that of the resistance.

Reducing spike leakage often requires a more complex limiter circuit. Fast-switching PIN diodes exhibit thin I-region widths W and short carrier lifetimes τ; however, the relatively low reverse breakdown voltage of these thin diodes can cause them to fail at higher power levels. Thick I-region devices typically exhibit higher reverse breakdown voltages and can therefore handle more power, but are often slower to activate than their thin I-region counterparts. Practical microwave and RF limiters combine these two types of PIN diodes to handle both the switching speed and power by placing the thicker devices closer to the antenna and the thinner devices closer to the circuits to be protected, often spaced $\lambda/4$ apart [28]. In this configuration, the thinner PIN diode acts quickly to initially limit the power to the protected circuit, with the thicker PIN diode eventually activating and reflecting the bulk of the incident RF power, thus protecting not only the receiver but also the thinner PIN diode.

PIN diode limiting circuits can be found beyond traditional communications and radar applications. One such application is in the MRI field for medical diagnostics. From an RF perspective, the MRI process resembles the classic radar problem, where a signal is transmitted with the receiver protected, and then the receiver listens for the return signal. In the case of MRI, the return signal originates with the relaxing spins of the desired nuclei [35–37]. Rather than using traditional antennas, clinical MRI systems use arrays of coils to receive the weak MR signal. These coils are tuned to the same frequency as the transmitter and so can couple RF power during transmit to the sensitive receiver or cause undesired heating that can cause burns in the patient being imaged because of the close receive coil proximity. PIN diodes are used to decouple receiver coils during transmit using the self-activating capabilities of the device.

This section will present two simulation examples of limiter operation: a single anti-parallel PIN diode limiter pair's response to a 1000 MHz RF pulse and a simple MOSFET limiter. The physical PIN diode and BSIM MOSFET models will be used. The first simulation will show the ability of the physical PIN diode model to accurately simulate all phases of limiter operation: spike leakage, flat leakage, and the RF pulse recovery period. The second MOSFET simulation will show the flat leakage dependence on the incident RF voltage. Simulation setup and interpretation of other more complex limiter circuits can be extended using the techniques based on these example cases.

8.3.1 PIN Diode Limiter Pair

The antiparallel PIN diode microwave limiter pair is a widely used circuit configuration, and it originates from traditional electronic clamping circuits. In this configuration, an antiparallel pair of PIN diodes is placed in shunt across the RF path with an inherent dc return for one diode through the complementary diode (Figure 8.17(a)). Two transient simulations at 1000 MHz have been performed; the first to determine the flat leakage for the pair and the second to observe the turn-on behavior of the PIN diode pair due to self-rectification. The time delay for the PIN diodes to become fully conducting is of special interest, since this time delay is the origin of spike leakage in this limiter configuration. For the simulation, the physical PIN diode model will be used, with a full list of device parameters as shown in Table 7.4 ('netlist8-14.net').

Flat leakage results from the first simulation (Figure 8.17(b)) show little impact of the dual-PIN diode limiter on the load voltage for low-source voltages, but once the source voltage exceeds the PIN diode turn-on voltage (approximately 1.0V), the PIN diodes start to conduct and reduce the shunt impedance of the pair, causing the load RF voltage to lower due to the reflection by the limiter stage.

The second PIN diode limiter example illustrates the spike and flat leakage phenomena as well as recovery (Figure 8.18). In this simulation, a 100V source at 1000 MHz (power available from the source of +44 dBm) is activated at 10 nS with

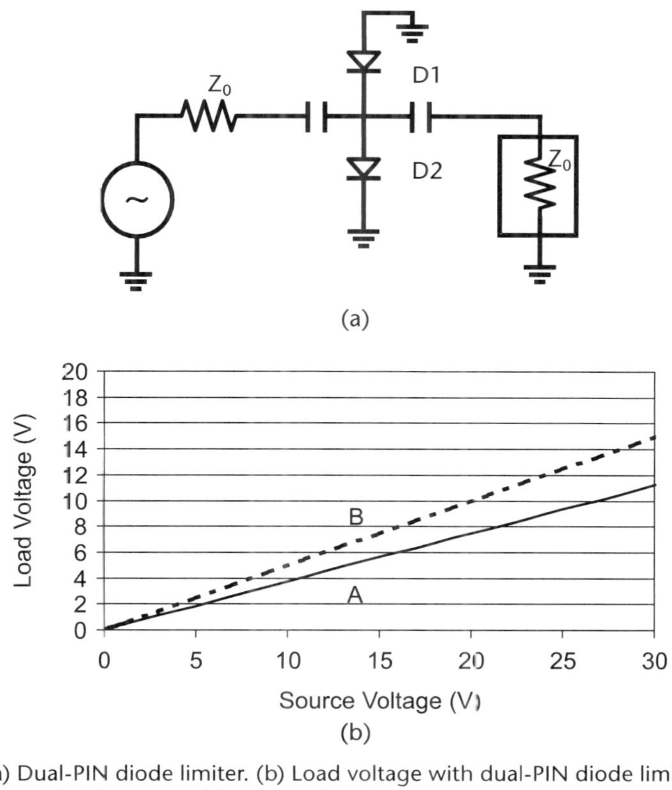

Figure 8.17 (a) Dual-PIN diode limiter. (b) Load voltage with dual-PIN diode limiter (A) and without the limiter (B). The onset of limiting takes place near the turn-on voltage of the PIN diodes.

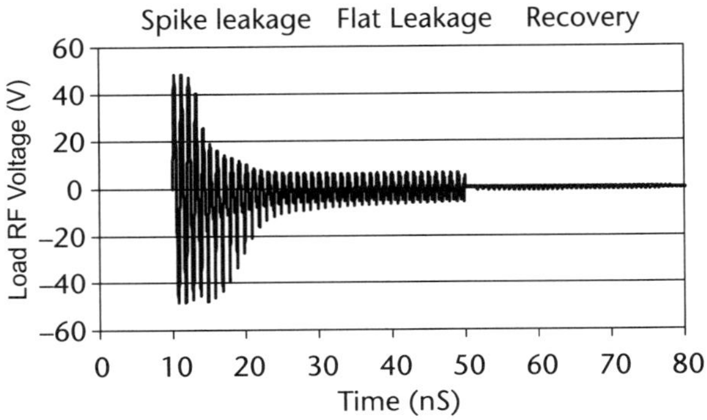

Figure 8.18 Dual-PIN diode limiter response to a 40 nS 1000 MHz RF pulse showing the three phases of limiting: spike leakage, flat leakage, and recovery.

pulse duration of 40 nS; the load voltage is monitored for 30 nS after the removal of the RF power. Figure 8.18 shows both spike and flat leakage occurring in this limiter during the active RF pulse, with the spike leakage lasting for approximately 5 nS as the PIN diodes go through their turn-on process. The spike leakage energy can be estimated as the product of the spike leakage duration and average power during the spike. For this example, the 50Ω load sees a peak voltage of approximately 50V lasting for 5 nS, with the associated spike leakage energy of $(5 \cdot 10^{-9}\text{S})(25\text{V}) = 125$ nJ.

8.3.2 MOSFET Limiter

For this example, dual diode-connected MOSFETs as illustrated in Figure 8.19(a) will be employed as a simple microwave limiter at 10 GHz. The 180 nm MOSFETs will be the same as those used in previous discussions with parameters shown in Table 5.3 and the example BSIM parameters used ('netlist8-15.net'). A series of transient simulations can be performed with varying source voltages to observe the limiting characteristics of the transistor pair. Flat leakage results from these simulations (Figure 8.19(b)) show little impact of the dual-MOSFET limiter at the load for low source voltages, but once the source voltage exceeds the MOSFET threshold voltage (approximately 0.35V), the diode-connected MOSFETs begin to conduct, reducing the shunting impedance of the pair, with RF energy being reflected back to the source and limiting the microwave energy at the load.

The dual-MOSFET limiter of this example exhibits approximately 3 dB insertion loss at low signal levels. This relatively high insertion loss at 10 GHz is due to the capacitance exhibited by the MOSFETs. The origin of this capacitance is from the drain and source connections of both M1 and M2 as well as the gate-source capacitance of M2, all in shunt with the RF path to the load.

8.4 Phase Shifters

Electronic beam steering of an array of spatially fixed antennas requires the use of phase shifters to set the proper phase at each antenna. The main lobe of the antenna

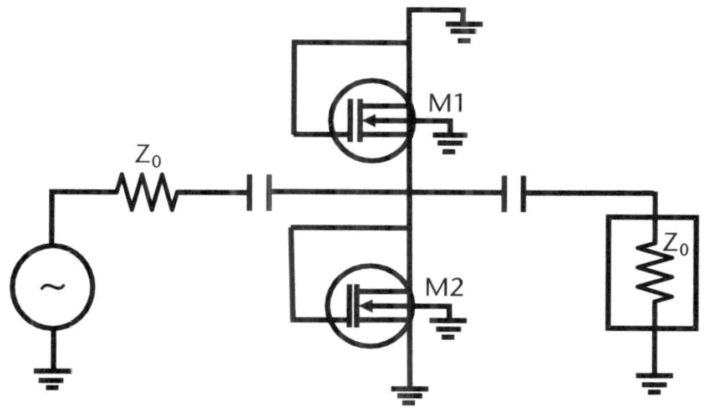

(a)

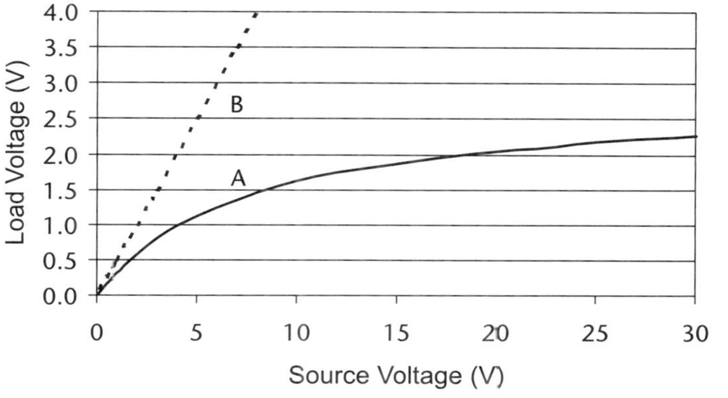

(b)

Figure 8.19 (a) Dual diode-connected MOSFET limiter. (b) Load voltage with dual-MOSFET limiter (A) and without the limiter (B). The onset of limiting takes place near the threshold voltage of the MOSFETs.

can then be physically swept in azimuth, elevation, or both by careful control of the phase and amplitude of the signal at the antenna feed points. Traditional phase shift circuits such as simple loaded line and switched transmission line were introduced in Chapter 1 along with other basic concepts of phase shifter circuits [38]. Like their digital stepped attenuator counterparts, most electronic phase shifters are designed to reconfigure fixed phase shifters as combinations digitally toggled in and out of the circuit to provide the required overall phase shift. The phase shifter step size is a function of the desired phase shift range Θ_{PS} and the number of bits N, with a corresponding phase shift per state of $\Delta\Theta = \Theta_{PS}/2^{N-1}$. Most popular fixed phase

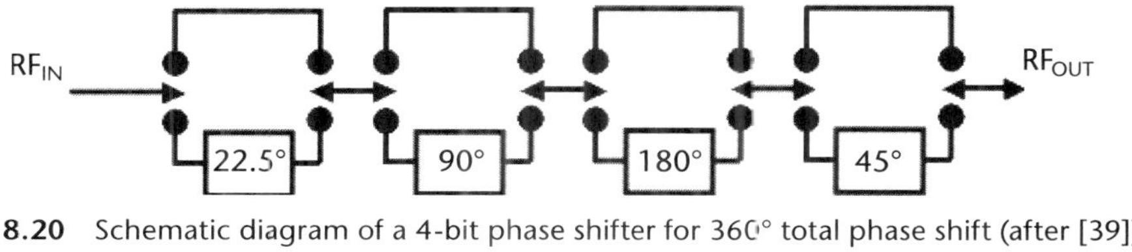

Figure 8.20 Schematic diagram of a 4-bit phase shifter for 360° total phase shift (after [39]).

shifters that cover a full 360° of phase shift exhibit step sizes of 22.5° for 4-bit and 11.25° for 5-bit phase shifters (Figure 8.20) [39]. Both PIN diodes and FET-based switches have been used in phased array antennas, with PIN diodes used in higher power pulsed systems and FET-based switches used in integrated circuit solutions. The order of the phase shifting elements (Figure 8.20) has been shown to be important, with the smaller phase shift stages placed between larger phase shift stages to reduce loading effects and increase phase linearity [40].

In phase shifters, the designer is concerned not only with obtaining the required phase shift but also for the insertion loss introduced by the phase shifter, the usable bandwidth where the phase does not vary considerably, and the phase error. For switched transmission line phase shifters, Garver has shown that off-state switch elements impact the phase error and that the shortest line in a switched line phase shifter should be limited to 20° to 50° degrees to minimize the error; the low on-state resistance of the switch elements helps to minimize insertion loss [38]. The use of coupled lines as discussed by Schiffman can be used to increase the bandwidth [38, 41]. A minimum of 10% bandwidth is possible with the simple phase shifters and an octave or more using Schiffman's approach [41].

In addition to these two fundamental phase shift circuits, other phase-shifting circuit topologies have been studied, with filter-based phase-shifting circuits seeing increased use [38, 39, 42]. These filter-based circuits have a space advantage over the loaded and switched line phase shifters because they are lumped element-based and therefore more space efficient, making them ideal candidates for use in integrated circuit phase shifters [39]. The simplest of these phase shifting circuits uses control elements to switch between a low-pass and a high-pass filter section (Figure 8.21). The filter component values are determined by the design frequency and the desired phase shift:

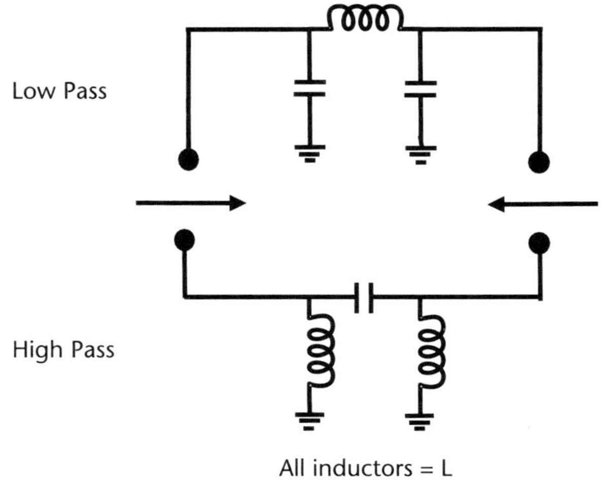

Figure 8.21 Fundamental low-pass/high-pass filter-based electronic phase shifter with L and C values given in (8.5) (after [38)]).

$$L = \frac{1}{\omega}\tan\left(\frac{\Delta\phi}{4}\right); \quad C = \frac{1}{\omega}\sin\left(\frac{\Delta\phi}{2}\right) \tag{8.5}$$

where ω is the radian frequency and $\Delta\phi$ is the desired phase shift.

This section will present one simulation case, that of the loaded line phase shifter using a PIN diode. Only a single phase shift will be studied using this example because the control elements are acting as switches in their fully on- or off-state rather than a continuously variable impedance as seen in the reflective attenuator examples. Simulation setup and interpretation for multiple phase shifts, and more complex phase shift circuits are readily extended using the techniques described in this example case.

The loaded line phase shifter is one of the simplest phase-shifting circuits because it consists of a single reactive element, usually shunt-connected, that is switched in and out of the circuit to give the differential phase shift. For a single shunt element, the total phase shift was shown in Chapter 1 to be (1.34)

$$\Delta\Theta = \pi/2 - \tan^{-1}\left(2X/Z_0\right) \tag{8.6}$$

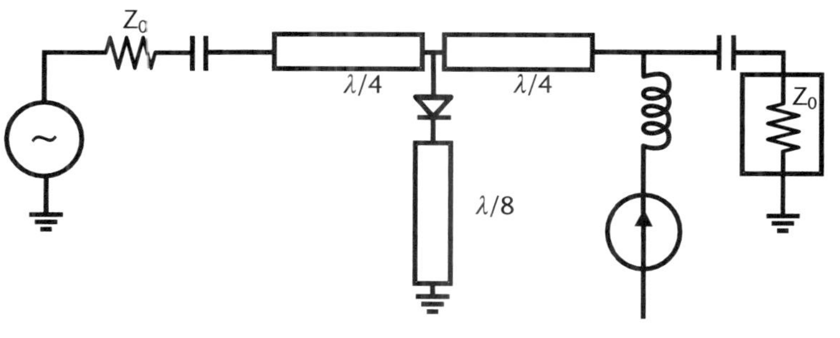

(a)

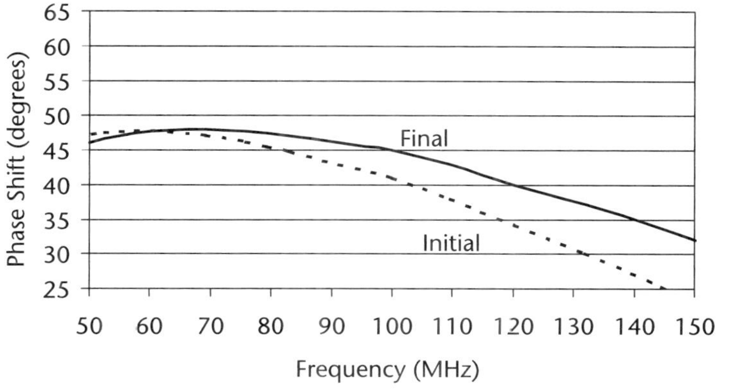

(b)

Figure 8.22 (a) Loaded line phase shifter using a single shorted λ/8 100 MHz transmission line stub and PIN diode control element. (b) Differential phase shift for the initial and final design of the λ/8 loaded line phase shifter.

where X is the shunt element reactance and Z_0 is the system impedance. Any reactance introduced by the switching element must be included in the definition for the reactance X, and so for real switching elements, the off-state reactance needs to be fully investigated. The loading by X on the line impacts the insertion loss as well; studies have shown an insertion phase of 45° is a typical upper limit for the use of this technique [43].

The simulation outlined in this section will use a shorted 25Ω shunt $\lambda/8$ transmission line stub ($X = +25\Omega$) at 100 MHz with a PIN diode used to toggle the stub in and out of the circuit (Figure 8.22(a)), with (8.6) predicting a phase shift of 45°. For the simulation example, the physical PIN diode model will be used, with a full list of parameters shown in Table 7.4 and biased at 10 mA dc forward current for the on-state, and zero bias in the off-state. The full SPICE subcircuit implementation of this model has been discussed previously and can be found in Table 4.3 ('netlist8-16.net').

The initial phase shift simulation results are shown in Figure 8.22(b) and show 41° of phase shift at 100 MHz, a phase error of 4°. This phase error is primarily due to the off-state capacitance of the PIN diode introducing a small amount of capacitive reactance that slightly reduces the overall value of the required reactance $X = +25\Omega$ when the PIN diode is toggled to its on-state. A slight reduction of stub length ($\lambda/8$ at 119 MHz) moves the differential phase shift to the required 45° at 100 MHz (Figure 8.22(b)). The phase shift varies from 48° at 70 MHz up to 40° at 120 MHz in the final design.

This example shows the importance of carefully investigating the impact of the switching element on phase shift performance. Significant phase errors in the final design can occur if the loading effects of the switching element itself are not considered in the design. Even though this example used a PIN diode switch, both MOSFET- and MESFET-based control elements exhibit capacitive off-state loading that can influence the final phase shift and must be considered as part of the design flow as well.

References

[1] LT-SPICE IV, Linear Technology Corporation, 2015. [http://www.linear.com/designtools/software/#Ltspice].

[2] McElroy, P. K., "Designing Resistive Attenuating Networks," *IRE Proc.*, Vol. 23, No. 3, March 1935, pp. 213–233.

[3] Hunton, J. K., and A. G. Ryals, "Microwave Variable Attenuators and Modulators Using PIN Diodes," *IRE Trans. on Microwave Theory and Techniques*, Vol. 10, No. 4, 1962, pp. 262–273.

[4] Mann, S., et al., "61 GHz Millimeter Wave Voltage Variable Attenuator Based on Flip-Chip Mounted PIN-Diodes," *2014 20th International Conference on Microwaves, Radar, and Wireless Communication (MIKON)*, 2014, pp. 1–4.

[5] Dai, Y-S., D-G. Fang, and Y. Guo, "A Novel UWB (0.045–50 GHz) Digital/Analog Compatible MMIC Variable Attenuator With Low Insertion Phase Shift and Large Dynamic Range," *IEEE Microwave and Wireless Components Lett.*, Vol. 17, No. 1, 2007, pp. 61–63.

[6] Peppiatt, H. J., J. A. Hall, and A. J. Fuller, "Broadband AGC Circuits," *IEEE Proc.*, Vol. 55, No. 2, 1967, pp. 220–220.

[7] Moradi, G., A. Abdipour, and A. Shabani, "Analysis and Design of a FET-PIN-FET Attenuator," *IEEE International RF and Microwave Conference, 2008. RFM 2008*, 2008, pp. 475–477.

[8] Franzwa, E., et al., "A 50MHz–16GHz Low Distortion SOI Voltage Controlled Attenuator IC with IIP3 > +38dBm and Control Range of > 25dB," *2012 IEEE MTT-S International Microwave Symposium Digest (MTT)*, 2012, pp. 1–3.

[9] Huang; Y-Y., et al., "Highly Linear RF CMOS Variable Attenuators with Adaptive Body Biasing," *IEEE Journal of Solid-State Circuits*, Vol. 46, No. 5, 2011, pp. 1023–1033.

[10] Paek, J.-S., H. Lee, and S. Hong, "Analysis and Design of CMOS Amplitude Modulator with Digitally Controlled Variable Attenuator," *IEEE Trans. on Microwave Theory and Techniques*, Vol. 59, No. 3, 2011, pp. 727–740.

[11] Parris, W. J., "P-I-N Variable Attenuator with Low Phase Shift (Short Papers)," *IEEE Trans. on Microwave Theory and Techniques*, Vol. 20, No. 9, 1972, pp. 618–620.

[12] Stukach, O. V., "Variable Attenuator with Low Phase Shift," *The 9th European Conference on Wireless Technology*, 2006, pp. 241–244.

[13] Hiller, G., *Design with PIN Diodes*, M/A-COM Application Note AG312, 2010 (see also *Design with PIN Diodes*, Skyworks Application Note 200312 Rev. C, April 2009).

[14] *Application Note: A Wideband General Purpose PIN Diode Attenuator*, Skyworks Solutions Inc, 2011.

[15] Ku, B-H., and Songcheol Hong, "6-bit CMOS Digital Attenuators with Low Phase Variations for X-Band Phased-Array Systems," *IEEE Trans. on Microwave Theory and Techniques*, Vol. 58, No. 7, 2010, pp. 1651–1663.

[16] Zhao, L. et al., "An Integrated Q-Band 6-Bit Digital Attenuator with Low Insertion Loss," *2014 Asia-Pacific Microwave Conference (APMC)*, 2014, pp. 1196–1198.

[17] Ali, F., S. Mitchell, and A. Podell, "Low-Loss, High-Power, Broadband GaAs MMIC Multi-Bit Digital Attenuators with On-Chip TTL Drivers," *Digest 1991 Gallium Arsenide Integrated Circuit (GaAs IC) Symposium*, 1991, pp. 243–246.

[18] Gupta, R. K., J. H. Reynolds, and P. J. McNally, "Modeling and CAD of an Ultra-Broadband Monolithic 5-bit Digital Attenuator," *18th European Microwave Conference*, 1988, pp. 151–155.

[19] Mikul, A., S. Zhu, P. Sun, Y. You, S. P. Sah, et al., "Compact low phase imbalance broadband attenuator based on SiGe PIN diodes," *2012 IEEE MTT-S Intl. Microwave Symp. Dig.*, June 2012, pp. 1-3.

[20] Eom, H. K. Yang, "A 6-20 GHz Compact Multi-Bit Digital Attenuator Using InP/InGaAs PIN Diodes," *20th International Conference on Indium Phosphide and Related Materials, 2008 (IPRM 2008)*, 2008, pp.1–3.

[21] Caverly, R., and G. Hiller, "Distortion in p-i-n Diode Control Circuits, with Gerald Hiller," *IEEE Trans. on Microwave Theory and Techniques*, Vol. MTT-35 (5), 1987, pp. 492–501.

[22] BSIM4 Official Release, BSIM Research Group at Univ. Calif. Berkeley, 2007. http://www-device.eecs.berkeley.edu/bsim/?page=BSIM4_Arc.

[23] *MicroSim PSpice A/D Reference Manual*, MicroSim Corp., 1997 (see also http://bwrcs.eecs.berkeley.edu/Classes/IcBook/SPICE/).

[24] Tsividis, Y. P., *Operation and Modeling of the MOS Transistor*, New York: McGraw-Hill, 1987.

[25] Mtimat, M., and A. B. Kouki, "Design and Characterization of Variable Attenuators and Phase Shifters with Linearized Bias Control Voltages," *Canadian Conference on Electrical and Computer Engineering*, Vol. 2, May 2004, pp. 1085–1088.

[26] Beyragh, D. S., H. Pahlevaninezhad, and S. R. Motahari, "A Broadband Low Reflection Electronically Variable PIN Diode-Based Attenuator," *IEEE International Conference on Ultra-Wideband, (ICUWB 2007)*, 2007, pp. 800–804.

[27] Leenov, D., "The Silicon PIN Diode as a Microwave Radar Protector at Megawatt Levels," *IEEE Trans. Electron Devices*, Vol. 11, No. 2, 1964, pp. 53–61.

[28] Brown, N. J., "Design Concepts for High-Power PIN Diode Limiting," *IEEE Trans. on Microwave Theory and Techniques*, Vol. 15, No. 12, 1967, pp. 732–742.

[29] Smith, D. G., D. D. Heston, and D. L. Allen, "Designing High-Power Limiter Circuits with GaAs PIN Diodes," *Digest, 1999 IEEE MTT-S International Microwave Symposium*, Vol. 1, 1999, pp. 329–332.

[30] Ward, A. L., R. J. Tan, and R. Kaul, "Spike Leakage of Thin Si PIN Limiters," *IEEE Trans. Microwave Theory and Techniques*, Vol. 42, No. 10, 1994, pp. 1879–1885.

[31] Seymour, D. J., D. D. Heston, and R. E. Lehmann, "Monolithic MBE GaAs Pin Diode Limiter," *Microwave and Millimeter-Wave Monolithic Circuits*, Vol. 87, No. 1, 1987, pp. 35–37.

[32] Garver, R. V., and J. A. Rosado, "Broad-Band TEM Diode Limiting," *IRE Trans. on Microwave Theory and Techniques*, Vol. 10, No. 5, 1962, pp. 302–310.

[33] Leenov, D., J. Forster, and N. Cranna, "PIN Diodes for Protective Limiter Applications," *Digest 1961 IEEE International Solid-State Circuits Conference*, Vol. IV, 1961, pp. 84–85.

[34] Zoscher, L., et al., "RF Voltage Limiters for Passive Differential UHF RFID Front-Ends in a 40 nm CMOS technology," *2015 IEEE MTT-S International Microwave Symposium (IMS)*, 2015, pp.1–4.

[35] Caverly, R., R. Watkins, and W. E. Doherty Jr., "Modeling PIN Diode Temperature Rise in High Induced Current MR Receive Coils," *Proc. 2014 Intl. Soc. Magnetic Resonance in Medicine Symp.*, May 2014, pp. 1–3.

[36] Caverly, R., "Electrothermal Modeling of PIN Diode Protection Circuits in MRI Surface Coils," *2014 IEEE Topical Conference on Biomedical Wireless Technologies, Networks, and Sensing Systems (BioWireleSS2014)*, January 2014, pp. 1–4.

[37] Caverly, R., "MRI Fundamentals: RF Aspects of Magnetic Resonance Imaging (MRI)," *IEEE Microwave Magazine*, Vol. 16, No. 6, 2015, pp. 20–33.

[38] Garver, R. V., "Broad-Band Diode Phase Shifters," *IEEE Trans. on Microwave Theory and Techniques*, Vol. 20, No. 5, May 1972, pp. 314–323.

[39] Tang, X, and K. Mouthaan, "Design Considerations for Octave-Band Phase Shifters Using Discrete Components," *IEEE Trans. on Microwave Theory and Techniques*, Vol. 58, No. 12, 2010, pp. 3459–3466.

[40] Li, W-T., et al., "60-GHz 5-Bit Phase Shifter with Integrated VGA Phase-Error Compensation," *IEEE Trans. on Microwave Theory and Techniques*, Vol. 61, No. 3, 2013, pp. 1224–1235.

[41] Schiffman, B. M., "A New Class of Broad-Band Microwave 90-Degree Phase Shifters," *IRE Trans. on Microwave Theory and Techniques*, Vol. 6, No. 2, 1958, pp. 232–237.

[42] Hayashi, H., T. Nakagawa, and K. Araki, "A Miniaturized MMIC Analog Phase Shifter Using Two Quarter-Wave-Length Transmission Lines," *IEEE Trans. on Microwave Theory and Techniques*, Vol. 50, No. 1, 2002, pp. 150–154.

[43] Pozar, D., *Microwave Engineering*, 2nd ed., New York: John Wiley, 1998.

Author Biography

Robert H. Caverly received his Ph.D. degree in electrical engineering from The Johns Hopkins University, Baltimore, in 1983. He has been a faculty member at Villanova University in the Department of Electrical and Computer Engineering since 1997 and is a full professor. Previously, he was a professor for more than 14 years at the University of Massachusetts Dartmouth, also in the Department of Electrical and Computer Engineering. Dr. Caverly's research interests are focused on the characterization of semiconductor devices, such as PIN diodes and FETs in the microwave and RF control environment. He has published more than 100 journal and conference papers and is the author of the book *CMOS RFIC Design Principles* from Artech House. An IEEE Fellow (2013), Dr. Caverly was an associate editor of the *IEEE Microwave Magazine* and an IEEE MTT-S Distinguished Microwave Lecturer at the time of the writing of this book.

Index

H

I

J

L

Q

R

Artech House Microwave Library

Behavioral Modeling and Linearization of RF Power Amplifiers, John Wood

Chipless RFID Reader Architecture, Nemai Chandra Karmakar, Prasanna Kalansuriya, Randika Koswatta, and Rubayet E-Azim

Control Components Using Si, GaAs, and GaN Technologies, Inder J. Bahl

Design of Linear RF Outphasing Power Amplifiers, Xuejun Zhang, Lawrence E. Larson, and Peter M. Asbeck

Design Methodology for RF CMOS Phase Locked Loops, Carlos Quemada, Guillermo Bistué, and Iñigo Adin

Design of CMOS Operational Amplifiers, Rasoul Dehghani

Design of RF and Microwave Amplifiers and Oscillators, Second Edition, Pieter L. D. Abrie

Digital Filter Design Solutions, Jolyon M. De Freitas

Discrete Oscillator Design Linear, Nonlinear, Transient, and Noise Domains, Randall W. Rhea

Distortion in RF Power Amplifiers, Joel Vuolevi and Timo Rahkonen

Distributed Power Amplifiers for RF and Microwave Communications, Narendra Kumar and Andrei Grebennikov

EMPLAN: Electromagnetic Analysis of Printed Structures in Planarly Layered Media, Software and User's Manual, Noyan Kinayman and M. I. Aksun

An Engineer's Guide to Automated Testing of High-Speed Interfaces, José Moreira and Hubert Werkmann

Envelope Tracking Power Amplifiers for Wireless Communications, Zhancang Wang

Essentials of RF and Microwave Grounding, Eric Holzman

FAST: Fast Amplifier Synthesis Tool—Software and User's Guide, Dale D. Henkes

Feedforward Linear Power Amplifiers, Nick Pothecary

Filter Synthesis Using Genesys S/Filter, Randall W. Rhea

Foundations of Oscillator Circuit Design, Guillermo Gonzalez

Frequency Synthesizers: Concept to Product, Alexander Chenakin

Fundamentals of Nonlinear Behavioral Modeling for RF and Microwave Design, John Wood and David E. Root, editors

Generalized Filter Design by Computer Optimization, Djuradj Budimir

Handbook of Dielectric and Thermal Properties of Materials at Microwave Frequencies, Vyacheslav V. Komarov

Handbook of RF, Microwave, and Millimeter-Wave Components, Leonid A. Belov, Sergey M. Smolskiy, and Victor N. Kochemasov

High-Linearity RF Amplifier Design, Peter B. Kenington

High-Speed Circuit Board Signal Integrity, Stephen C. Thierauf

Integrated Microwave Front-Ends with Avionics Applications, Leo G. Maloratsky

Intermodulation Distortion in Microwave and Wireless Circuits, José Carlos Pedro and Nuno Borges Carvalho

Introduction to Modeling HBTs, Matthias Rudolph

Introduction to RF Design Using EM Simulators, Hiroaki Kogure, Yoshie Kogure, and James C. Rautio

Introduction to RF and Microwave Passive Components, Richard Wallace and Krister Andreasson

Klystrons, Traveling Wave Tubes, Magnetrons, Crossed-Field Amplifiers, and Gyrotrons, A. S. Gilmour, Jr.

Lumped Elements for RF and Microwave Circuits, Inder Bahl

Lumped Element Quadrature Hybrids, David Andrews

Microstrip Lines and Slotlines, Third Edition, Ramesh Garg, Inder Bahl, and Maurizio Bozzi

Microwave Circuit Modeling Using Electromagnetic Field Simulation, Daniel G. Swanson, Jr. and Wolfgang J. R. Hoefer

Microwave Component Mechanics, Harri Eskelinen and Pekka Eskelinen

Microwave Differential Circuit Design Using Mixed-Mode S-Parameters, William R. Eisenstadt, Robert Stengel, and Bruce M. Thompson

Microwave Engineers' Handbook, Two Volumes, Theodore Saad, editor

Microwave Filters, Impedance-Matching Networks, and Coupling Structures, George L. Matthaei, Leo Young, and E. M. T. Jones

Microwave Materials and Fabrication Techniques, Second Edition, Thomas S. Laverghetta

Microwave Materials for Wireless Applications, David B. Cruickshank

Microwave Mixer Technology and Applications, Bert Henderson and Edmar Camargo

Microwave Mixers, Second Edition, Stephen A. Maas

Microwave Network Design Using the Scattering Matrix, Janusz A. Dobrowolski

Microwave Radio Transmission Design Guide, Second Edition, Trevor Manning

Microwave and RF Semiconductor Control Device Modeling, Robert H. Caverly

Microwave Transmission Line Circuits, William T. Joines, W. Devereux Palmer, and Jennifer T. Bernhard

Microwaves and Wireless Simplified, Third Edition, Thomas S. Laverghetta

Modern Microwave Circuits, Noyan Kinayman and M. I. Aksun

Modern Microwave Measurements and Techniques, Second Edition, Thomas S. Laverghetta

Neural Networks for RF and Microwave Design, Q. J. Zhang and K. C. Gupta

Noise in Linear and Nonlinear Circuits, Stephen A. Maas

Nonlinear Microwave and RF Circuits, Second Edition, Stephen A. Maas

Q Factor Measurements Using MATLAB®, Darko Kajfez

QMATCH: Lumped-Element Impedance Matching, Software and User's Guide, Pieter L. D. Abrie

Passive RF Component Technology: Materials, Techniques, and Applications, Guoan Wang and Bo Pan, editors

Practical Analog and Digital Filter Design, Les Thede

Practical Microstrip Design and Applications, Günter Kompa

Practical Microwave Circuits, Stephen Maas

Practical RF Circuit Design for Modern Wireless Systems, Volume I: Passive Circuits and Systems, Les Besser and Rowan Gilmore

Practical RF Circuit Design for Modern Wireless Systems, Volume II: Active Circuits and Systems, Rowan Gilmore and Les Besser

Production Testing of RF and System-on-a-Chip Devices for Wireless Communications, Keith B. Schaub and Joe Kelly

Radio Frequency Integrated Circuit Design, Second Edition, John W. M. Rogers and Calvin Plett

RF Bulk Acoustic Wave Filters for Communications, Ken-ya Hashimoto

RF Design Guide: Systems, Circuits, and Equations, Peter Vizmuller

RF Linear Accelerators for Medical and Industrial Applications, Samy Hanna

RF Measurements of Die and Packages, Scott A. Wartenberg

The RF and Microwave Circuit Design Handbook, Stephen A. Maas

RF and Microwave Coupled-Line Circuits, Rajesh Mongia, Inder Bahl, and Prakash Bhartia

RF and Microwave Oscillator Design, Michal Odyniec, editor

RF Power Amplifiers for Wireless Communications, Second Edition, Steve C. Cripps

RF Systems, Components, and Circuits Handbook, Ferril A. Losee

The Six-Port Technique with Microwave and Wireless Applications, Fadhel M. Ghannouchi and Abbas Mohammadi

Solid-State Microwave High-Power Amplifiers, Franco Sechi and Marina Bujatti

Stability Analysis of Nonlinear Microwave Circuits, Almudena Suárez and Raymond Quéré

Substrate Noise Coupling in Analog/RF Circuits, Stephane Bronckers, Geert Van der Plas, Gerd Vandersteen, and Yves Rolain

System-in-Package RF Design and Applications, Michael P. Gaynor

Terahertz Metrology, Mira Naftaly, editor

TRAVIS 2.0: Transmission Line Visualization Software and User's Guide, Version 2.0, Robert G. Kaires and Barton T. Hickman

Understanding Microwave Heating Cavities, Tse V. Chow Ting Chan and Howard C. Reader

Understanding Quartz Crystals and Oscillators, Ramón M. Cerda

For further information on these and other Artech House titles, including previously considered out-of-print books now available through our In-Print-Forever® (IPF®) program, contact:

Artech House Publishers
685 Canton Street
Norwood, MA 02062
Phone: 781-769-9750
Fax: 781-769-6334
e-mail: artech@artechhouse.com

Artech House Books
16 Sussex Street
London SW1V 4RW UK
Phone: +44 (0)20 7596 8750
Fax: +44 (0)20 7630 0166
e-mail: artech-uk@artechhouse.com

Find us on the World Wide Web at: www.artechhouse.com